Biogas Technology

Liangwei Deng · Yi Liu ·
Wenguo Wang

Biogas Technology

Springer

Liangwei Deng
Biogas Institute of Ministry of Agriculture
and Rural Affairs
Chengdu, China

Yi Liu
Biogas Institute of Ministry of Agriculture
and Rural Affairs
Chengdu, China

Wenguo Wang
Biogas Institute of Ministry of Agriculture
and Rural Affairs
Chengdu, China

ISBN 978-981-15-4939-7 ISBN 978-981-15-4940-3 (eBook)
https://doi.org/10.1007/978-981-15-4940-3

This Springer imprint is published by the registered company Springer Nature Singapore Pte Ltd.
The registered company address is: 152 Beach Road, #21-01/04 Gateway East, Singapore 189721, Singapore

Preface

Biogas technology has won universal attention to its active role in waste disposal, renewable energy recovery and greenhouse gas emission reduction. In 1989, Biogas Institute of Ministry of Agriculture (BIOMA) organized a group of experts for the edition of Biogas Technology in China, which was a reference book of the international training courses on biogas technology held at BIOMA where more than 2,700 technical and managerial personnel from around 130 countries were trained. After over three decades, China, along with other countries, has witnessed great progress in biogas technology in basic theory, process technology, device and equipment, development mode and operation management. In the era of the technical renewal, the knowledge needs to be reviewed and summarized. To this end, Biogas Institute of Ministry of Agriculture and Rural Affairs (BIOMA) re-compiled the book Biogas Technology, jointly contributed by its talent pool of rich experience in scientific research and technical extension.

This book begins with the chapter of anaerobic digestion microorganisms, which is a brief introduction of anaerobic digestion microorganisms, biochemical processes and influencing factors of digestion process. The following chapters were devoted, respectively, to the principles and technical characteristics of three types of digestion systems, i.e. rural household biogas digesters, biogas digester for domestic sewage treatment and biogas plants, and then a comprehensive introduction and assertion of the common basic knowledge and technical keys of the three digestion systems, including building materials and structures of digesters, biogas cleaning and upgrading, biogas storage, utilization of biogas and digestate.

The authors of this book are professionals engaged in research, design, construction and commissioning of digestion systems for years, with solid and rich experience in the design, commissioning and operation of household digesters, biogas digester for domestic sewage treatment and biogas plants. Fruitful achievements of professors, engineers and managers in China and other countries were cited as references. The completion of this book benefits from long-term research accumulation of predecessors of BIOMA and the summary of engineering experience of engineers and managers. I would like to thank the predecessors and the relevant authors for their contributions. After the completion of the manuscript, I made a revision on all the chapters of the book, and the authors of similar majors checked each other's contents. Miss Ruijjie Yu revised the English language of the book. I would like to express my heartfelt thanks.

Chengdu, China Liangwei Deng

Contents

Contributor to Chapters

Chapter I

Assis. Prof. Yansheng Xu Biogas Institute of Ministry of Agriculture and Rural Affairs, Chengdu, China

Prof. Lei Cheng Biogas Institute of Ministry of Agriculture and Rural Affairs, Chengdu, China

Assoc. Prof. Yan Long Biogas Institute of Ministry of Agriculture and Rural Affairs, Chengdu, China

Chapter II

Assoc. Prof. Yi Liu Biogas Institute of Ministry of Agriculture and Rural Affairs, Chengdu, China

Assis. Prof. Jingsi Cheng Biogas Institute of Ministry of Agriculture and Rural Affairs, Chengdu, China

Chapter III

Seni. Engir. Ke Pan Biogas Institute of Ministry of Agriculture and Rural Affairs, Chengdu, China

Prof. Guozhong Shi Biogas Institute of Ministry of Agriculture and Rural Affairs, Chengdu, China

Assis. Prof. Jingsi Cheng Biogas Institute of Ministry of Agriculture and Rural Affairs, Chengdu, China

Chapter IV

Prof. Liangwei Deng Biogas Institute of Ministry of Agriculture and Rural Affairs, Chengdu, China

Assoc. Prof. Yan Long Biogas Institute of Ministry of Agriculture and Rural Affairs, Chengdu, China

Chapter V

Seni. Engir. Li Song Biogas Institute of Ministry of Agriculture and Rural Affairs, Chengdu, China

Assoc. Prof. Yan Long Biogas Institute of Ministry of Agriculture and Rural Affairs, Chengdu, China

Chapter VI

Seni. Engir. Xiaodong Pu Biogas Institute of Ministry of Agriculture and Rural Affairs, Chengdu, China

Assis. Prof. Wen Zhang Biogas Institute of Ministry of Agriculture and Rural Affairs, Chengdu, China

Chapter VII

Seni. Engir. Zhiyong Wang Biogas Institute of Ministry of Agriculture and Rural Affairs, Chengdu, China

Assis. Prof. Wen Zhang Biogas Institute of Ministry of Agriculture and Rural Affairs, Chengdu, China

Chapter VIII

Seni. Engir. Chuixue Kong Biogas Institute of Ministry of Agriculture and Rural Affairs, Chengdu, China

Assis. Prof. Wen Zhang Biogas Institute of Ministry of Agriculture and Rural Affairs, Chengdu, China

Chapter IX

Prof. Wenguo Wang Biogas Institute of Ministry of Agriculture and Rural Affairs, Chengdu, China

Assis. Prof. Jingsi Cheng Biogas Institute of Ministry of Agriculture and Rural Affairs, Chengdu, China

Chapter 1
Anaerobic Digestion Microorganisms

Anaerobic digestion is a common process in nature. Biogas is generated from many places such as the bottom sediments of fresh water and of the sea, rice field, wetland, marsh and rumen of some animals (Lin et al. 2014). People use artificial reactors to process solid waste and wastewater to generate and utilize biogas (Deublein and Steinhauser 2008). Generation of biogas is a bio-chemical process with cooperative action of multiple microorganisms, involving various obligate or facultative anaerobic microorganisms. Microorganisms play a decisive role for the efficiency of biogas production.

1.1 Characteristics of microorganisms

Anaerobic digestion microorganisms are a kind of organisms related to biogas generation and have some basic characteristics of common microorganisms. Therefore, prior to illustrating anaerobic digestion microorganisms, the generic characteristics of all microorganisms will be introduced.

1.1.1 Definition of microorganisms

Microorganisms (or microbes) refer to all lower organisms that have minimal bodies and simple structures and can hardly be observed with naked eyes. With a large and complex group and various categories, microorganisms can be divided into cellular and noncellular microorganisms. Cellular microorganisms refer to those with cellular morphology, including bacteria, archaea and eukaryotic microorganisms, while noncellular microorganisms refer to those without cellular morphology, including virus and viroid. Cellular microorganisms can also be divided into prokaryotic and eukaryotic microorganisms. Prokaryotic microorganisms indicate bacteria in a broad sense,

L. Deng et al., *Biogas Technology*,
https://doi.org/10.1007/978-981-15-4940-3_1

that is, a large category of primitive unicellular microorganisms whose cell nucleus is not wrapped by nuclear membrane and whose DNA is exposed in nucleus zone, including bacteria and archaea. Bacteria (narrow sense), actinomycetes, cyanobacteria, mycoplasma, chlamydia and rickettsia belong to the category of bacteria. The nucleus of eukaryotic microorganisms is wrapped by nuclear membrane. The cells can perform mitosis, and mitochondria or multiple organelles such as chloroplast which exists in the cytoplasm. Fungi (yeasts, molds, and fungi) in Holomycota, microscopic algae in plantage and protozoa in animalia are all eukaryotic microorganisms.

Due to the minimal size and inconspicuous appearance of microorganisms, people had recognition for microorganisms far later than they had for animal and plant in a very long time. Since the first time of observing microorganisms by Antoni van Leeuwenhoek (1632–1723), people have known microorganisms for merely 300 years. Since its foundation, microbiology has only been developed for more than 100 years. However, microorganisms possess a decisive position in the entire life science field due to their rich diversity, unique biological characteristics and functions, and they play a significant role in natural material circulation and in many fields such as industrial, agricultural, health care and environment protection fields.

1.1.2 Key events in microbiology development

In the long history of human social activity, our recognition for animal and plant can date back to the emergence of mankind, but the recognition for microorganisms had been inadequate for a long term. Robert Hooke (1635–1703) described the structure of mold through observation by microscope in his book, *Micrographia,* and that was the first known description for microorganisms. In 1676, Antoni van Leeuwenhoek (1632–1723), the "Microbiology Pioneer", observed bacteria for the first time using his self-made microscope. However, during the following 200 years, microorganisms did not draw enough attention from people. In mid-19th century, Ferdinand Cohn (1828–1898) found that bacteria generates endospore when he was doing research on heat resistance of bacteria. After that, Robert Koch (1843–1910) achieved rapid progress in microorganism isolation, and he had isolated and described multiple pathogenic bacteria. The culture preparation technology and basic operation technology concerning isolation based on solid culture medium and purification technology created by Koch are still used today. In 1857, Louis Pasteur (1822–1895) overturned the prevailing theory of autogenesis using the curving bottle experiment and established the germ theory of life. Therefore, a new subject—microbiology was gradually established.

1.1.3 Characteristics and distribution of microorganisms

There are many categories of microorganisms with various shapes and extremely small sizes. Microorganisms belong to lower organisms with simple structures, so they share some common characteristics.

(1) Minimal sizes and large specific surface area

Microorganisms are measured by micrometer while virus by nanometer. In recent years, some microorganisms that can be seen by naked eyes were found, such as *Epulopicsium fishelsoni* found in 1993 and *Thiomargarita namibiensis* reported in 1998. The specific surface area of microorganisms is large due to their minimal sizes.

(2) A wide distribution and great variety

Microorganisms are widely distributed in nature, including inside and outside of animals and plants, soils, air, oceans, glaciers, salt lakes, deserts, oil wells and under rock stratum. The exact number of species on earth is not known yet. Up to now, recorded species has nearly 2 million, out of which the number of species of microorganisms sums up to 200,000. Isolated microorganism species merely account for 3%–5% of the total number. Therefore, there are a large number of microorganisms remaining to be discovered.

(3) Rapid growth rate and strong metabolism

Quick conversion, vigorous metabolism and rapid reproduction of microorganisms come from their large specific surface areas that contribute to material, energy and information exchange with surrounding environment. Reproduction of one generation can be done within tens of minutes or several hours under suitable conditions, and microorganisms can decompose substances that are thousands times heavier than themselves.

(4) Strong environment adaptability and easy variation

Microorganisms exist in various environments and can adapt to any severe condition unlike animals and plants. However, most microorganisms are unicellular organisms, so they contact directly with various environments and their DNA is prone to change.

The above characteristics of microorganisms provide a foundation for human to exploit and utilize microorganism resources.

1.1.4 Function of microorganisms in ecological circulation

Organisms in the ecological system include producers, consumers and decomposers, of which the producers and decomposers are indispensable, and the absence of either can cause collapse of the ecological system. Microorganisms play an important role in the ecological system. They act as both primary producers and decomposers, both being an essential member of material circulation and key conservators of materials and energy. Their major function, however, is to act as decomposers, since producers such as plants and consumers such as animals are eventually decomposed by heterotrophic microorganisms in order to accomplish a complete material circulation and energy. Microorganisms play a significant role in circulation of elements such as C, N and S. If there is no action of microorganism, various elements and substances cannot be cyclically utilized and ecological balance will be broken. Even in the biology development history, microorganisms are always the major decomposers and the promoters for element circulation.

1.2 Anaerobic digestion microorganisms

Anaerobic digestion refers to the process of converting organic matters into methane and CO_2 under anaerobic conditions by obligate or facultative anaerobic microorganisms. The mechanism is also known as the methane formation action, and the microorganisms participating in such processes are called anaerobic digestion microorganisms (Ma et al. 2017). With great varieties, the anaerobic digestion microorganisms can be generally divided into the methane-producing group and the non-methane-producing group.

1.2.1 Biochemical process of anaerobic digestion

Theoretically, all substrates containing organic matters such as carbohydrate, protein and fat can be used as feedstock for anaerobic digestion. Conversion of organic matters into biogas can be expressed by the following equation:

$$C_cH_hO_oN_nS_s + yH_2O \rightarrow xCH_4 + nNH_3 + sH_2S + (c - x)CO_2 \quad (1.1)$$

$x = 1/8\ (4c + h - 2o - 3n + 2\ s)$
$y = 1/4\ (4c - h - 2o + 3n + 3\ s)$

The reaction equations for conversion of carbohydrate, protein and fat into biogas are as follows:

$$\text{Carbohydrate: } C_6H_{12}O_6 \rightarrow 3CH_4 + 3CO_2 \quad (1.2)$$

$$\text{Fat: } C_{12}H_{24}O_6 + 3H_2O \rightarrow 7.5CH_4 + 4.5CO_2 \quad (1.3)$$

$$\text{Protein: } C_{13}H_{25}O_7N_3S + 6H_2O \rightarrow 6.5CH_4 + 6.5CO_2 + 3NH_3 + H_2S \quad (1.4)$$

The proportion of CO_2 and CH_4 generated depend on feedstock composition and degradation level. When simply using carbohydrates, fats, or proteins as feedstock, theoretically the methane content in biogas is respectively 50%, 62.5% and 38.2%.

In studies on anaerobic digestion, different researchers proposed different theories about anaerobic digestion stages, i.e., two-stage, three-stage and four-stage theories of anaerobic digestion.

1.2.1.1 Two-stage theory of anaerobic digestion

In the early 20th century, Thumn and Reichie (1914), Imhoff (1916) respectively discovered that anaerobic digestion include acid and alkaline fermentation stages. After that, Buswell and Neave (1930) verified and confirmed this idea. In later years, it was widely believed that anaerobic digestion consists of the above two stages. In the acid stage, macromolecular organic matters (carbohydrate, fat and protein) are decomposed into fatty acid, alcohol, CO_2 and H_2O by fermentative bacteria. Due to accumulation of fatty acids, pH of the mixed liquid decreases, so the process is called acid fermentation or an acidogenesis stage, and accordingly the microorganisms participating in the process are called fermentative bacteria or acidogenic bacteria. In the alkaline fermentation stage, methanogens further convert the metabolite produced in the first stage into CH_4 and CO_2. The organic acid is gradually consumed, and pH continues to increase, so this stage is called an alkaline fermentation stage or a methanogenesis stage.

1.2.1.2 Three-stage theory of anaerobic digestion

By continuously researching on anaerobic microorganisms, especially methanogenic microorganisms, researchers have found that methanogenic microorganisms can only utilize some simple-structured organisms such as acetic acid, formic acid, methanol, methylamine, H_2/CO_2 to produce methane rather than using fatty acid containing more than two carbon atoms. Therefore, it is hard to use the two-stage theory to explain how those heavier alcohols and organic acids are decomposed in the second stage.

Based on their research of anaerobic microorganisms and biochemical reaction of anaerobic digestions, Lawrence and McMarty (1967), Bryant (1979) respectively proposed the "Three-stage theory" of anaerobic digestion. This theory states that anaerobic digestions include three stages: the hydrolysis stage, the hydrogen-producing acetogenesis stage, and the methanogenesis stage (Fig. 1 1), each of which is achieved by different microorganisms. In this process, the complex organics such as carbohydrates, fats, and proteins are hydrolyzed into soluble matters that can be absorbed and decomposed by microorganisms into simpler sugar, fatty acid and amino acid. Next, fermentation bacteria degrade the soluble organic matters into micro molecules such as volatile acids and ethyl alcohol, which are converted into

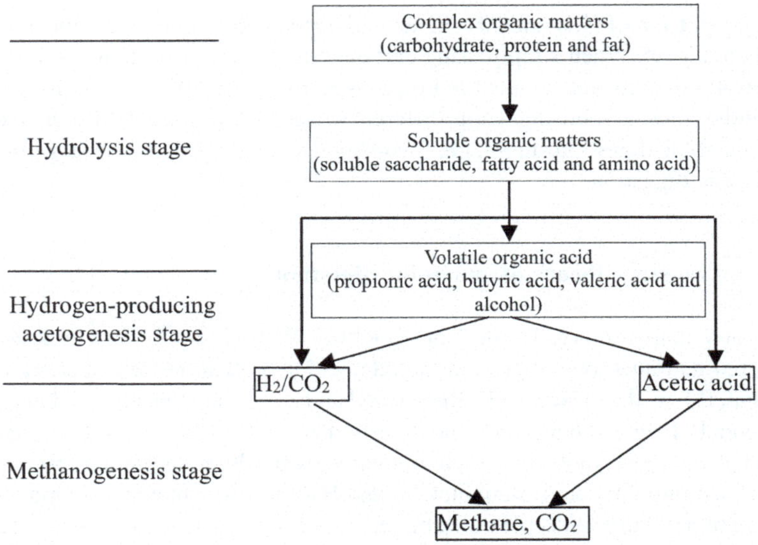

Fig. 1.1 Three stages of anaerobic digestion

smaller molecules (such as acetic acid, H_2, and CO_2) by hydrogen-producing acetogenic bacteria in the second stage. In the third stage, methanogens convert acetic acid, H_2/CO_2 into CH_4. Microorganisms participating in such process are divided into three categories, i.e., fermentative bacteria, hydrogen-producing acetogenic bacteria and methanogens.

1.2.1.3 Four-stage theory of anaerobic digestion

After the three-stage theory of anaerobic digestion was proposed, Zeikus (1979) found that during anaerobic digestion, there are also some microorganisms that transform H_2 and CO_2 into acetic acid and can be used by methanogens. These microorganisms are called homoacetogenic bacteria and the process is called homoacetogenesis (Fig. 1.2). Based on these researches, Zeikus (1979) proposed a four-stage theory of anaerobic digestion.

It is believed that anaerobic digestion is a complicated biochemical process in the four-stage theory. Based on different substrates and products of degradation, anaerobic digestion is composed of a hydrolysis stage, an acidogenesis stage, a H_2-producing acetogenesis stage and a methanogenesis stage (Fig. 1.2), each of which is accomplished by different groups of microorganisms. To some extent, there is syntrophic and symbiotic relationship among these different microorganisms which have different requirements for growth.

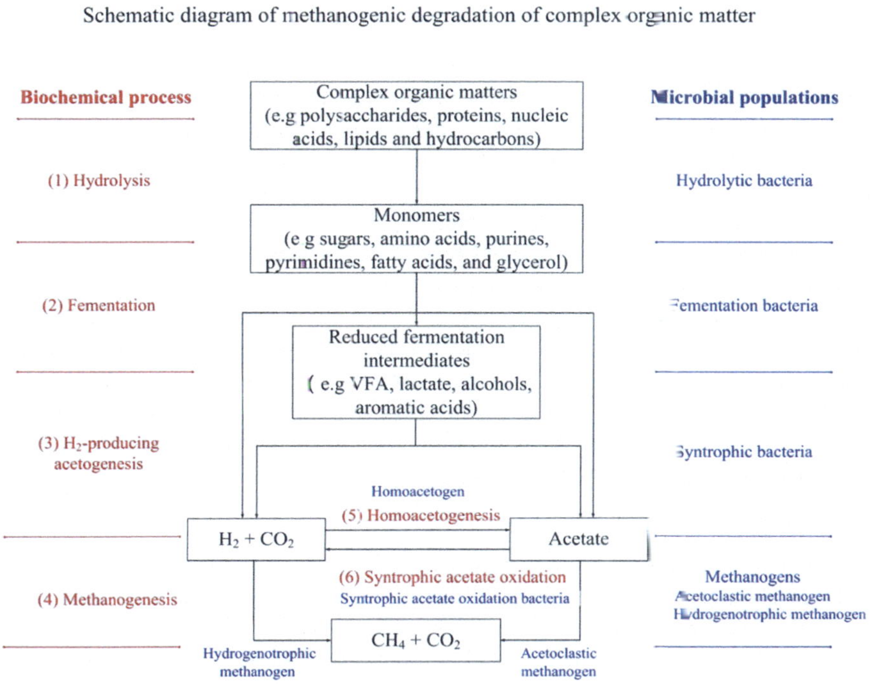

Fig. 1.2 Four stages of anaerobic digestion

(1) Hydrolysis stage

In the hydrolysis stage, insoluble macromolecular compounds such as carbohydrates, fats and proteins cannot penetrate through cell membranes. Therefore, they need to be hydrolyzed into soluble compounds such as soluble sugar, fatty acid, glycerinum and amino acid by catalysis of extracellular enzymes (such as protease, cellulose, hemicellulose, amylase and lipase) secreted by hydrolytic bacteria.

In this reaction, the covalent bonds of macromolecular substance are broken. Generally, the reaction can be expressed by the following equation:

$$R - X + H_2O \rightarrow R - OH + X^- + H^+ \tag{1.5}$$

The time hydrolysis takes varies with different feedstocks. Hydrolysis of carbohydrates can be completed within a few minutes while that of proteins and fats take several days. In this process, facultative anaerobic microorganisms will consume oxygen dissolved in water and decrease oxidation-reduction potential in anaerobic reactors to provide favorable growing conditions for obligate anaerobic microorganisms.

(2) Acidogenesis stage

In the second stage of anaerobic digestion, which is also called the acidogenesis stage or acidification stage, fermentative bacteria use hydrolysates as substrates to further convert the substrates into acetic acid, propionic acid, butyric acid and micro molecules such as ethanol, H_2 and CO_2. Category of substrates and microorganisms participating in acidification and concentration of H^+ produced will influence the type of products during acidogenesis stage.

Acidogenesis process is completed by a large number of various fermentative bacteria. For example, *Acetobacter* degrade fatty acids through β-oxidation. During degradation, fatty acids firstly combine with coenzyme A and then get gradually oxidized. Two carbon atoms broken in β-oxidation process are released as acetate. Amino acid is degraded by *Clostridium botulinum* through Stickland reaction. *Clostridium botulinum* absorbs 2 amino acid molecules (one as hydrogen donor and the other as hydrogen acceptor) simultaneously to produce acetic acid, ammonia and CO_2. H_2S is produced during degradation of cysteine.

(3) Hydrogen-producing acetogenesis stage

Intermediate metabolites, such as propionic acid, butyric acid and lactic acid produced in acidogenesis stage can be utilized by mathenogenic archaea only after being converted into simpler micromolecules. The acetic acid producing process is called acidogenesis. Part of acetic acid is produced by hydrogen-producing acetogenic bacteria, another part by bacteria that can both utilize organic acid, and H_2 and CO_2. Such bacteria are called homoacetogenic bacteria. Acetic acid production from several substrates are listed in Table 1.1.

1 Acetic acid produced through fermentation

During this process, hydrogen-producing acetogenic bacteria use the product from the acidogenesis stage as substrates to produce acetic acid. Short chain fatty acids such as propionic acid and butyric acid, together with ethanol, are converted into acetic acid, H_2 and CO_2 by catalysis of hydrogen-producing acetogenic bacteria. Methanogens can then use these simple C_1 and C_2 compounds to produce methane.

Table 1.1 Acetic acid production and free energy change of some substrates under standard conditions (Dolfing 1988; Khanal 2008)

Substrate	Reaction	$\Delta G^{0\prime}$(kJ/mol)
Ethanol	$CH_3CH_2OH + H_2O \rightarrow CH_3COO^- + H^+ + 2H_2$	+9.6
Propionic acid	$CH_3CH_2COO^- + 2H_2O \rightarrow CH_3COO^- + HCO_3^- + H^+ + 3H_2$	+76.1
Butyric acid	$CH_3CH_2CH_2COO^- + 2H_2O \rightarrow 2CH_3COO^- + H^+ + 2H_2$	+48.1
Benzoic acid	$\mathbf{C_6H_5COO^- + 7H_2O \rightarrow 3CH_3COO^- + 3H^+ + HCO_3^- + 3H_2}$	**+53**

25°C; H_2 is in gas state and others are in liquid state

Hydrogen and acetate production is an endoergic reaction. For example, the oxidation and degradation of propionic acid need to absorb 76.1 kJ/mol of energy and those reactions of ethanol need 9.6 kJ/mol of energy (Table 1.1). Hydrogen partial pressure needs to be maintained at a very low level in the acidogenesis stage or the reaction cannot proceed. This is due to the fact that the reaction is possible only under extremely low hydrogen partial pressure (PH_2) from the viewpoint of thermodynamics, and with a low hydrogen partial pressure, hydrogen-producing acetogenic bacteria can get necessary energy for growth and reproduction. If H_2 is not consumed, continuous accumulation of H_2 is not favorable for the growth of hydrogen-producing acetogenic bacteria and production of hydrogen and acetic acid. Therefore, hydrogen-producing acetogenic bacteria must live with hydrogen consuming microorganisms (such as hydrogenotrophic methanogens) to grow well.

2 Homoacetogenesis

During hydrogen and acetic acid production, homoacetogenic bacteria can continuously utilize H_2 to reduce CO_2, produce acetic acid, and release energy (Eq. 1.6), and the acetic acid is further utilized by methanogens. The whole process can both reduce hydrogen partial pressure and provide substrates for methanogens.

$$2CO_2 + 4H_2 \leftrightarrow CH_3COOH + 2H_2O \ \Delta G^{0'} = -104.6(kJ/mol) \quad (1.6)$$

Nearly 30% of methane during anaerobic digestion is produced by reduction of CO_2 by H_2. But only 5–6% of methane is produced by reduction of CO_2 dissolved in water by H_2. This can be explained by "interspecies hydrogen transfer", that H_2 produced by hydrogen-producing acetogenic bacteria is directly transferred to methanogens without the process of hydrogen dissolving.

(4) Methanogenesis stage

Methane production takes place in the fourth and also the last stage of anaerobic digestion which is catalyzed by methanogens. Basically, the substrates that can be used by methanogens are the simplest C_1 and C_2 compounds. Metabolism of methane production is shown in Fig. 1.3.

The methane production process can normally proceed under syntrophism of hydrogen-producing acetogenic bacteria with hydrogenotrophic methanogens, but the process can be influenced under the syntrophism of hydrogen-producing acetogenic bacteria with other microorganisms utilizing hydrogen. Other microorganisms utilizing hydrogen such as sulfate reducing bacteria can compete with hydrogenotrophic methanogens for hydrogen, which may reduce the substrates that can be utilized by methanogens and influence the production of methane. In addition, H_2S produced by sulfate reducing bacteria can poison methanogens. Different methane production processes will produce different amounts of energy. Compared with reduction of CO_2 to produce methane, acetic acid fermentation can only release a few energy (Eqs. 1.7 and 1.8).

$$CH_3COOH \leftrightarrow CH_4 + CO_2 \Delta G^{0'} = -31 kJ/mol \quad (1.7)$$

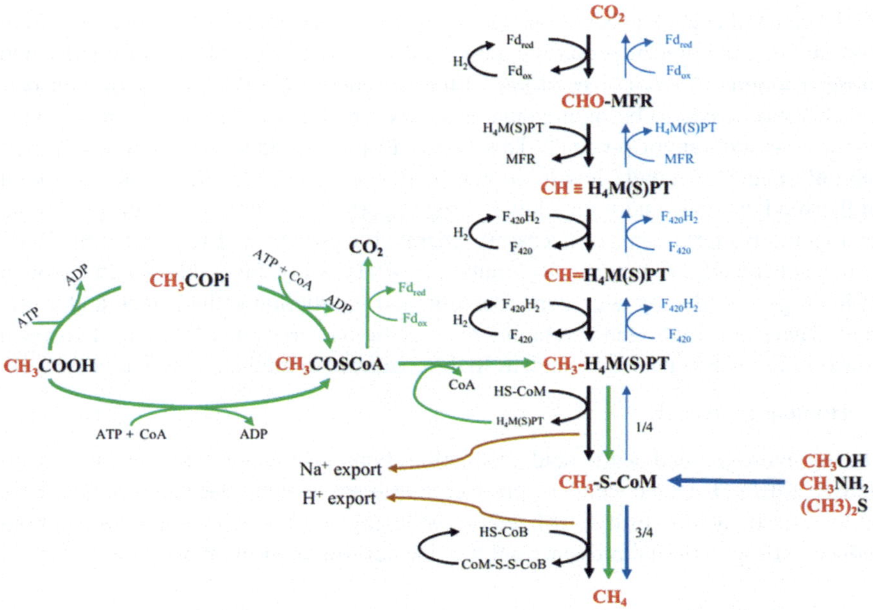

Fig. 1.3 Metabolism of methane production (Cheng and Zheng et al. 2016; Costa and Leigh. 2014)

$$CO_2 + 4NADH/H^+ \leftrightarrow CH_4 + 2H_2O + 4NAD^+ \Delta G^{0'} = -136kJ/mol \quad (1.8)$$

Based on thermodynamic condition of the above reaction, despite that reduction of CO_2 to produce methane can release more energy and is easier to proceed thermodynamically, 70% of methane is produced through acetate fermentation other than some produced by reduction of CO_2 (about 30%).

1.2.2 Microorganisms of anaerobic digestion

Anaerobic digestion is a biochemical reaction completed by cooperation of different microorganisms. Success of anaerobic digestion is closely related to such microorganisms. If any one category of these microorganisms is inhibited, the anaerobic digestion may fail, especially the inhibition of methanogens that has severer impacts. Methanogens can only utilize simple substrates such as acetic acid, methyl compound, H_2 and CO_2. They grow slowly and are relatively sensitive to external environment. If activity of methanogenic archaea is inhibited or the archaea dies, anaerobic digestion cannot proceed or even fail.

Researches on microorganisms of anaerobic digestion contribute to an in-depth understanding of anaerobic digestion as well as startup and operation of biogas plants.

Microorganisms of anaerobic digestion can be divided into non-menthanogenic bacteria and methanogen, or they can be divided into hydrolytic bacteria, acetogenic bacteria, homoacetogenic bacteria and methanogens by function (Fig. 1.2).

1.2.2.1 Fermentative bacteria

Fermentative bacteria mainly functions at the hydrolysis stage and the acidogenesis stage of anaerobic digestion. Extracellular enzyme produced by fermentative bacteria can hydrolyze insoluble organic matters into soluble organic matters such as polysaccharides, long chain fatty acids, and amino acids. Acidogenic bacteria further decompose the soluble organic matters into volatile organic acid (acetic acid, propionic acid and butyric acid), alcohol, ketone, CO_2, H_2, NH_3 and H_2S. Generally, fermentative bacteria convert the organic matters into pyruvic acid and then degrade them into different products as per different conditions. For example, the fermentative bacteria type, substrates, and environmental conditions (pH, hydrogen partial pressure and temperature) can influence the production of metabolites. Accumulation of metabolites can influence reactions, especially when production and accumulation of H_2 influence hydrogen and acetic acid production. If hydrogen pressure is too high, propionic acid and other organic acids will be produced, and acid accumulation or acid inhibition will occur. Therefore, it is essential to maintain the balance of metabolism between fermentative bacteria and other microorganisms.

Fermentative bacteria cannot completely degrade organic matters such as carbohydrates and fats, and byproducts such as organic acid, solvent, vitamins, hydrogen will be generated and utilized by other microorganisms. In anaerobic digestion, fermentative bacteria is an extremely complicated mixed bacteria group, which mainly belong to *Streptococcus*, *Enterobacteraceae*, *Fusobacterium*, *Bacteroides*, *Butyrivibrio*, *Bifidobacterium*, *Lactobacillus*, etc. The fermentative bacteria can be categorized into cellulose degradation bacteria, starch degradation bacteria, protein degradation bacteria, fat degradation bacteria, and so on by different degradation products. They can also be categorized into obligate and facultative anaerobic bacteria. They are mainly heterotrophic bacteria and have strong adaptability for environmental changes. But the dominant categories vary with environmental conditions and fermentation substrates.

1.2.2.2 Acetogens

Hydrogen-producing acetogenic bacteria were found by Bryant (1967), who found that the so-called "*Methanobacterium omelianskii*" is in fact comprised of two categories of bacteria, one fermenting ethanol to produce hydrogen while the other utilizing hydrogen to produce methane. He called this coupling of hydrogen production and utilization by all different types of microbes the "interspecies hydrogen transfer."

(1) Hydrogen-producing acetogenic bacteria

The major function of hydrogen-producing acetogenic bacteria is to decompose various short chain fatty acids (propionic acid, butyric acid, etc.), ethanol and specific aromatic compounds (benzoate) into acetic acid, H_2 and CO_2 and provide substrates for acetotrophic methanogens and hydrogenotrophic methanogens (Table 1.1).

Hydrogen and acetic acid production is an endoergic reaction (Table 1.1) and cannot proceed spontaneously but to proceed under extremely low hydrogen partial pressure. For example, propionic acid can be oxidized into acetic acid only under 10^{-4} atm of hydrogen partial pressure and the reaction of hydrogen-producing acetogenic bacteria can proceed. Hydrogen partial pressure for butyric acid oxidization must be lower than 10^{-3} atm. When acetogenic bacteria and hydrogenotrophic methanogens are co-cultured, the reaction can successfully proceed. Hydrogenotrophic methanogens can rapidly utilize hydrogen to maintain an extremely low level of hydrogen partial pressure. During the process, $\Delta G^{0\prime}$ is a negative value (Eq. 1.8) and it is an exergonic reaction, which provides favorable thermodynamics for hydrogen and acetic acid production bacteria growth while ensuring successful oxidization of propionic acid, butyric acid and ethanol. A syntrophic relationship exists between the two categories of microorganisms. In addition to hydrogenotrophic methanogens, sulfate reducing bacteria or homoacetogenic bacteria can also consume hydrogen during this process.

It needs to be mentioned that during anaerobic digestion, up to 30% electrons are used by propionic acid oxidization to produce acetic acid and hydrogen. Therefore, the oxidization of propionate seems to be more important than that of other organic acids. Accumulation of propionic acid under high temperature is more severe than that under low temperature. Structure of the anaerobic digester, provision of nutrition, characteristics of substrates, and microorganism categories have important influences on facilitation of propionate degradation. Most of hydrogen-producing acetogenic bacteria are microorganisms such as *Syntrophomonas*, *Syntrophobacter* and *Clostridium*.

(2) Homoacetogenic bacteria

Homoacetogenic bacteria can both utilize organic matters to produce acetic acid and convert H_2 and CO_2 into acetic acid. Regardless of the kind of substrate used, the product is always acetate, so the bacteria is called homoacetogenic bacteria. The homoacetogenic bacteria can reduce hydrogen partial pressure when H_2 and CO_2 is used to produce acetate, which not only contributes to continuous hydrogen and acetate production but also provides more substrates for methanogens. Both autotrophic and heterotrophic bacteria can perform homoacetogenesis. Autotrophic homoacetogenic bacteria use H_2 and CO_2 to produce acetic acid, and CO_2 is used as a carbon source for cell growth (Eq. 1.9). Some homoacetogenic bacteria can use CO as the carbon source (Eq. 1.10). On the other hand, in addition to using hydrogen to reduce CO_2, some homoacetogenic bacteria can also use other C_1 compound as carbon sources and electron donors of acetic acidogenesis, such as methanol, ethanol, methylated aromatic compounds, sugars, organic acids, amino acids and alcohols (Eqs. 1.11 and 1.12). Many homoacetogenic bacteria can also reduce nitrate (NO_3^-) and thiosulfate ($S_2O_3^{2-}$). The major process of homoacetogenesis, however, is reduction of CO_2 to acetate using acetyl coenzyme A.

$$2CO_2 + 4H_2 \rightarrow CH_3COOH + 2H_2O \tag{1.9}$$

$$4CO + 2H_2O \rightarrow CH_3COOH + 2CO_2 \tag{1.10}$$

$$4HCOOH \rightarrow CH_3COOH + 2CO_2 + 2H_2O \tag{1.11}$$

$$4CH_3OH + 2CO_2 \rightarrow 3CH_3COOH + 2H_2O \tag{1.12}$$

Clostridium aceticum and *Acetobacterium woodii* are two species of mesophilic homoacetogenic bacteria isolated from wastewater. Homoacetogenic bacteria have high thermodynamic efficiency and therefore will not cause accumulation of H_2 and CO_2 when using poly-carbon compound to grow and produce acetic acid. By comparing the Gibbs free energy of methane production by hydrogenotrophic methanogens (Eq. 1.8) and that of acetate production by homoacetogenic bacteria (Eq. 1.6), it is found that reduction of CO_2 to produce methane can release more energy (Eq. 1.8), which seems more likely to occur thermodynamically and can make hydrogenotrophic methanogens acquire more hydrogen as the electron donor. However, further research on the mechanism of competition between homoacetogenic bacteria and hydrogenotrophic methanogen is still needed.

During anaerobic digestion, there is a syntrophic relationship between hydrogen-producing acetogenic bacteria and hydrogenotrophic methanogens. The syntrophic relationship is essential for stability of anaerobic digestion. One of the indicators of this balance is the concentration of volatile fatty acids in the mixed liquid. Volatile fatty acids refer to short chain organic acids that are the intermediate products of degradation of complex organic matters. When the syntrophic relationship is destroyed (e.g. when feedstock is excessive, hazardous materials exist, nutrition is in shortage, or there is biomass loss), volatile acid will accumulate and its concentration will continuously increase, which may cause dramatic decrease of pH in the anaerobic digester and give rise to acidification. The anaerobic digestion is likely to terminate or fail eventually in case of no prompt adjustment.

1.2.2.3 Syntrophic bacteria

There are many examples of syntrophism in microbiology. Syntrophism means that some kinds of microorganisms cannot perform degradation by their own kinds but need cooperation with some other kinds of microorganism to achieve the goal. Syntrophism is generally a secondary fermentation action and uses the metabolites of

Table 1.2 Major syntrophic bacteria and their characteristics (Madigan et al. 2014)

Species	Quantity of known species	Phylogenetic analysis	Substrate used for co-culture
Syntrophobacter	4	δ-Proteobacteria	Propionic acid, lactic acid and some alcohols
Syntrophomonas	9	Firmicutes	C_{14}–C_{18} saturated or unsaturated fatty acids and some alcohols
Pelotomaculum	2	Firmicutes	Propionic acid, lactic acid, some alcohols and some aromatic compounds
Syntrophus	3	δ-Proteobacteria	benzoic acid, some aromatic compounds, some fatty acids and alcohol

other microorganisms as growing substrate. During methane production, the syntrophic metabolism is sometimes a main process. Major categories of syntrophic bacteria and the substrates used for degradation are listed in Table 1.2. All syntrophic bacteria are obligate anaerobic bacteria. Many organic compounds can be degraded by syntrophism, including aromatic compounds and aliphatic hydrocarbons, but the major compounds used in syntrophic actions are fatty acids and alcohols.

During anaerobic digestion, the core of syntrophism is interspecies hydrogen transfer, which means that the hydrogen produced by one microorganism is used by another microorganism through syntrophism, and a syntrophic relationship is said to exist between these two groups of microorganisms. Hydrogen consuming microorganisms can also be classified into different categories including denitrifying bacteria, iron bacteria, sulfate reducing bacteria, homoacetogenic bacteria and hydrogenotrophic methanogens. In case of syntrophic behavior between hydrogen-producing acetogenic bacteria and hydrogenotrophic methanogens, hydrogen-producing acetogenic bacteria provide hydrogen for methanogens, and in return the hydrogenotrophic methanogens can consume hydrogen for successful hydrogen-producing acetogenesis. For example, *Pelotomaculum* ferments alcohol to produce acetic acid and hydrogen. Through co-culture of methanogens, *Pelotomaculum* uses hydrogen as the electron donor to reduce CO_2 and produce methane. It can be known from the equation below that the standard free energy change ($\Delta G^{0\prime}$) of ethanol fermentation is a positive value (Eq. 1.12) and therefore the fermentation alone cannot proceed thermodynamically.

Ethanol fermentation:

$$2CH_3CH_2OH + 2H_2O \rightarrow 4H_2 + 2CH_3COO^- + 2H^+ \Delta G^{0\prime} = +19.4\,\text{kJ/mol} \tag{1.13}$$

Methane production:

$$4H_2 + CO_2 \rightarrow CH_4 + 2H_2O \; \Delta G^{0'} = -130.7 kJ/mol \qquad (1.14)$$

Coupling reaction:

$$2CH_3CH_2OH + CO_2 \rightarrow 2CH_3COO^- + CH_4 + 2H^+ \Delta G^{0'} = -111.3 kJ/mol \qquad (1.15)$$

Under standard conditions, *Pelotomaculum* cannot use ethanol to grow, but when hydrogen produced by *Pelotomaculum* is used by hydrogenotrophic methanogens, the reaction is exergonic (Eq. 1.14) and the hydrogen will be consumed rapidly. When the two reactions couple, $\Delta G^{0'}$ of the entire reaction turns into a negative value (Eq. 1.15) and the reaction can thermodynamically. That means that syntrophic metabolism needs the participation of two groups of microorganism so as to form a close cooperation relationship through energy distribution and substance transfer. Table 1.3 shows the syntrophic relationship between some hydrogen-producing acetogenic bacteria and methanogens.

From the perspective of ecology, syntrophism is key to carbon circulation in anaerobic digestion. In anaerobic digestion, syntrophic bacteria can consume a large number of reductive substrate and produce hydrogen. There are two categories of obligate anaerobic microorganisms using CO_2 as electron acceptors and store energy within. One is homoacetogenic bacteria and the other is methanogens. They both use H_2 as the major electron donor. In the meantime, the number of electron acceptors is limited in anaerobic condition, which (in addition to CO_2) can inhibit anaerobic digestion without syntrophic bacteria. In contrast, the syntrophic relationship will be of little importance under aerobic condition or when there is enough electron acceptors. Therefore, syntrophic relationship is the characteristic reaction of anaerobic metabolism. In an anaerobic system, methanogenesis or acidogenesis is the major final reaction and there is a close connection between hydrogen-producing and hydrogen-consuming microorganisms in the environment.

1.2.2.4 Methanogens

Methanogens are obligate anaerobic archaea. Researches on methanogen did not receive much progress in a long time before Hungate invented anaerobic microorganism culture technique. The name of methanogens was proposed by Bryant in 1974 to differentiate it with methane-oxidizing bacteria. Methanogen contains coenzyme F_{420}, the key cofactor participating in methane metabolism. F_{420} in oxidization state can absorb UV-light of 420 nm and can stimulate blue-green fluorescent light of 470 nm. Methanogens and non-menthanogens can be differentiated under fluorescent/visible light microscope using the spectroscopy characteristics. Methanogens can be divided into three categories: hydrogenotrophic

Table 1.3 Syntrophic relationship between hydrogen-producing acetogenic bacteria and hydrogenotrophic methanogens (Zhang et al. 2016)

Substrate	Syntrophic microorganism	Reaction	$\Delta G^{0\prime}$(kJ/mol)
Ethanol	"S" bacteria	$2CH_3CH_2OH + 2H_2O \rightarrow 2CH_3COO^- + 2H^+ + 4H_2$	+19.3
	Methanogens	$4H_2 + HCO_3^- + H^+ \rightarrow 3H_2O + CH_4$	−135.6
	Overall reaction	$2CH_3CH_2OH + HCO_3^- \rightarrow 2CH_3COO^- + H^+ + H_2O + CH_4$	+116.3
Acetic acid	*Clostridium aceticum*	$CH_3COO^- + 4H_2O \rightarrow 4H_2 + 2HCO_3^- + H^+$	+104.6
	Methanogens	$4H_2 + HCO_3^- + H^+ \rightarrow 3H_2O + CH_4$	−135.6
	Overall reaction	$CH_3COO^- + H_2O \rightarrow HCO_3^- + CH_4$	−31
Propionic acid	*Syntrophobacter wolinii*	$4CH_3CH_2COO^- + 12H_2O \rightarrow 4CH_3COO^- + 4HCO_3^- + 4H^+ + 12H_2$	+304.6
	Methanogens	$3HCO_3^- + 3H^+ + 12H_2 \rightarrow 9H_2O + 3CH_4$	−406.6
	Overall reaction	$4CH_3CH_2OO^- + 3H_2O \rightarrow 4CH_3COO^- + HCO_3^- + H^+ + 3CH_4$	−102
Butyric acid	Syntrophomonas wolfei	$2CH_3(CH_2)_2COO^- + 4H_2O \rightarrow 4CH_3COO^- + 2H^+ + 4H_2$	+96.2
	Methanogens	$4H_2 + HCO_3^- + H^+ \rightarrow 3H_2O + CH_4$	−135.6
	Overall reaction	$2CH_3(CH_2)_2COO^- + HCO_3^- + H_2O \rightarrow 4CH_3COO^- + H^+ + CH_4$	−39.4
Benzoic acid	*Syntrophus buswelii*	$4C_6H_5COO^- + 28H_2O \rightarrow 12CH_3COO^- + 4HCO_3^- + 12H^+ + 12H_2$	+359
	Methanogens	$3HCO_3^- + 3H^+ + 12H_2 \rightarrow 9H_2O + 3CH_4$	−406.6
	Overall reaction	$4C_6H_5COO^- + 19H_2O \rightarrow 12CH_3COO^- + HCO_3^- + 9H^+ + 3CH_4$	−47.6

Table 1.4 Type of methanogens and methane production reaction (Oremland 1988; Khanal 2008)

Nutrition type	Methane production reaction
Hydrogenotrophic methanogens	(A) $4H_2 + CO_2 \rightarrow CH_4 + 2H_2O$
	(B) $4HCOO^- + 2H^+ \rightarrow CH_4 + CO_2 + 2HCO_3^-$
Acetotrophic methanogens	(C) $CH_3COO^- + H_2O \rightarrow CH_4 + HCO_3^-$
Methylotrophic methanogens	(D) $4CH_3OH \rightarrow 3CH_4 + CO_2 + 2H_2O$
	(E) $4CH_3NH_2 + 2H_2O \rightarrow 3CH_4 + CO_2 + 4NH_3$
	(F) $(CH_3)_2S + H_2O \rightarrow 1.5CH_4 + 0.5CO_2 + H_2S$

methanogens, acetoctrophic methanogens and methylotrophic methanogens. Correspondingly, there are three ways to produce methane, i.e., CO_2 reduction, methyl cracking and acetate fermentation (Fig. 1.3 and Table 1.4).

A large number of methanogens exists in anaerobic environment containing rich organic matters, such as marsh, wetland, pond, bottom of lake, bottom sediments in the sea, anaerobic digesters and intestine of animals. Though a few methanogens can live under extreme temperature (up to 122 °C and down to below 0 °C), most of them are mesophilic microorganisms.

(1) Hydrogenotrophic methanogens

Hydrogenotrophic methanogens use H_2 as electron donors to reduce CO_2 to CH_4 (Table 1.4). Hydrogen is produced by other microorganisms, such as hydrogen fermentative bacteria, especially *Clostridia* and hydrogen-producing acetogenic bacteria. In type strain of methanogens, more than 3/4 can grow using H_2 and CO_2, such as *Methanobacter* and *Methanoculleus* that are common in biogas plants. Hydrogenotrophic methanogens can also use formate as the electron donor to reduce CO_2 to CH_4 (reaction B, Table 1.4). A few methanogens can also obtain electron donors through oxidizing secondary alcohol or obtain electron donors from pyruvate to reduce CO_2 to CH_4. For example, *Methanofollis ethanolicus* can directly use ethanol, 2-propyl alcohol or 2-butanol to produce CH_4. Some *Methanococcus* spp. can use pyruvate as the electron donor to reduce CO_2 to CH_4. A few hydrogenotrophic methanogens can also grow with CO as the substrate, although the growth rate of methanogens is very slow in this case, and this circumstance is not common in nature. For example, *Methanothermobacter thermoautotrophicus* can use CO of low concentration (< 60%) to produce CH_4.

(2) Acetotrophic methanogens

The way acetotrophic methanogens produce methane is the main process of methane production process. Acetic acid is converted into CH_4 and CO_2 (reaction C, Table 1.4), and 70% CH_4 is produced by this pathway. *Methanosarcina* and *Methanothrix* (used to be called *Methanosaeta*) are two important acetoctrophic methanogens. *Methanosarcina* forms a sarcina shape when growing in order to use multiple substrates including methanol (reaction D, Table 1.4), methylamine (reaction E, Table 1.4), H_2/CO_2 (reaction A, Table 1.4) and pyruvic acid to produce methane. For example, *Methanosarcina barkeri* can use pyruvate as carbon and energy sources to produce CH_4 (López et al. 2015). *M. barkeri* can use CO (100%) and methanol (50 mM) to grow and can also grow in CO of concentration below 50% to produce methane and hydrogen. In addition to using acetate, *Methanosaeta harudinacea* can directly use electrons to reduce CO_2 to produce CH_4. The typical multiplication time taken by *Methanosarcina* is 1–2 day (s) when acetate is used for its growth, while the multiplication time of *Methanosaeta* is 4–9 days. If hydraulic retention time (HRT) of a continuous stirred-tank reactor is short, generally *Methanosarcina* will become dominant in the digester.

(3) Methylotrophic methanogens

Methylotrophic methanogens can use compounds containing methyl group to produce methane, such as methanol (reaction D, Table 1.4), methylamine compounds (including methylamine, dimethylamine and trimethylamine) (reaction E, Table 1.4), and methyl sulfide (including methyl mercaptan and dimethylsulfide) (reaction F, Table 1.4). Methylotrophic methanogens mainly exist in *Methanosarcinaceae*, *Methanomassiliicoccus* and *Methanosphaera*. In methylotrophic methane production, methyl group is converted to methyl carrier and reduced to methane. Methylotrophic methanogens can obtain electrons to reduce CO_2 through oxidizing a part of methyl groups or use H_2 as electron donors. However, some researchers have found that *Methanococcoides* spp. in *Methanomassiliicoccus* can also use methyl compounds such as N, N-dimethylethanolamine, choline and glycine betaine to grow and produce CH_4 (L'Haridon et al. 2014). Methyl mercaptan is used as the precursor of methane in liquid phase but no methane is produced in pure culture.

(4) Phyletic classification of methanogenic archaea

Microorganisms were classified by morphology and physiological and biochemical characteristics in the past, so methanogens are classified as bacteria. However, with the development of microbial diversity, traditional methods for microorganism classification become increasingly limited, and some even believe that a scientific microorganism classification is impossible. As the emergence and development of molecular biochemical technique, Phyletic classification has gradually become the standard of microorganism classification. In 1977, Woese et al. proposed the phyletic classification method based on homology of nucleotide sequences of 16S rRNA and 18S rRNA to classify organisms into three types ("Three Domains Theory"). Based on this method, methanogens are different from bacteria and eucaryon and belong to a unique branch that is named as archaea, in which methanogens are included.

Up to now, there have been 6 valid orders of methanogens published, i.e., *Methanobacteriales*, *Methanococcales*, *Methanocellales*, *Methanomicrobiales*, *Methanopyrales* and *Methanosarcinales*. They are classified into 15 families, 35 genus and more than 150 valid species (Table 1.5). The *Thermoplasmata* that are discovered recently represent the 7th order of methanogens, and Iino et al (2013) renamed it as *Methanomassiliicoccales* as per bacteria nomenclature. All

Table 1.5 Phylogenic classification of methanogens (Liu and Whitman 2008; Cheng and Zheng et al. 2016)

Order	Family	Genus	Number of valid species	Separation resource
Methanobacteriales	Methanobacteriaceae	*Methanobacterium*	24	Anaerobic digesters, rice fields, tundra, fresh water and sentiments in the sea
		Methanobrevibacter	15	Excrement of mammal, intestine of termite and anaerobic digesters
		Methanosphaera	2	Human excrement and intestine of rabbit
		Methanothermobacter	3	Anaerobic digesters and oil fields
	Methanothermaceae	*Methanothermus*	2	Sulfur-containing hot springs and mud
Methanococcales	Methanocaldococcaceae	*Methanocaldococcus*	7	Hot liquid and sediments in deep sea
		Methanotorris	2	Hot liquid and sediments in deep sea
	Methanococcaceae	*Methanococcus*	4	Sediments and mud at seabed
		Methanothermococcus	2	Sediments at seabed and hot sea water
Methanocellales	Methanocellaceae	*Methanocella*	3	Rice soils
Methanomicrobiales	Methanocalculaceae	*Methanocalculus*	6	Oil field, salt lakes, maricultural farm, and waste plant
	Methanocorpusculaceae	*Methanocorpusculum*	4	Anaerobic reactors and sediments
	Methanomicrobiaceae	*Methanoculleus*	11	Oil fields, deep sediment, wetlands and anaerobic reactors
		Methanofollis	5	Aquafarms, waste processing plants, lotus ponds and hot sulfur liquid mud

(continued)

Table 1.5 (continued)

Order	Family	Genus	Number of valid species	Separation resource
		Methanogenium	4	Sediments in the sea and lakes in Antarctic
		Methanolacinia	2	Oil fields and sediments in the sea
		Methanomicrobium	1	Rumen of cow
		Methanoplanus	2	Drilling mud and ciliate in the sea
	Methanoregulaceae	*Methanolinea*	2	Anaerobic reactors, rice fields, tundra, fresh water and sentiments in the sea
		Methanoregula	2	Peat bog and anaerobic reactors
		Methanosphaerula	1	Mineral peat bog
	Methanospirillaceae	*Methanospirillum*	4	Anaerobic reactors, sewage sludge and wetlands
Methanosarcinales	Methanosarcinaceae	*Halomethanococcus*	1	Salt lakes
		Methanimicrococcus	1	Intestine of cockroach
		Methanococcoides	4	Sediments in the sea and lake in Antarctic
		Methanohalobium	1	Sediments in the sea
		Methanohalophilus	4	Sediments, blue-green algae meadow, deep underground water and salt marsh
		Methanolobus	7	Sediments in the sea and fresh water, water of coal bed and deep underground water
		Methanomethylovorans	3	Anaerobic reactors, sediments in the lake and wetland
		Methanosalsum	2	Salt lakes and alkaline water lakes

(continued)

Table 1.5 (continued)

Order	Family	Genus	Number of valid species	Separation resource
		Methanosarcina	13	Anaerobic reactors, deep underground water, sediments in the sea, and fresh water and tundra
	Methanotrichaceae	*Methanothrix*	2	Anaerobic reactors and sediments at lakebed
	Methermicoccaceae	*Methermicoccus*	1	Oil fields
Methanopyrales	Methanopyraceae	*Methanopyrus*	1	Sediments in hot springs at seabed
Methanomassiliicoccales	Methanomassiliicoccaceae	*Methanomassiliicoccus*	1	Human excrement

methanogens that are currently obtained through separation belong to Euryarchaeota, but the latest research shows that methanogens may also exist in other branches including Bathyarchaeota.

1.3 Factors influencing anaerobic digestion

Anaerobic digestion of organic matters can be influenced by multiple factors, which can be divided into design operation factors and environment factors, the latter of which mainly influence anaerobic digestion microorganisms. Any factor that may influence the growth of microorganisms can influence normal operations of anaerobic digestion. Metabolism of microorganisms relies on many factors. Nutritional requirements, substrates, temperature, pH and oxygen will influence anaerobic digestion. In addition, the environmental requirements of fermentative bacteria and methanogens are different, so different needs of the two categaries microbes at different digestion stages should be considered.

If the entire anaerobic digestion (hydrolysis/acidification stage and hydrogen-producing acetogenesis/methanogenesis stage) takes place in a single reaction system (single phase anaerobic digestion), the environment for growth and the conditions of methanogens should be preferentially considered, for the growth rate of methanogens is slower and they are more sensitive to the environment. Otherwise, the methanogens may lose activity or die, causing failure of the entire anaerobic digestion.

1.3.1 Oxidation-reduction potential

Redox potential indicates the relative strength of oxidizing and reducing agents, and in anaerobic digestion it can be influenced by dissolved oxygen content as well as oxidization and reduction characteristics of intermediate metabolites. For different microorganisms, the optimal oxidization and reduction characteristics vary. Anaerobic digestion microorganisms are mainly obligate or facultative anaerobic microbes, the former of which have no superoxide dismutase or catalase and are therefore sensitive to oxidization or oxides, and they cannot grow or even survive under high redox potential. Therefore, the redox potential of anaerobic digestion needs to be maintained at a relatively low level. Relevant researches show that the oxidization reduction potential needs to be maintained at −300 to +400 mV during hydrolysis and acidification and lower than −250 mV during methanogenesis. The optimal oxidization reduction potential is −330 to −300 mV when methanogens are cultured alone. During anaerobic digestion, the entrance of oxides such as oxygen may inhibit the growth of microorganisms, resulting to their death and further causing extension of the startup stage or failure. Therefore, entrance of oxides should be prevented during anaerobic digestion.

1.3.2 Temperature

Temperature is an important factor that can remarkably influence the growth and metabolism of microorganisms. During anaerobic digestion, different categories of microorganisms favor different temperatures. For acidogenic bacteria, the optimal temperature for mesophilic bacteria is 25–35°C; for methanogens, the optimal temperature for mesophilic bacteria is 32–42°C and that for thermophilic bacteria is 48–55°C. During methanogenesis, most methanogens are mesophilic and only a few are thermophilic microorganisms. But some methanogens can also grow under low temperature (0.6–1.2°C). Compared with mesophilic methanogens, thermophilic methanogens are more sensitive to temperature changes and will lose activity significantly with temperature variations. Researches show that temperature changes during anaerobic digestion should not exceed 2°C, otherwise the biogas production will decrease. Compared with low temperatures, high temperatures can cause more troubles to microorganisms, for high temperatures will inactivate enzymes of microorganisms and cause irreversible damage.

However, in thermophilic anaerobic digestion, metabolism of microorganisms is more vigorous. The degradation rate of substrates can increase by 50% on average, especially in reactors having fat-containing feedstocks. In this case, more usable materials are provided and biogas productivity increases. Anaerobic digestion in thermophilic condition can inactivate the epidemic germs and pathogenic microorganisms. When temperature is above 55°C and retention duration of feedstocks is more than 23 h, additional hygienic and disease prevention measures are no longer necessary. In many two-phase anaerobic digestion plants, different temperatures can be applied to different anaerobic digestion stages to achieve an optimized fermentation effect. Undoubtedly, a feasible measure to increase anaerobic digestion efficiency is to adopt mesophilic fermentation in the hydrolysis stage and thermophilic fermentation in the methanogenesis stage. But it is not absolutely true for different feedstocks, and temperature of the hydrolysis stage can also be higher than that of the methanogenesis stage.

1.3.3 pH

pH is an important parameter influencing growth and metabolism of microorganisms. A suitable pH is a necessary premise for growth of microorganisms. When pH is not within a suitable range, microorganisms cannot grow well. Therefore, a specific pH should be maintained, and excessively high or low pHs will influence the activity of microorganisms and cause cell deactivation or death, which is detrimental to a normal anaerobic digestion.

Different microorganisms need different optimal pHs. Generally, acidogenic bacteria can adapt to a wide range of pH while methanogens can only adapt to a narrow range. Acidogenic bacteria can grow with a pH of 4.5–8.0. For methanogens, the optimal pH is 6.7–7.5, which is close to neutrality. Only *Methanosarcina* can grow well with a pH ≤ 6.5. Therefore, pH of hydrogen-producing acetogenesis and methanogenesis stages is higher than that of hydrolysis and acidification stages and it is generally maintained at around 6.5–7.5 in a single phase anaerobic digestion.

Organic acid produced by carbohydrates will decrease the pH of the mixed liquid. Carbohydrates are more prone to acidification and the hydrogen partial pressure is easier to increase along with the production of intermediate metabolites. In addition, some CO_2 generated during anaerobic digestion is discharged as a part of biogas and the rest is dissolved in liquid, increasing the concentration of H^- and decreasing the pH.

$$CO_2 + H_2O \leftrightarrow H_2CO_3 \quad (1.16)$$

$$H_2CO_3 \leftrightarrow H^+ + HCO_3^- \quad (1.17)$$

$$HCO_3^- \leftrightarrow H^+ + CO_3^{2-} \quad (1.18)$$

However, during degradation of nitrogen-containing compounds, the pH will increase; for example, ammonia will be generated in protein degradation.

$$NH_3 + H_2O \leftrightarrow NH_4^+ + OH^- \quad (1.19)$$

$$NH_3 + H^+ \leftrightarrow NH_4^+ \quad (1.20)$$

The above two buffer systems can basically maintain the pH within the optimal scope and avoid excessive acidification or alkalization, but this is related to the characteristics of feedstock.

Generally, attention should be paid to any decrease of pH. Volatile fatty acids such as acetic acid, propionic acid, butyric acid, and dissolved carbonic acid can decrease the pH. Decrease of pH or increase of concentration of CO_2 in fermentation broth indicates acidification of digestion. The acidification can be solved by stopping feeding, decreasing loading rate, adding alkaline and reseeding sludge.

1.3.4 Hydrogen partial pressure

Hydrogen will be generated in both acidogenesis and acetate production stages. Because hydrogen and acetic acid production takes place only under extremely low hydrogen partial pressure, the existence and accumulation of hydrogen can seriously influence acidogenesis and acetate production. Slow degradation and accumulation of volatile fatty acids such as propionic acid and butyric acid is resulted, causing pH decreases, acidification, and possible termination of digestion. In the growth of anaerobic microorganisms, tolerable hydrogen partial pressure is related to the microorganism and substrate types. In terms of conversion of benzoate, the range of suitable hydrogen partial pressures is narrow no matter the acetate process or the H_2/CO_2 process is used to generate methane. The degradation of propionic acid can be used as a measurable standard of biogas-producing efficiency, because it is often the limiting factor of digestion in practice.

Interspecies hydrogen transfer can greatly balance the concentration of hydrogen. On one hand, methanogens need enough hydrogen as the reducing agent to reduce CO_2 to methane. On the other hand, hydrogen partial pressure must be very low to ensure acetate production from oxidation of organic matters such as propionate and butyrate. In order to maintain the interspecies hydrogen transfer and the syntrophic relationship between hydrogen-producing acetogen and hydrogenotrophic methanogens, the physical distance between these substances must be very short.

1.3.5 Types of substrates

For the growth of microorganisms, macro-elements (nitrogen and phosphorus) and micro-elements (micro mineral substance) are essential. In addition to nitrogen and phosphorus, many micro-elements are necessary for anaerobic microorganisms, including cobalt, nickel, molybdenum etc. Nickel is particularly important to methanogens, because it is a structural component of F_{430} factor that only exists in methanogens. Cobalt is also important to methanogens, because it is a structural component of vitamin B_{12} which can promote methane production.

The type of substrates determines the rate of degradation, so the influence of substrates must be considered in anaerobic digestion and process control. For anaerobic digestion, microorganisms can stop growing if an essential substrate is completely consumed with no refill. Therefore, continuous addition of feedstock (carbohydrate, fat and protein) and, minerals and micro-elements that are necessary for microorganism growth is important for normal operation of anaerobic digestion.

Sugars such as starch and monosaccharides can be hydrolyzed in a very short time and is easy to be acidified while the degradation of cellulose is extremely slow. Metabolites produced during hydrolytic processes will also limit or inhibit degradation. For example, the accumulation of fatty acids will cause a decrease of pH and influence the activity of microorganisms. Hydrogen will influence the oxidization rate of volatile fatty acids such as propionate and butyrate. Free ammonia and H_2S produced in protein degradation will inhibit methanogenesis.

1.3.6 Inhibiting substances

During anaerobic digestion, some feedstocks contain poisonous and hazardous substances that decrease the anaerobic digestion efficiency and even cause failure, especially when processing industrial organic feedstocks. Attention should be paid to avoid intoxication by poisonous and hazardous substances on microorganisms. The most common inhibiting substances are ammoniacal nitrogen (NH_3 and NH_4^+), sulfides, salts, heavy metals and some artificially synthetic substances.

(1) NH_3 and NH_4^+

Ammoniacal nitrogen in anaerobic digestion, including NH_3 and NH_4^+, comes from degradation of organic compounds containing nitrogen. Free ammonia (NH_3) of concentration above 80 mg/L can inhibit anaerobic digestion, and when the concentration is above 150 mg/L, the digestion process will be poisoned. Free ammonia is the most important form of ammoniacal nitrogen that poisons methanogens, because free ammonia can penetrate through cell membrane. Methanogens can use NH_4^+ as a nitrogen resource. NH_4^+–N of low concentration (50–100 mg/L) is helpful for anaerobic digestion, but a concentration up to 1500–3000 mg/L can inhibit anaerobic digestion and that above 3000 mg/L can poison microorganisms. It is also reported, however, that no intoxication of microorganisms was caused by NH_4^+ around 7000–9000 mg/L. pH has a significant influence on concentration of NH_3 and NH_4^+. As pH increases, concentration of NH_3 will also increase, so does the inhibition and toxicity for anaerobic digestion. Temperature also has important influence on formation of ammonia. When temperature increases, more NH_3 are produced. Therefore, increasing the temperature can enhance ammonia inhibition.

(2) Sulfides

Sulfur is a macro-element of bacteria cells and plays an important role in cell synthesis. Adequate sulfur in feedstock can promote the growth of bacteria, but excessive sulfur will inhibit anaerobic digestion. Sulfides including H_2S will be produced in anaerobic digestion of tannery wastewater, petrochemical refining wastewater, citric acid production wastewater and protein-rich wastewater. During acidogenesis, sulfides below 300 mg/L will not influence the acid production process, while sulfides in methanogenesis of above 80 mg/L will inhibit the methanogenesis process and of around 150–200 mg/L will cause remarkable inhibition. Sulfides are produced by reduction of sulfate under anaerobic condition. So other forms of compounds of sulfur can also inhibit anaerobic digestion, one such example is SO_4^{2-} above 5000 mg/L. Among sulfides, non-ionized H_2S is more poisonous to methanogens than ionized HS^-. Although sulfides will poison microorganisms, the microorganisms can also adapt to sulfides of different concentration after domestication and can survive under circumstances of $Na_2S<600$ mg/L and $H_2S<1000$ mg/L.

(3) Salts

High-concentration salts exist in kitchen wastes and food or industrial wastewater such as pickle processing and casings processing wastewater. When concentration is extremely high, salts can be poisonous. The toxicity of salts to microorganisms mainly comes from positive ions, and the influence of negative ion is relatively small. Monovalent ions (Na^+ and K^+) are less poisonous than divalent ions (Ca^{2+} and Mg^{2+}). Under slightly alkaline conditions, the solubility of divalent ions present as bicarbonate salts is very small, so the actual influence by divalent ions is small. The influence of NaCl on microorganisms is mainly expressed by inhibiti on of Na^+, so there are many researches on toxicity of Na^+. The toxicity of Na^+ is reversible and adaption of microorganisms to high concentration of Na^+ can be improved through domestication. When concentration of Na^+ is 10000 mg/L, relative production rate of methane is 4% for undomesticated digestion system and 29% after domestication (Lei 2000). In biogas plants treating starch wastes, there is no remarkable influence on anaerobic digestion when concentration of NaCl increases to 10 g/L. Anaerobic digestion will be severely inhibited when the concentration increases to 20 g/L, and biogas production rate decreases by 30.96% compared with concentration of 0 g/L while methane production rate decreases by 42.14% (Hou and Su 2015). In an anaerobic digestion test for the treatment of kitchen waste, addition of 2 g/L NaCl causes no influence for the biogas production; addition of 5 g/L increase gas production by 7.5%, and addition of 10, 20 and 30 g/L respectively decreases gas production by 17%, 24% and 51% compared with the control group (Tao et al. 2013).

(4) Heavy metals

As a micro-element, heavy metals can increase the activity of microorganisms at low concentration and will be poisonous at high concentration, especially lead, cadmium, copper, zinc, nickel and chromium that can cause abnormal operation of biogas plants. The toxicity ranking is generally Ni > Cu > Pb > Cr > Zn. Compared with non-dissolved heavy metal, dissolved heavy metal is more likely to cause anaerobic digestion failure. But sulfides such as H_2S produced during anaerobic digestion can react with dissolved heavy metals and turn it into non-dissolved compound so as to alleviate the toxicity of heavy metals.

(5) Organic acids

Organic acids are produced from complex organic matter by hydrolysis during anaerobic degradation. Such acids include short chain fatty acids (SCFAs) (C-2-C-6), long chain fatty acids (LCFAs) and amino acid. The short chain fatty acids include acetic acid, propionic acid, butyric acid and a few isobutyric acid, pentanoic acid, isopentanoic acid and caproic acid. The common forms of organic acids include the free state and the non-free state. Generally, free state organic acids can inhibit microorganisms, because they can penetrate into cells and cause protein denaturation.

Under normal anaerobic digestion operations, the concentration of volatile fatty acids is low. If too much organic acid is added at once, organic acid degradation is blocked, or syntrophic relationship between acidogenic bacteria and methanogens is imbalanced in a short time, organic acids will accumulate, and the mixed liquid will be acidified. Accumulation of acids will decrease pH of the mixed liquid and further aggravate the inhibition of acid. When pH < 7, acetic acid that reach 1000 mg/L, isobutyric acid or isoamyl acid that reach 50 mg/L, and propionic acid of around 5 mg/L will all cause remarkable inhibition. However, a research shows that, whether the acetic acid or the butyric acid, only when its concentration exceeds 10000 mg/L can cause inhibitory effect, and only when the propionic acid reaches 6000 mg/L, the inhibitory effect is noticeable under neutral pH. Therefore, organic acids are usually used as monitoring indicators of anaerobic digestion.

(6) Artificially synthetic organic matters

Artificially synthetic organic matters are difficult to be degraded during anaerobic degradation, which is closely related to intoxication or inhibition. The common poisonous synthetic organic matters include halogens, aldehydes and aromatic compounds. However, after adaption of microorganisms to these substances, the tolerance of microorganisms can be improved by 50 times and these compounds can then be degraded.

In addition, nonspecific inhibition will be caused by disinfectants, herbicides, insecticides, surfactants and antibiotics entering anaerobic digestion reactors.

References

Bryant, M.P. 1967. https://doi.org/10.1007/BF00406313.

Bryant, M.P. 1979. Microbial methane production—theoretical aspects. *Journal of animal science* 48: 193–201.

Cheng, L., Z.Z. Zheng, C. Wang, and H. Zhang. 2016. Recent advances in methanogens. *Microbilogy China* 05: 1143–1164. (in Chinese).

Costa, K.C., and J.A. Leigh. 2014. Metabolic versatility in methanogens. *Current Opinion in Biotechnology* 29(Supplement C): 70–75.
Deublein, D., and A. Steinhauser. 2008. *Biogas from waste and renewable resources*. Strauss GmbH, Mörlenbach, WILEY-VCH Verlag GmbH & Co. KGaA, Weinheim.
Dolfing, J. 1988. Acetogenesis. In *Biology of Anaerobic Microorganisms*, ed. J.B. Alexander Zehnder, 417–468. New York, USA, Wiley, Inc.
Hou, X.C., and H.J. Su. 2015. Effect of high salt concentration on methane production by anaerobic fermentation. Annual meeting of China Chemical Engineering Association. (in Chinese).
Iino, T., H. Tamaki, S. Tamazawa, Y. Ueno, M. Ohkuma, K. Suzuki, Y. Igarashi, and S. Haruta. 2013. *Candidatus* Methanogranum caenicola: a Novel Methanogen from the Anaerobic Digested Sludge, and Proposal of *Methanomassiliicoccaceae* fam. nov. and *Methanomassiliicoccales* ord. nov., for a Methanogenic Lineage of the Class Thermoplasmata. *Microbes and Environments* 28 (2): 244–250.
Khanal S K. 2008. Anaerobic biotechnology for bioenergy production: principles and applications, Wiley.
Lei, Z.F. 2000. The effect of high concentration of sodium on the instability of wastewater biological treatment system is reviewed. *Industrial Water Treatment* 20 (4): 6–10. (in Chinese).
L'Haridon, S., M. Chalopin, D. Colombo, and L. Toffin. 2014. *Methanococcoides vulcani* sp. nov., a marine methylotrophic methanogen that uses betaine, choline and N, N-dimethylethanolamine for methanogenesis, isolated from a mud volcano, and emended description of the genus *Methanococcoides*. *International Journal of Systematic and Evolutionary Microbiology* 64 (6): 1978–1983.
Lin, H.L., Q.Y. Li, Y.F. Li, and D.Z. Li. 2014. *Microbiology in aerobic environment*. Harbin: Harbin University of Technology Press. (in Chinese).
Liu, Y., W.B. Whitman. 2008. Metabolic, phylogenetic, and ecological diversity of the methanogenic archaea. *Annals of the New York Academy of Sciences* 1125 (1): 171–189.
López, M., M. Madeline, P. Schönheit, W.W. Metcalf, and G.A. O"Toole. 2015. Genetic, genomic, and transcriptomic studies of pyruvate metabolism in *methanosarcina barkeri* fusaro. *Journal of Bacteriology* 197 (22): 3592–3600.
Ma, X.P., C.B. Xu, and B.R. Fu. 2017. *Anaerobic microbiology and wastewater treatment*. Beijing: Chemical Industry Press. (in Chinese).
Madigan, M.T, J.M. Martinko, K.S. Bender, D.H. Buckley, and D.A Stahl. 2014. Brock biology of microorganisms.14edn.
O'Brien, J.M., R.H. Wolkin, T.T. Moench, J.B. Morgan, and J.G. Zeikus. 1984. Association of hydrogen metabolism with unitrophic or mixotrophic growth of Methanosarcina barkeri on carbon monoxide. *Journal of Bacteriology* 158 (1): 373–375.
Offre, P., A. Spang, and C. Schleper. 2013. Archaea in biogeochemical circulations. *Annual Review of Microbiology* 67 (1): 437–457.
Oremland, R.S. 1988. Biogeochemistry of methanogenic bacteria. New York, USA, Wiley, Inc.
Schink, B. 2006. Syntrophic associations in methanogenic degradation. *Progress in Molecular and Subcellular Biology* 41: 1–19.
Shen, P., and X.D. Chen. 2016. *Microbiology*. Tianjin: Higher Education Press. (in Chinese).
Tao, Z.P., X. Zhao, and W.Q. Ruan. 2013. Effect of sodium chloride on biogas production by anaerobic fermentation of kitchen waste. *Journal of Food and Biotechnology* 32 (6): 596–602. (in Chinese).
Wan, S., Y.F. Li, and T.M. Yin. 2013. Anaerobic biological treatment of wastewater. Harbin University of Technology Press. (in Chinese).
Wang, J.L, S.P. Li, and Z. Huang. 2004. Environmental microbiology Beijing. Higher Education Press. (in Chinese).
Watkins, A.J., E.G. Roussel, R.J. Parkes, and H. Sass. 2013. Glycine betaine as a direct substrate for methanogens (Methanococcoides spp.). *Applied and Environmental Microbiology* 80: 289–293.

Watkins, A.J., E.G. Roussel, G. Webster, R.J. Parkes, and H. Sass. 2012. Choline and N, N-Dimethylethanolamine as direct substrates for methanogens. *Applied and Environmental Microbiology* 78 (23): 8298–8303.

Woese, C.R., and G.E. Fox. 1977. Phylogenetic structure of the prokaryotic domain: the primary kingdoms. *Proceedings of the National Academy of Sciences* 74 (11): 5088–5090.

Wu, W., Z.Z. Zheng, T.T. Ma, L. Cheng, and H. Zhang. 2016. Recent advances on syntrophic acetate oxidation bacteria. *China Biogas* 02: 3–8. (in Chinese).

Yu, F.B., X.P. Luo, L.B. Guan, M.X. Zhang, and S.D. Shan. 2008. Research advances in biogas fermentation microorganism. *Journal of Anhui Agriculrual Science* 36 (35): 15658–15660. (in Chinese).

Zehnder, A.J.B. 1988. Biology of anaerobic microorganisms, ed. A.J.B Zehnder. Wiley, New York, USA.

Zeikus, J.G. 1977. The biology of methanogenic bacteria. *Bacteriological reviews* 41 (2): 514–541.

Zeikus, J.G. 1979. https://doi.org/10.1016/0141-0229(79)90043-7.

Zhang, W., F. Yin, and I. Monroe. 2016. *Anaerobic digestion technology and engineering*. Beijing: Chemical Industry Press.

Zhao, Y.Z., H. Zhang, Y. Deng, X.B. Yin, G.Y. Liu, G.Q. Hu, L.W. Lian, and Y.B. Qian. 1997. *Methanogenic bacteria and research methods*. Chengdu: Chengdu University of Science and Technology Press. (in Chinese).

Zhou, D.Q. 2011. *Microbiology course*. Beijing: Higher Education Press. (in Chinese).

Chapter 2
Rural Household Digesters

A rural household digester refers to an anaerobic digestion device that satisfies the domestic energy needs of one household or several households. This type of digester is suitable for farmers to treat and utilize human and animal excrements and other organic wastes for producing biogas as a living energy. China has documented the use of biogas since the late 20th century, when biogas was produced by fermenting agricultural wastes in pits. Over the past one hundred years, household digesters in China have greatly developed, taking a leading position in the world in terms of quantity and technical level. Under the joint efforts of a vast number of scientific and technological stakeholders and technical extension personnel, new digester types such as the rotational flow distribution digesters, strong reflux digesters, and plug flow digesters have been developed successively, and new digester materials such as glass fiber reinforced plastics and red mud plastics have been promoted. With household digesters as the core, China has developed various rural energy and ecological models such as "pig—biogas—orchard model" in the south, the "four-in-one model" in the north, and the "five supplementary elements model" in the northwest. Although the role of rural household digesters in rural energy becomes less significant because of economic growth, household digesters in remote or economically underdeveloped areas is still a suitable technology to treat agricultural wastes and organic rubbishes. It still plays an important role in energy security, transformation of agricultural development modes and promotion of ecological civilization.

2.1 Feedstock of Household Digesters

In the process of anaerobic digestion, anaerobic digestion microorganisms need to absorb adequate nutrition so as to carry out normal living activities, exuberant and uninterrupted production of biogas. Therefore, sufficient feedstock is the foundation of biogas production. Theoretically, all biomass can be decomposed by the anaerobic

L. Deng et al., *Biogas Technology*,
https://doi.org/10.1007/978-981-15-4940-3_2

digestion process, but different kinds of feedstock have different physical and chemical properties and gas production performance, so it is necessary to understand and master the characteristics of various feedstocks and their gas production potentials.

2.1.1 *Types and Characteristics of Feedstocks for Household Digesters*

Due to their small scale, rural household digesters serve to satisfy domestic energy needs of one or several families, so the collection of feedstocks of rural household digesters should be conducted as close to the digesters as possible under the premise of ensuing biogas output. According to different sources, digester feedstocks can be categorized into three types: crop waste, animal waste, and domestic organic waste. This part section only introduces the types and the main characteristics of the feedstocks suitable for rural household digesters. The output and biogas production potential of feedstock can be referred to in the fourth chapter of this book.

(1) Crop waste

Crop waste mainly refers to all kinds of waste from crop production and processing, mainly crop straws. One of its characteristics is that the carbon content in raw materials is high, and the carbon nitrogen ratio is generally more than 30:1. In addition, the cellulose and lignin content in crop straws is high, and the straws is of a large volume and cannot be destroyed easily. Based on these characteristics, the anaerobic degradation rate of crop waste is low, thus the biogas production duration is long. In order to increase its biogas production rate, crop waste is usually pretreated before entering a digester. Although the gas production of crop waste is slow, the biogas production potential is generally greater than that of breeding waste and domestic organic waste given long enough digestion time. The main problem of crop straw as feedstock for household digester is its high content of cellulose and lignin, so it is difficult to feed and discharge. Pretreatment measures such as composting and acidification are often required.

(2) Animal waste

Animal waste mainly refers to manures, urine and sewage produced by breeding livestock and poultry, usually referred to as livestock and poultry manures. On one hand, the nitrogen content of such feedstock is high, with carbon nitrogen ratio generally lower than 25:1. On the other hand, its particles are relatively fine and contain many low-molecular weight compounds, so there is no need to carry out physical pretreatment such as crushing, and the decomposition and gas production rates are relatively high. The characteristics of this kind of feedstock include a short digestion time, fast gas production, and the total gas production per unit feedstock is lower than that of the crop waste. Livestock and poultry manure is the main feedstock source of household biogas plants in China. The output of livestock and poultry manure can refer to the data in Table 2.4.

(3) Domestic organic waste

Domestic organic waste refers to feces, urine, sewage and household garbage produced in daily life of rural residents. Its most prominent characteristic is the diversity of the components. It contains substances that are harmful to anaerobic digestion, such as detergents and plastic bags. This kind of organic waste has a high content of protein and lipid and a high biogas production rate. However, with the improvement of rural living standards, a large amount of flushing sewage increases, leading to a decrease in the concentration and retention time of feedstock in digester.

2.1.2 Parameters of Feedstock for Anaerobic Digestion

(1) Total Solid (TS)

Total Solid (TS) refers to the amount of residues remaining after evaporation and dehydration of feedstock at a certain temperature. It is also called Dry Matter (DM) in some other countries. TS includes suspended solids (SS) and dissolved solids (DS) Available drying temperatures are 103–105 and 180 °C. Drying at 103–105 °C will retain water of crystallization and the attached water, while drying at 180 °C can remove all attached water. 103–105 °C is preferable in most cases. The disadvantage of the parameter of total solids is that it cannot accurately reflect the content of organic substances. However, for feedstocks with similar physical and chemical properties, the proportion of organic substances in TS is similar, and the test method of TS is relatively simple. Therefore, the content of organic matters in feedstock of rural household digesters is usually roughly represented by TS (Table 2.1). TS is usually expressed in percentage, and DM is usually expressed in g/L or mg/L. TS is calculated as follows:

Table 2.1 Total solid content of common feedstocks in rural household digester (approximate value)

Feedstock	Total solid content(%)	Water content(%)
Dry rice straw	85	15
Dry wheat straw	85	15
Dry Corn straw	82	18
Grass	24	76
Human feces	20	80
Pig manure	18	82
Cow manure	17	83
Human urine	0.4	99.6
Pig urine	0.4	99.6
Cow urine	0.6	99.4

$$TS(\%) = \frac{W_2}{W_1} \times 100\% \quad (2.1)$$

where

W_1 weight of feedstock before drying (g)
W_2 weight of feedstock after drying (g)

Example: 10 g of pig manure is dried at 105 °C in oven to a constant weight of 1.8 g, what is the percentage content of total solid m_0?

$$m_0 = \frac{W_2}{W_1} \times 100\% = \frac{1.8}{10} \times 100\% = 18\%$$

(2) Suspended solid (SS)

It refers to the feedstock solid substances that cannot pass through the filter membrane with a certain pore size, expressed by g/L or mg/L. It is the remained suspended solid after passing through a filter membrane with a pore size of 0.45 μm and dried at 103–105 °C.

(3) Volatile solid (VS)

The remaining solid after burning total solids for one hour at 550 ± 50 °C is ash content. TS subtracts ash content equals to VS, also known as Organic Dry Matter (Organic Dry Matter, ODM). Volatile solids mainly include suspended matter, colloids and dissolved organic matters in feedstock. VS is usually expressed by percentage, while ODM is usually expressed by g/L or mg/L. VS is calculated in two ways. One is the ratio of volatile solids to total solids with the following equation:

$$VS/TS(\%) = \frac{W_2 - W_3}{W_2} \times 100\% \quad (2.2)$$

where

W_3 ash content (g)

The other is the ratio of VS to feedstock with the following equation:

$$VS(\%) = \frac{W_2 - W_3}{W_1} \times 100\% \quad (2.3)$$

(4) Volatile Suspended Solid (VSS)

Burn suspended solid at 550 ± 50 °C for one hour, and the burnable volatile matter is VSS, usually expressed by g/L or mg/L.

(5) Chemical oxygen demand (COD)

It refers to the amount of oxygen consumed by oxidizing feedstock under heating condition with chemical oxidants, usually expressed in terms of oxygen consumption in mg/L or g/L. The most commonly used oxidant is potassium dichromate or potassium permanganate. The COD value measured when potassium dichromate is the oxidant is called COD_{Cr} or COD for short, and the value measured when potassium permanganate is the oxidant is called COD_{Mn} or potassium permanganate index. COD represents the amount of organic substances that can be chemically oxidized and a small amount of inorganic substances in feedstock, as well as the maximum chemical energy in the feedstock. Common test methods include potassium dichromate method, coulomb method, fast closed catalytic digestion method (including photometric method), energy saving heating method, chlorine gas correction method, etc.

(6) Biochemical oxygen demand (BOD)

It refers to the amount of oxygen consumed by microorganisms to decompose organic matters in feedstock under aerobic conditions, expressed by mg/L or g/L. Since the microbial and biochemical process is slow, BOD is commonly represented by the amount of oxygen consumption of feedstock during five days' cultivation at 20 °C, expressed as BOD_5. BOD_5 represents the amount of organic matter in feedstock that can be decomposed by microorganisms. Common test methods include dilution inoculation, microbiological sensor rapid determination, activated sludge aeration degradation, etc.

(7) Total organic carbon (TOC)

It is a comprehensive index that uses carbon content to indicate the total amount of organic matters in feedstock. TOC is more accurate than COD or BOD_5 in representing the total amount of organic matters because it can oxidize all organic matters by combustion. The main test methods are oxidative combustion—non-dispersive infrared absorption method, conductance method, gas chromatography, wet oxidation-non-dispersive infrared absorption method.

(8) Digestibility

It refers to the ability of feedstock to be degraded by microorganisms in anaerobic digestion. In a wastewater treatment plant, it is often referred to as biodegradability, which is an important parameter to characterize feedstocks. For some specific feedstocks, the digestibility depends on the content of degradable organic matter contained. At the same time, feedstock also contains some substances that are difficult to be degraded such as lignin, phenols etc. In addition, the degradation of different components also takes different time durations. Under anaerobic conditions, micromolecule carbohydrates, volatile fatty acids, and ethanol degrade within a few hours, while it takes days to degrade proteins, hemicellulose, and lipids and weeks to degrade cellulose. Some feedstocks contain lipids, which have a high biogas output and a long degradation time. Compared with easily degradable feedstocks, this kind of feedstocks requires a digester with a larger volume. However, in practical design, to optimize construction cost, maximum biogas production and shorter retention time should be taken into account simultaneously.

The ratio of BOD_5 to COD can also represent the digestibility of feedstocks. COD and BOD_5 are two commonly used indirect indices to indicate the organic content in feedstock. COD is determined by chemical oxidation while BOD_5 by microbial oxidation. A part of COD is difficult to be oxidized and degraded by microorganisms, so COD is larger than BOD_5. For the same kind of feedstock, the ratio of BOD_5 to COD is usually used to indicate the degree of biochemical decomposition of it. The higher the ratio, the better the anaerobic digestion effect. It is generally believed that when BOD_5/COD is less than 0.3, the feedstock is difficult to be biodegraded, thus not suitable for anaerobic digestion.

(9) Carbon nitrogen ratio (C/N)

It refers to the ratio of carbon and nitrogen content in feedstock denoted as C/N. Carbon provides energy for microbial activities and is the main element in formation of methane. Nitrogen is the main constituent of microbial cells. Generally, the C/N of feedstock for anaerobic digestion should be 20–30:1. It should be noted that not all carbon and nitrogen elements can be utilized by microorganisms. For example, lignin, cellulose, hemicellulose and glucose are all sources of carbon, but the ability of anaerobic microorganisms to degrade and utilize lignin and glucose is quite different. When C/N is too high, the anaerobic digestion process is not easy to start, and the gas production efficiency is not good. When C/N is too low, excess of nitrogen will be converted into free ammonia, causing ammonia poisoning and inhibiting normal anaerobic digestion. The carbon-nitrogen ratios of feedstocks commonly used in rural areas are shown in Table 2.2.

Table 2.2 Carbon nitrogen ratios (approximate value) of common feedstocks in rural household digester

Feedstock	Proportion of carbon in feedstock (%)	Proportion of nitrogen in feedstock (%)	C/N
Dry wheat straw	46	0.53	87:1
Dry rice straw	42	0.63	67:1
Dry corn straw	40	0.75	53:1
Fallen leaves	41	1.00	41:1
Soybean stem	41	1.30	32:1
Wild grass	14	0.54	26:1
Peanut stem	11	0.59	19:1
Fresh sheep manure	16	0.55	29:1
Fresh cow manure	7.3	0.29	25:1
Fresh horse manure	10	0.42	24:1
Fresh pig manure	7.8	0.60	13:1
Fresh human feces	2.5	0.85	2.9:1

Note Due to the different sources of feedstocks, the components may vary. Figures in this table are only approximations

As can be seen from Table 2.2, the carbon to nitrogen ratios of crop waste (crop straw) are relatively high. For example, the carbon and nitrogen ratio of dry wheat straw is as high as 87:1. The C/N of breeding wastes (livestock and poultry manure) and domestic organic wastes are relatively low. For example, the C/N of human feces is only 2.9:1. In feedstock preparation, feedstocks with high carbon content such as crop straw need to be mixed with those of high nitrogen content (such as human feces and animal manure) in order to achieve the appropriate C/N and a high anaerobic digestion efficiency. In particular, during the start-up stage of a digester, the appropriate C/N ratios can speed up the start-up process. If the amount of human and animal excrement is not enough, ammonium bicarbonate, urea and other nitrogen fertilizers can be added to supplement nitrogen when crop waste is used as the main feedstock.

(10) pH

It refers to the negative logarithm of hydrogen ion concentration in liquid feedstock. Testing methods include glass electrode, portable pH meter and test paper. For testing of semi-solid and solid feedstocks, the sample can be mixed with a certain amount of distilled water and get tested. Due to the high buffering capacity of the anaerobic digestion system, pH of the feedstock can have a wide range. In general, the buffering capacity of anaerobic digestion system is determined by the concentration of CO_2 in gas, the concentration of ammonia in the mixed liquid, and the water content. If pH of the feedstock is too high or too low that exceeds the buffering capacity of the digestion system, it should be neutralized before feeding. If there is slight acidification in the digestion process, feeding should be stopped. If the acidification is serious, methanogenic bacteria should be added to increase the amount of methanogens, or plant ash, hydrated lime should be added to neutralize the mixed liquid. If the acidification is even stronger, the mixed liquid should be changed.

(11) Biochemical methane potential—BM

Biochemical Methane Potential (BMP) was proposed by the McCarty research group of Stanford University. It represents the number of organic matters in feedstock that can be degraded by anaerobic microorganisms. Since most organic matters can be degraded under both anaerobic and aerobic conditions, the values of BMP and BOD are similar in most cases. Only when the biodegradability is poor, the values of BMP and BOD are drastically different.

In China, BMP is often used to indicate biogas output per unit quantity of feedstock. Feedstock with high suspended matter content is usually expressed by (m^3/kg TS), or by the gas production rate of volatile solid per unit (m^3/kg VS). Feedstock with low suspended substance content is usually expressed by gas production of COD (m^3/kg COD). BMP is divided into gas production rate of feeding and gas production rate of removal. BMP is usually measured by batch digestion with a single feedstock.

For the same type of digestion, the BMPs introduced in different literatures vary greatly, mainly because BMPs usually have theoretical values, experimental values, and full scale values. Some literature does not distinguish between the different types when introducing BMPs. The theoretical value of BMP of carbohydrates is

Table 2.3 BMP of feedstocks in rural household digesters and feedstock consumption of producing 1 m^3 of biogas

Feedstock	Water content (%)	BMP (m^3biogas/kg DM)	Feedstock consumption of producing 1 m^3 of biogas	
			Dry matter (kg DM)	Fresh matter (kg FM)
Pig manure	82	0.30	3.33	18.52
Cow manure	83	0.19	5.26	30.96
Chicken manure	70	0.33	3.03	10.10
Human feces	80	0.31	3.23	16.13
Sheep manure	75	0.22	4.55	18.18
Straw	15	0.31	3.23	3.80
Wheat straw	15	0.30	3.33	3.92
Corn straw	18	0.33	3.03	3.70
Water hyacinth	93	0.31	3.23	46.08
alligator weed	90	0.29	3.45	34.48

about 0.7 m^3/kg TS. When COD is used as the unit of feedstock concentration, the theoretical gas production rate is 0.35 m^3CH_4/kg COD. The experimental value is the maximum BMP measured by a certain method in a laboratory, which is generally about 70% of the theoretical value. The full scale value is the BMP value obtained through engineering application. See Table 2.3 for the BMP of common feedstock used in rural household digesters and the feedstock consumption of producing 1 m^3 of biogas.

2.2 Digestion Process of Household Digester

Digester is the place where microorganisms carry out anaerobic digestion processes. First, the digestion chamber of a household digester should be closed to provide a strictly anaerobic environment for the anaerobic digestion process. Second, the digester should facilitate feedstock and digestate to enter and exit the digestion chamber. Third, the digester should have a buffering capacity for biogas storage. In terms of technique, digesters for rural households worldwide can be mainly divided into two categories: the first is the fixed dome type digester developed and promoted in China, which is usually called a hydraulic digester in China; the second is the floating roof type digester which is popular in India and other South Asia regions, also known as a Gobar digester.

2.2.1 Fixed Dome Digester

A fixed dome digester is usually called a hydraulic digester in China or abroad where Chinese digesters are used (Fig. 2.1). Hydraulic digesters are the most widely used type in rural China.

A hydraulic digester mainly consists of several parts, including an inlet chamber, a digestion chamber, a gas storage chamber and a hydraulic chamber. When there is no biogas generation at the beginning of the digestion process or when the storage chamber is connected with the atmosphere, the biogas pressure in the digester is equal to the atmospheric pressure, and the mixed liquid of the inlet chamber, the digestion chamber and the hydraulic chamber is at the same level. The liquid level in this case is the zero pressure liquid level, also known as the "zero pressure line" When the upper part of storage chamber is completely closed, biogas generated from the degradation of organic matters by microbe rises to the storage chamber. With the increase of biogas volume, the pressure of biogas in the digester will increase. When biogas pressure is higher than the atmospheric pressure, biogas will push mixed liquid in the digestion chamber into the inlet chamber and the hydraulic chamber. Liquid level in digestion chamber drops, while that in inlet pipe and hydraulic chamber rises, leading to a level difference that maintains the biogas pressure in gas chamber. This process is called "gas presses liquid". This process ends when the pressure inside and outside the digester arrives at an equilibrium. When the produced biogas is being consumed, it is discharged from a biogas discharge pipe, and the pressure in the digester decreases so that the mixed liquid in the inlet chamber and the hydraulic chamber returns to the digestion chamber. The liquid level of inlet pipe and the hydraulic chamber drops, while the liquid level in the digestion chamber rises and the liquid level difference decreases, resulting in a decrease in biogas pressure. If the biogas produced during digestion is less than the consumption, liquid level in the digestion chamber will gradually keep up with the level of the inlet pipe and the

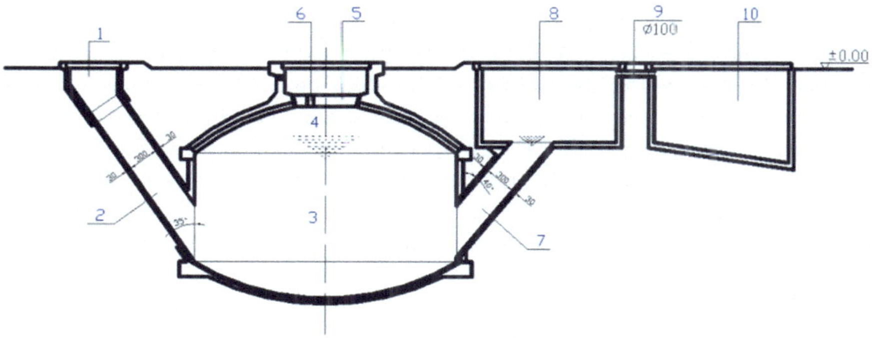

Fig. 2.1 A fixed dome (hydraulic) biogas digester
1. inlet chamber, 2. inlet pipe, 3. digestion chamber, 4. biogas storage chamber, 5. movable cover, 6. biogas pipe, 7. outlet pipe, 8. hydraulic chamber, 9. overflow pipe, 10. digestate storage tank. The components and function of hydraulic digester are introduced as follows

hydraulic chamber. Finally, the pressure difference disappears, and the biogas output is terminated. At this point, the liquid level will again return to the "zero pressure line" to maintain a new pressure balance. This process is called "liquid presses gas". As a result of ceaselessly gas production and consumption, the liquid level difference inside and outside the digester is continually changing, but the internal and external pressure balance is always maintained. This is the working principle of hydraulic digester. Biogas pressure of the hydraulic digester changes with the difference of liquid level in the inlet pipe, the hydraulic chamber, and the digestion chamber. Therefore, pressure is unstable when using gas product.

The advantages of a hydraulic digester are as follows:

(1) Rational structure, even force, convenient construction, sturdy and durable, simplicity of uses;
(2) Generally constructed by bricks or reinforced-concrete, so building materials are widely available.
(3) Low cost, making it more suitable for the economic development level of rural areas;
(4) Generally built underground with a good resistance to forces; conducive to heat preservation in winter.
(1) Inlet chamber

The inlet chamber is a space for gathering fresh feedstock and the entrance of the feedstock to the digester. Generally, it can be covered by a cover plate. In many cases, an inclined pipe is used to connect the inlet chamber and the digestion chamber to facilitate feeding. The lower end of the inlet is at 1/2 from the digester bottom to digester top cover. Too high a location will reduce the capacity of the gas chamber, while too low a location will inhibit new feedstock from entering the center of the digestion chamber. The size of the inlet pipe is determined according to the characteristics of the feedstock and the volume of the digester.

(2) Digestion chamber and biogas storage chamber

The digestion chamber and the biogas storage chamber, as the core of the digester, are integrated. The lower part is the digestion chamber and the upper part is the biogas chamber. The function of the former is to produce biogas through conversion of organic matters in the feedstock by microorganisms. After the biogas produced during digestion escapes, it rises to the upper biogas storage chamber, whose function is to store biogas. In normal operation, the liquid level in digestion chamber is variable, so is the volume of the digestion chamber and the biogas storage chamber. The total volume of the two is generally called the effective volume of a household digester.

(3) Hydraulic chamber

The hydraulic chamber, also known as the outlet chamber, is connected with the digestion chamber at the bottom. The roles of a hydraulic chamber is to discharge digestate and to store the mixed liquid produced from the digestion chamber during gas production. When using gas, the hydraulic chamber provides refluxing mixed liquid and provide pressure. The hydraulic digester stores gas through the process of "gas presses liquid" so the hydraulic chamber needs large volume to store mixed liquid in order to let the gas chamber store more biogas.

(4) Movable cover

The movable cover is on top of the digester cover, and a gas pipe is inserted in it. Generally, it is flat and cork-shaped, and circular reverse cover plates are used most often. The function of a movable cover is to seal the digester and to discharge biogas from the gas pipe for uses by farmers. Some other functions include a manhole or an access opening for feedstocks during reloading, as well as a depressurization cover that can be ejected in case of high gas pressure inside the digester.

2.2.2 *Floating Dome Digester*

Most overseas literatures show that floating dome digesters were developed in India in 1962. This type of digester is mainly composed of several parts including an inlet chamber, a digestion chamber, a floating dome, and an outlet chamber. The function of an inlet chamber and a digestion chamber is the same as that of a hydraulic digester. An outlet chamber is only for discharges of digestate, without the function of a hydraulic chamber. The body of this type of digester is a digestion chamber, the top of which is a gas storage device similar to floating domes of water-sealed gasholders (wet gasholders) in a biogas plant (Fig. 2.2). The upward and downward movement of the floating dome functions to store biogas, and the weight of the floating dome provides a certain pressure to the biogas to ensure a smooth transportation and use of the biogas. When there is no biogas generation or when the floating dome is connected to the atmosphere, the inner surface of the floating dome is closely attached to the upper liquid level during digestion. In this case, the pressure of biogas is 0.00 Pa.

Biogas is produced along with degradation of organic matters by microbe in the digestion process. The volume of biogas product increases continuously, and biogas pressure rises accordingly. The biogas pressure pushes mixed liquid inside the floating dome to the outside, and the liquid level in the floating dome decreases. When the sum of the biogas pressure and the buoyancy of the mixed liquid exerted onto the floating dome is greater than the gravity of the floating dome, the floating dome will rise until the upward and downward forces are equal. When biogas is in use, the biogas in the floating dome decreases and the floating dome drops. Compared with a fixed dome digester, the most prominent characteristic of a floating dome is that with the change of gas storage volume, the pressure of biogas is basically constant, which is conducive to stable operations of equipment powered by biogas. However, the disadvantages of this type of digester are also obvious: first, if the floating dome is made of metal, it is easy to be corroded even if there are anti-corrosion measures, thus reducing the service life of the digester. Second, if there are large substance blocks or a large number of fibrous substances in the feedstock, the overlapping parts of the floating dome and the digester body get easily blocked and movement of the floating dome is largely hindered; third, compared with a fixed dome digester that is

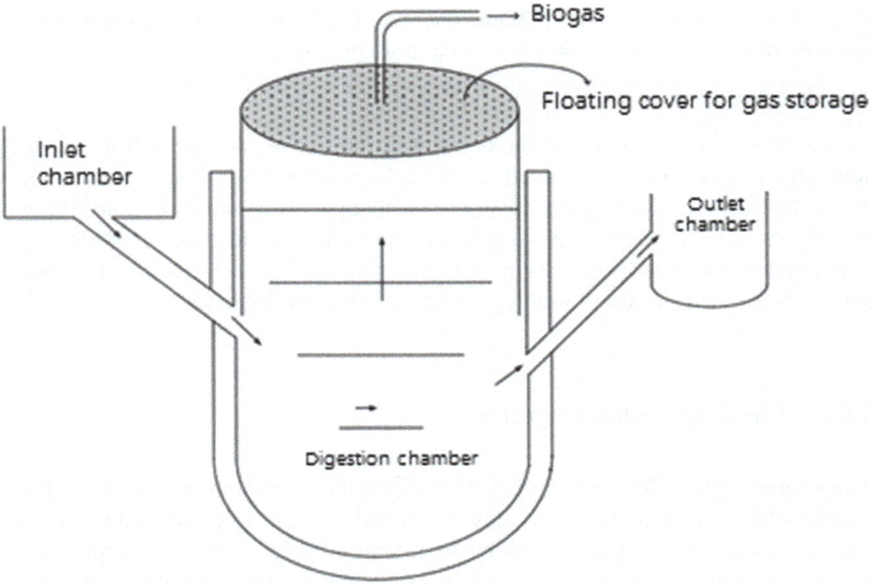

Fig. 2.2 A floating dome digester

completely built underground, this type of aboveground digester has poor insulation. This is one of the reasons why this type of digester can only be promoted in India, Pakistan and many other South Asian and Southeast Asian countries, as well as some regions in south China.

2.3 Derivative Types of Household Digesters

In using rural household digesters, through continuous research and practice, a variety of new digesters were developed based on traditional hydraulic digesters.

2.3.1 Bent Flow Distribution Digester

A bent flow distribution digester is a derivative of hydraulic digesters. Its characteristics include that its bottom tilts from the inlet to the hydraulic chamber, so the lower end of the bottom is at the bottom of the hydraulic chamber. It has three digester types A, B, and C. A-series type is traditional (Fig. 2.3), and there is no special structure in the digester. In type B, central inlet and outlet pipes and plug flow plates are added (Fig. 2.4). The central pipes are beneficial to sending feedstocks in and out from the central part of the digestion chamber, and the plug flow plate is beneficial

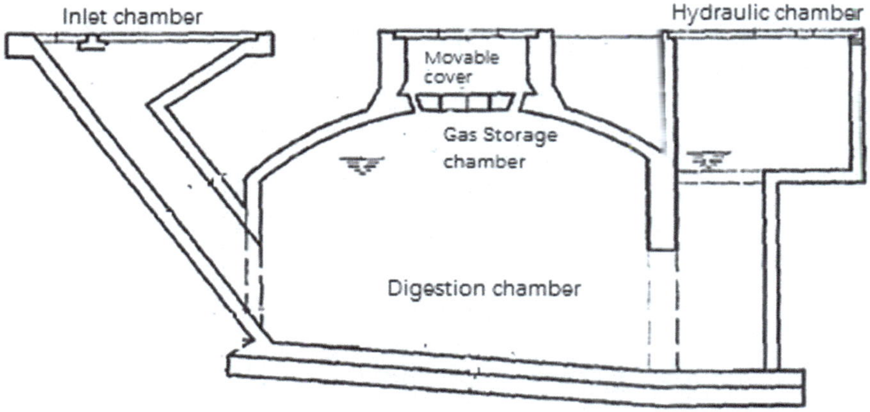

Fig. 2.3 Type A bent flow distribution digester

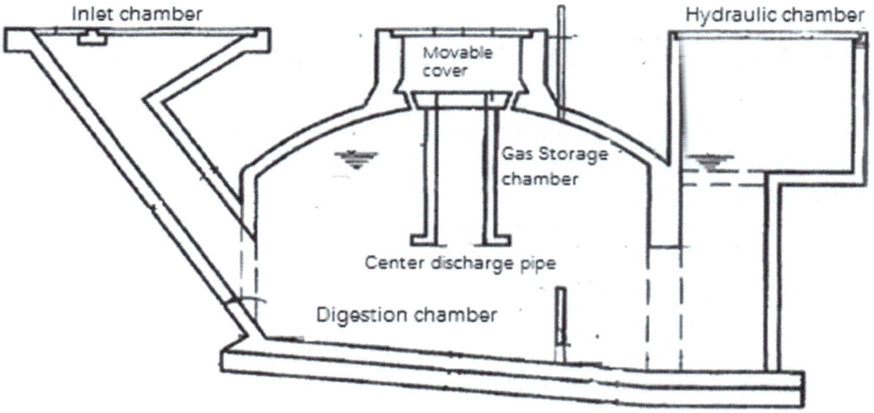

Fig. 2.4 Type B bent flow distribution digester

to controlling the flow rate of the feedstock at the bottom of the digester as well as controlling the solid retention time. In type C digesters, a distributing plate, a central shell breaking gas transmission hanging cage, and a feedstock pretreatment tank are added (Fig. 2.5). These new facilities can effectively improve the distribution of fresh feedstocks in the digester and improve load capacity and the efficiency of anaerobic digestion.

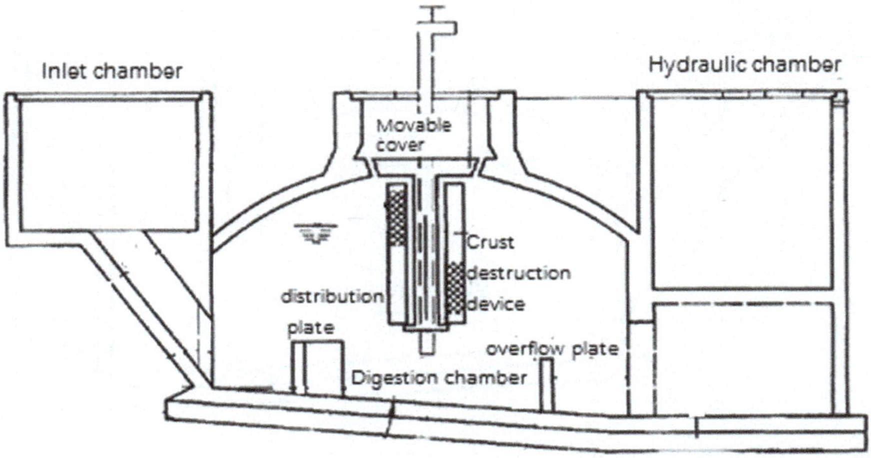

Fig. 2.5 Type C bent flow distribution digester

2.3.2 *Rotational Flow Distribution Digester*

Compared with traditional hydraulic digesters, a rotational flow distributing digester separates the hydraulic chamber from the outlet chamber and adds facilities such as the cyclone distributing wall and residue suction pipe. According to different ways of liquid circulation, rotational flow distribution digesters are categorized into the cyclone distributing automatic circulation type (Fig. 2.6) and the cyclone distributing compulsory circulation type (Fig. 2.7).

Rotational flow distribution digesters have the following characteristics: first, through reasonable distribution and coordination of a spiral bottom and an arc-shaped distribution wall, it can reduce short flows of slurry, extend the process and retention time of feedstock in the digestion chamber. Second, by means of the microorganism adhesion on the surface of the arc-shaped distribution wall, the highly active anaerobic microorganisms are fixed and enriched to avoid microorganism loss with liquid discharge. Third, by means of the anaerobic digestion and automatic circulation of the mixed liquid, the digester can achieve automatic stirring, circulation, and shell breaking, reducing the intensity of manpower management. Fourth, through the discharge agitator and mixed liquid reflux system, the digester can achieve the purpose of manually forced reflux stirring and residue discharge so as to ensure convenient management and sustainability.

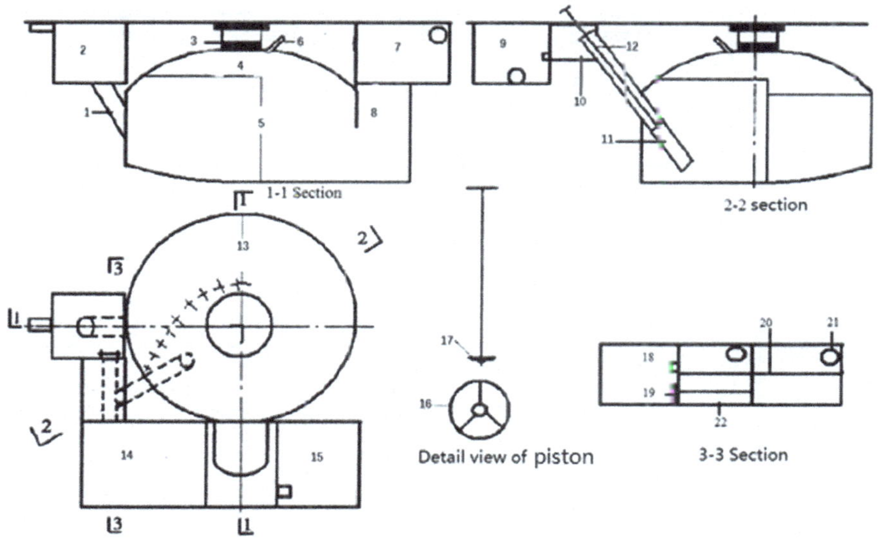

Fig. 2.6 A rotational flow distributing automatic circulation digester
1. Inlet pipe, 2. Inlet chamber, 3. Window cover, 4. Gas storage chamber, 5. Digestion chamber, 6 Gas-guide tube, 7. Hydraulic chamber, 8. Discharge access, 9. Inlet chamber, 10. Outlet chamber, 11. Residue suction pipe, 12. Piston, 13. Rotational flow distribution wall, 14. Acidification chamber, 15. Storage chamber, 16. Pistol plate, 17. Rubber cushion, 18. Inlet chamber, 19. One-directional valve, 20. Ultimate reflux level, 21. Overflow port, 22. Auto circulation pipe

2.3.3 Forced Reflux Digester

For a forced reflux digester, reflux pipes and pumping devices are installed to in the outlet chamber to forcefully pump out the mixed liquid from the digestion chamber, and the liquid is again sent to the digestion chamber via the feed inlet pipe or the acidification tank. This circulation results in a large amount of mixed liquid reflux, so purpose of complete mixing, shell breaking, and bacteria reflux (Fig. 2.8) can be achieved. The bottom of this type of digester is usually spherical, which is more benefits reflow of mixed liquid and mass transfer at the bottom. Forced reflux is intermittent, and it requires an interval of usually 1–2 months between forced refluxes.

2.3.4 Plug Flow Digester

Generally, the body of a plug flow digester is a slanting cylinder and is often horizontal, with its inlet and outlet locating at both ends of the digester body (Fig. 2.9). It can be built aboveground or underground, usually made of steel structure, PVC, fiber reinforced glasses, red mud plastic and so on. The characteristics of this type

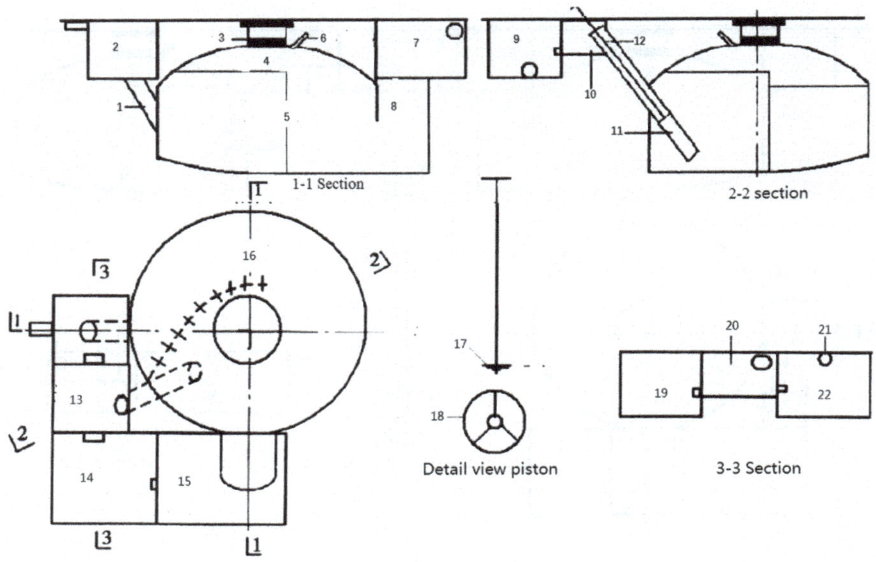

Fig. 2.7 A rotational flow distribution forced-circulation digester
1. Inlet pipe, 2. Inlet chamber, 3. Window cover, 4. Gas storage chamber, 5. Digestion chamber, 6. Gas-guide tube, 7. Hydraulic chamber, 8. Discharge access, 9. Inlet chamber, 10. Discharge chamber, 11. Residue suction pipe, 12. Piston, 13. Outlet chamber, 14. Fertilizer storage chamber, 15. Hydraulic chamber, 16. Rotational flow distribution wall, 17. Rubber cushion, 18. Piston plate, 19. Inlet chamber, 20. Outlet chamber, 21. Overflow port, 22. Fertilizer storage chamber

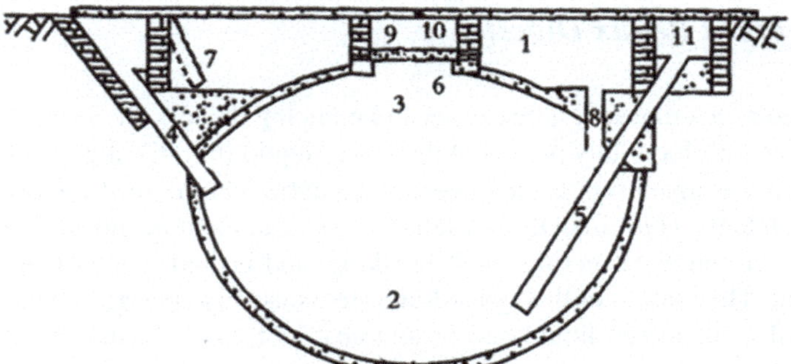

Fig. 2.8 A forced reflux digester
1. hydrolysis and acidification tank, 2. digestion chamber, 3. gas storage chamber, 4. inlet pipe, 5. outlet pipe, 6. movable cover, 7. backflow washing pipe, 8. pressure limiting backflow pipe, 9. water storage ring, 10. biogas pipe, 11. outlet chamber

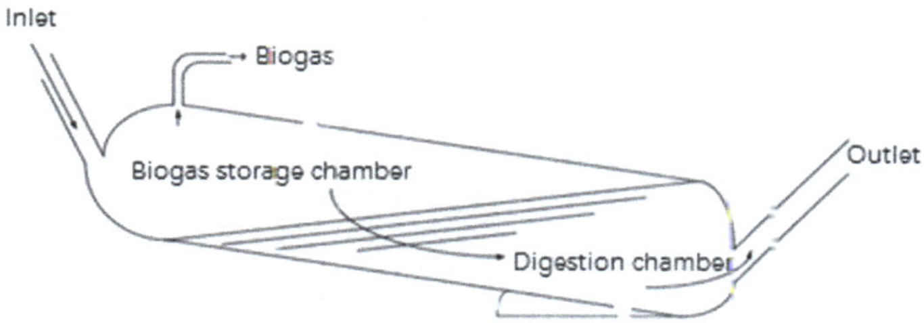

Fig. 2.9 A plug flow digester

of digester are as follows. First, the volume of the digester is constant, but the pressure of biogas changes with generation and consumption of biogas. Second, due to its long body, an acid phase can easily form at the inlet, and a methane phase can easily form at the outlet, thus forming a system similar to two-phase digestion in the digester. Third, if digester is built on ground, it can be moved in order to easily utilize biogas and digestate. Plug flow digesters have been widely used in Latin America, especially in the Andes of Peru. In addition, the Environmental Protection Agency of the United States estimated that more than half of the small biogas plants they built were the plug flow type.

2.3.5 Separate Floating Cover Digester

Traditional hydraulic digesters utilize a gas chamber on top of the digester to store biogas, while separate floating cover digesters store biogas by a water seal tank and a floating cover installed outside the digester (Fig. 2.10). As a result of this change in gas storage mode, this type does not belong to the category of hydraulic digesters, and its major advantages include the following aspects.

(1) Separation of the digestion chamber and the gas chamber allows without a hydraulic chamber, which helps to expand the volume of the digestion chamber, with the maximum feeding capacity accounting for 98% of the digester volume.
(2) Compared with a hydraulic digester, the gas pressure of the floating cover is relatively stable, which facilitates stable operations of equipment using biogas;
(3) Compared with a floating dome digester, the heat preservation performance of the digestion chamber of this type is better, which is conducive to normal digestion.

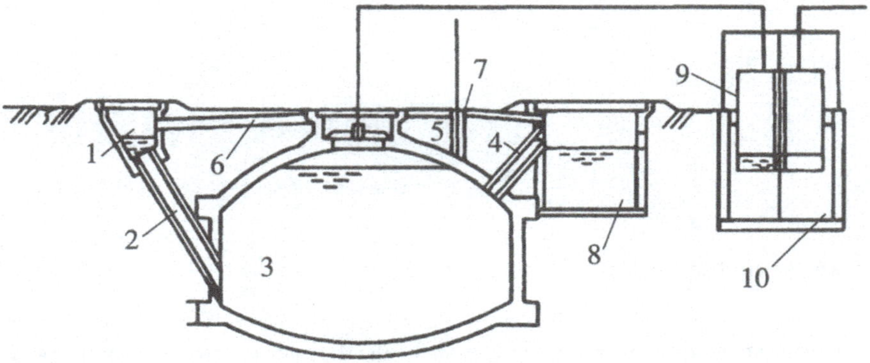

Fig. 2.10 A separate floating cover digester
1. inlet chamber, 2. inlet pipe, 3. digestion chamber, 4. overflow pipe, 5. discharge agitator. 6. sludge reflux ditch. 7. residue discharge ditch, 8. fertilizer storage tank, 9. floating cover, 10. water seal tank

2.3.6 Other Types of Household Digesters

In addition to the above several types of digesters, China has also developed various types of digesters, such as cylindrical digesters and ellipsoidal digesters. The advantages of different types of digesters include improved efficiency of anaerobic digestion, simplified operation management, and lower construction or operation costs.

2.4 Design of Household Digesters

If microbiology and biochemistry are the basis of anaerobic digestion, designs and construction of digesters are indispensable to meeting the requirements of anaerobic digestion process, utilization of biogas and digestate, environmental protection and sanitation, technical management, and so on. Comprehensive utilization of biogas and digestate is the purpose of anaerobic digestion. Therefore, anaerobic digestion facilities and equipment should be designed scientifically and rationally according to the requirements of digestion processes. Due to the vast variety of household digesters, different types of digesters do differ in their designs. This book introduces the designs of household digesters by taking hydraulic digesters, the most widely used in China, as an example.

2.4.1 Principles of Digester Design

(1) The "combined three" principle

The "Combined three" refers to the combination of the digester, the livestock and poultry breeding house, and the toilet. The digester site should be close to livestock and poultry breeding houses and toilets. For the construction of a digester, livestock and poultry breeding houses and toilets should be connected with each other, so that human and animal wastes can directly enter the digester. The "Combined three" not only saves labor, but also ensures continuous supply of fresh feedstocks, resulting in continuous gas production.

(2) The "round, small and shallow" principle

"Round" means that the top view of the digestion chamber in a hydraulic digester should be round, which allows even mixing of feedstock in the digester, saves construction materials, and ensures the structure stability. "Small" refers to the volume of the hydraulic digester is as small as possible (usually less than 10 m^3) as long as the biogas production requirement is met. 4, 6, 8 and 10 m^3 are most common in rural China. "Shallow" refers to the depth of a hydraulic digester that is usually less than 2.5 m, which is convenient for construction and easy to discharge underground water when the underground water level is high.

2.4.2 Design Parameters of Digesters

(1) Gas pressure

Biogas utilization facilities require relatively stable gas pressures. For a hydraulic digester, if the design gas pressure is too small, it cannot meet the pressure requirements of equipment fueled by biogas, and it also results in a large occupied area of the outlet chamber. If the design pressure of biogas is too high, the required strength of the digester increases, so does the thickness of the digester wall. Therefore, the design pressure of domestic hydraulic digester in rural China normally takes 4000–8000 Pa.

(2) Volumetric biogas production rate

Volumetric biogas production rate refers to the biogas production per cubic meter of digester per day, and this rate is affected by many factors such as temperature, the types and concentration of feedstock, stirring, pretreatment degree of feedstock, inoculant, technical management, digester type, hydraulic retention time, etc. Due to different conditions in different areas, the rate is not a fixed value, generally ranging from 0.15 to 0.30 m^3/m^3 d. The volumetric biogas production rate for different feedstock under different digestion temperature can be determined by tests.

(3) Gas storage capacity

A hydraulic digester functions to push mixed liquid into an outlet chamber (the majority of liquid) and an inlet chamber (the minority of liquid) to store biogas by difference in gas pressure. Therefore, the volume of the outlet chamber determines the storage capacity. Determination of gas storage volume is related to gas consumption. Biogas storage design capacity of household digesters in rural China is usually half of the daily biogas output.

(4) Volume

Determination of digester volume is the key to digester designs. If a digester is too small, it cannot make full use of feedstock and meet the production demand; If a digester is too large, and feedstock is inadequate, the volumetric utilization rate is low, wasting human and material resources. Therefore, volume of a digester should be reasonably determined based on the available feedstock, digestion process, gas consumption and many other factors. According to the current living standard and multi-energy complementarity, the average gas consumption per person per day is 0.2–0.3 m^3, and digesters are mostly available in volumes of 4, 6, 8, and 10 m^3. Table 2.4 shows the number of livestock and poultry on feed that can satisfy requirements of household digesters of different volume.

In the process of biogas production in a hydraulic digester, the space occupied by the digestion chamber, gas chamber and hydraulic chamber keeps changing interactively. When pressure in the digester is 0.00 Pa, the volume of mixed liquid in the digestion chamber reaches the maximum (the maximum design value), while the volume of biogas in the gas chamber and the volume of mixed liquid in the hydraulic chamber reaches the minimum (the minimum design value). In order to increase the volume of mixed liquid in the digestion chamber and increase biogas production, the volume of digestion chamber should be 85–95% of the total effective volume of the digester, whereas the volume of the gas chamber should be no more than 15%.

(5) Feedstock amount and their ratio

Table 2.4 Relationship between the volume of rural household digesters and the number of livestock and poultry on feed *

Items	Unit	Fattening pig	Layer chicken	Fattening cow	Dairy cow	Fattening sheep
Manure amount	kg/d	3.0	0.1	15	30	1.2
Total solid	%	18	30	18	17	30
BMP of manure	m^3 biogas/kg TS	0.30	0.33	0.19	0.19	0.22
6 m^3 digester	head	7	121	2	1	15
8 m^3 digester	head	10	162	3	2	20
10 m^3 digester	head	12	202	4	3	25

*The volumetric biogas production rate is calculated as 0.20 m^3/m^3 d

The feeding capacity of a digester is 80–90% of its working volume. Above the level of mixed liquid, sufficient space shall be left to prevent the gas pipe from blockage and to facilitate biogas collection. The minimum design feeding quantity is based on the principle of no overflow of biogas from the connection between the digestion chamber and the hydraulic chamber.

According to the total solid of the feedstock, the concentration of digestion liquid, the amount of water to be added to the feedstock, and the amount of ingredients can be calculated according to the requirements of the digestion process. The calculation is shown in Eq. 2.4.

a. Calculation of the total solid content in mixed feedstock

$$m_0 = \frac{\sum X_i m_i}{\sum X_i} \times 100\% \tag{2.4}$$

where

m_0 total solid content of the mixed feedstock;
X_i mass of feedstock i;
m_i total solid content of feedstock i

Assume: human excrement $X_1 = 100$ kg $m_1 = 20\%$

Pig manure $X_2 = 100$ kg $m_2 = 20\%$
Rice straws $X_3 = 98.9$ kg $m_3 = 90\%$

Calculate the total solid content of the mixture.

According to Eq. 2.4,

$$m_0 = \frac{100 \times 20\% + 100 \times 20\% + 98.9 \times 90\%}{100 + 100 + 98.9} \times 100\% = 43.2\%$$

That is, the total solid content of the mixed feedstock is 43.2%, and the water content is 56.8%.

b. Calculation of the amount water to be added to mixed feedstock

Make the mixed feedstock in the above example into a 10% mixed liquid, what is the amount of water needed, W?

$$M_0 = \frac{\text{weight of total solid}}{\text{weight of mixed liquid}} \times 100\% \tag{2.5}$$

$$10\% = \frac{\sum X_i m_i}{\sum X_i + W} = \frac{100 \times 20\% + 100 \times 20\% + 98.9 \times 90\%}{100 + 100 + 98.9 + W}$$

$$10\% = \frac{129.01}{298.9 + W}$$

$$W = 991.2\,\text{kg}$$

From the above calculation, it can be seen that 100 kg of human feces, 100 kg of pig manure, 98.9 kg of rice straws, and 991.2 kg of water can together prepare a mixed liquid with a solid concentration of 10%.

c. Calculation based on the carbon to nitrogen ratio and total solid concentration

The effective volume of an 8 m^3 rural household digester is 80% of the total volume, and the C/N ratio is required to be 25:1. The concentration of mixed liquid is 10%, and the total solid content of inoculants is 10%. The weight of added inoculants is 30% of the total weight of the feedstock. Now pig manure and rice straws are used as the feedstock. Take pig manure and rice straw as the feedstock and assume that the bulk density of the mixed liquid is 1, what is the weight of pig manure, rice straws, inoculant, and water that should be added (nitrogen and carbon contents of the inoculant are not considered).

The known variables:

pig manure $m_1 = 20\%$ rice straw $m_2 = 85\%$

$$C_1 = 7.8\% \qquad C_2 = 42\%$$

$$N_1 = 0.6\% \qquad N_2 = 0.63\%$$

Assume the required amount of pig manure is X_1; the required amount of rice straws is $X_{2;}$

Inoculum $m_3 = 10\%$, required amount of inoculum is X_3;

Calculate the required water amount, W (kg)

$$\frac{C}{N} = 25 = \frac{0.078X_1 + 0.42X_2}{0.006X_1 + 0.0063X_2}$$

Result: $X_1 = 3.646X_2$

$$m_0 = 10\% = \frac{X_1m_1 + X_2m_2 + X_3m_3}{8000 \times 80\%} = \frac{0.2X_1 + 0.2X_2 + 0.1X_3}{8000 \times 80\%}$$

$$6400 = 0.23X_1 + 0.88X_2 = 0.23\,(3.646X_2) + 0.88X_2$$

$$6400 = 0.838X_2 + 0.88X_2 \quad X_2 = 372\,(\text{kg})$$

$$X_1 = 3.646 \times 372 = 1358\,(\text{kg})$$

$$X_3 = (1358 + 372) \times 30\% = 519\,(\text{kg})$$

$$W = 6400 - 1358 - 519 - 372 = 4151\,(\text{kg})$$

According to the requirements of the feedstock ratio, the calculated results are 1358 kg of pig manure, 372 kg of straws, 519 kg of inoculum, and 4151 kg of added water.

2.4.3 Digester Design

(1) Inlet chamber

An inlet chamber is usually a brick structure, with a length of 400 mm, width of 500 mm, and depth of 300–500 mm. The size of the inlet chamber can also be adjusted according to local terrains.

(2) Digestion chamber and gas chamber

a. Determination of volume

The volume of a digestion chamber is determined by the amount of biogas consumption and the volumetric biogas production rate by the following equation:

$$V_{23} = \frac{Q}{R_p} \quad (2.6)$$

In the equation:

V_{23} volume of the digestion chamber (m^3)
Q daily biogas consumption (m^3/d)
R_p volumetric biogas production rate (m^3/m^3 d).

The volume for gas storage is half of the volume of daily biogas production.

b. Geometric shapes and symbols

The digestion chamber of a circular hydraulic digester is composed of three parts: a positive cut-off spherical cap, a cylindrical wall, and a reverse cut-off spherical cap. The geometric shape is shown in Fig. 2.11

D_0 inner diameter of the digester
R_0 inner radius of the digester
f_1 top spherical cap height
f_2 bottom spherical cap height
ρ_1 radius of the dome curvature
ρ_2 radius of the bottom curvature
H_0 height of the cylinder.

c. Calculate the geometric dimensions based on volume

Assume the internal volume of the digester dome is V_1, the internal volume of the digester bottom is V_2, and the internal volume of the cylinder is V_3, then by the volume equation of cylinders and spherical caps, we can get:

$$V_1 = \frac{\pi}{6} f_1 \left(3R_0^2 + f_1^2\right) \quad (2.7)$$

$$V_2 = \frac{\pi}{6} f_2 \left(3R_0^2 + f_2^2\right) \quad (2.8)$$

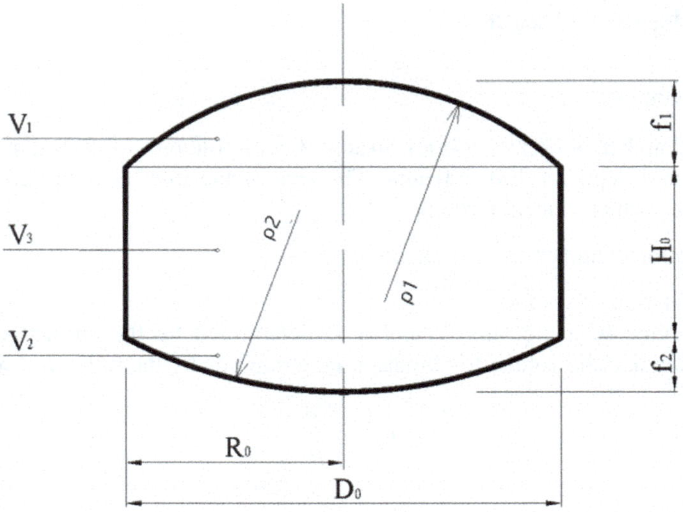

Fig. 2.11 Geometric shape of a circular digester

$$V_3 = \pi R_0^2 H_0 \tag{2.9}$$

$$V = V_1 + V_2 + V_3 \tag{2.10}$$

$$\rho_1 = \frac{R_0^2}{2f_1} + \frac{f_1}{2} \tag{2.11}$$

$$\rho_2 = \frac{R_0^2}{2f_2} + \frac{f_2}{2} \tag{2.12}$$

set:

the ratio of top spherical cap height f_1 to inner diameter of body as $\alpha_1 = \frac{f_1}{D_0}$;
the ratio of bottom spherical cap height f_2 to inner diameter of body D_0 as $\alpha_2 = \frac{f_2}{D_0}$;
the height of cylinder H_0 to inner diameter of body D_0 $\alpha_3 = \frac{H_0}{D_0}$.
Plug α_1, α_2, and α_3 into Eqs. 2.7–2.10, we can get:

$$V_1 = \frac{\pi \alpha_1}{6}(0.75 + \alpha_1^2)D_0^3 \tag{2.13}$$

$$V_2 = \frac{\pi \alpha_3}{6}(0.75 + \alpha_3^2)D_0^3 \tag{2.14}$$

$$V_3 = \frac{\pi \alpha_2}{4}D_0^3 \tag{2.15}$$

Generally, $\frac{1}{5}$, $\frac{1}{8}$ and $\frac{1}{2.5}$ are taken for α_1, α_2, and α_3 respectively. Equations 2.10–2.12 have now become:

$$V_1 = 0.8227D_0^3$$
$$V_2 = 0.0501D_0^3$$
$$V_3 = 0.3142D_0^3$$

Equation 2.7 becomes:

$$V = V_1 + V_2 + V_3 = 0.4470D_0^3 \tag{2.16}$$

According to Eq. 2.16, if we know the total volume V of a digestion chamber and a gas chamber, we can calculate the diameter D_0. Based on the relationships between α_1, α_2, and α_3, we can calculate f_1, f_2, H_0, and then according to the Eq. 2.8 and 2.9, ρ_1 and ρ_2 can be calculated.

d. Calculation of surface area

Surface area of a digester includes the inner surface, the outer surface, and the middle surface. Based on the surface area, the amount of building materials needed for digester construction can be calculated. Surface area F consists of the cover surface area F_1, the wall surface area F_2, and the bottom surface area F_3 All surface areas in the following equations refer to the inner surface area. For a digester with a large volume, thickness becomes not negligible, so the material consumption should be calculated based on its middle surface area.

$$F_1 = 2\pi\rho_1 f_1 = \pi\left(R_0^2 + f_1^2\right) \tag{2.17}$$

$$F_2 = 2\pi R_0 H_0 \tag{2.18}$$

$$F_3 = 2\pi\rho_2 f_2 = \pi\left(R_0^2 + f_2^2\right) \tag{2.19}$$

$$F = F_1 + F_2 + F_3 \tag{2.20}$$

(3) Hydraulic chamber

The volume of a hydraulic chamber is half of daily biogas production volume. Its shape is determined by topography, geological conditions, and some other factors.

2.4.4 Energy and Ecology Model with Biogas

With the development of household digesters, many places have developed a variety of agricultural biogas-based ecological models based on local natural and social environment. Such models include "pig-biogas-orchard" model in the south, "four in one" model in the north, and "five supporting facilities" in the northwest. These patterns have combined rural biogas production, courtyard economy, and ecological agriculture, as well as having transformed traditional production patterns, life styles, and ideology in rural areas. As a result, these patterns have realized efficient utilization of agricultural wastes as resources, improvement in agricultural production efficiency, sanitation of rural environment, and higher living standards in rural areas. Remarkable economic, ecological, and social benefits have been achieved.

2.4.5 "Pig-Biogas-Orchard" Model in the South

In the energy and ecology model of "pig-biogas-orchard", each household is the basic unit. Natural resources such as mountains, fields, water, and courtyards are used, and advanced technology is introduced. This model consists of livestock and poultry houses, digesters, and orchards. Moreover, digester construction is combined with that of livestock houses and toilets, forming a "trio" courtyard economy of breeding-biogas-planting (Fig. 2.12) and realizing ecological virtuous circle. Biogas is used as the energy supply for daily life of farmers in this operation model, while digestate is used as fertilizers for fruit trees and other crops. Vegetables and feed crops are

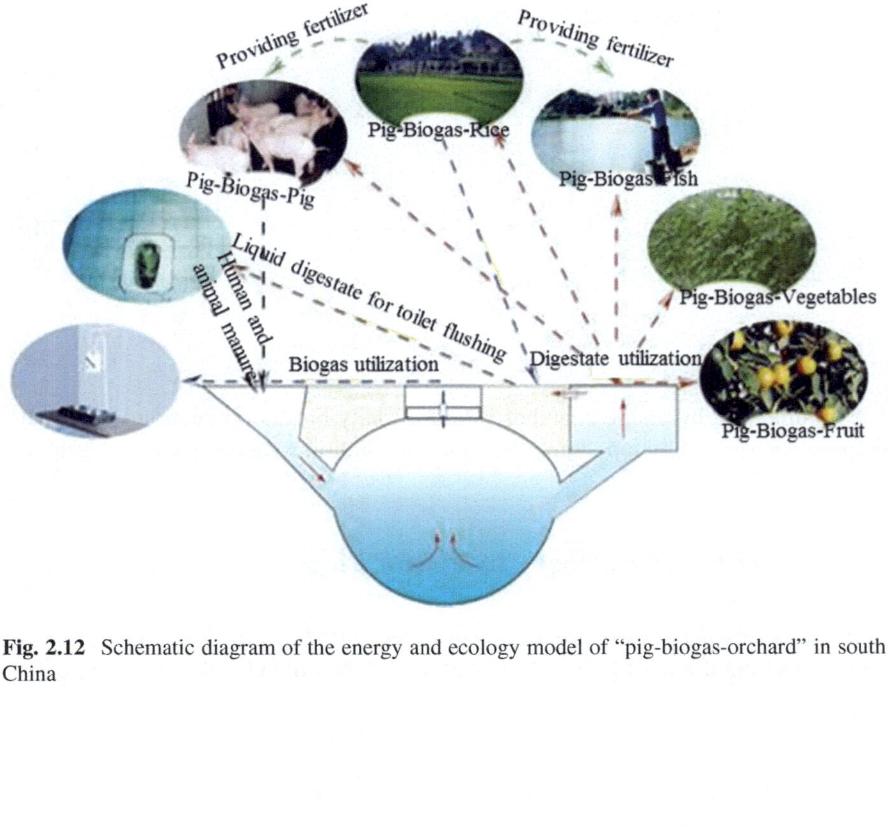

Fig. 2.12 Schematic diagram of the energy and ecology model of "pig-biogas-orchard" in south China

also grown in orchards to meet the feed demand of livestock and poultry. This model focuses on the agriculture-dominant industry and fully utilizes digestate. In addition to raising pigs, this model also allows cattle, sheep, and chickens. In addition to the combination with fruit industry, it also combines with grain, vegetables and cash crops to form derivative patterns such as "pig-biogas-fruit", "pig-biogas-vegetable", "pig-biogas-fish" and "pig-biogas-rice".

The planting area and breeding scale should take the digester capacity into careful consideration. Taking orchards and pigs as an example, a 5–8 m^3 rural household digester for household with four people should be supplemented with a 1.5 m^2 toilet and a 6–10 m^2 pig house with 4–6 pigs all year around, as well as an orchard covering about 0.26 hm^2.

2.4.6 *Energy and Ecology Model of "Four-in-One" in Northern*

"Four-in-one model" is a comprehensive utilization system built on farmer fields or yards. This model consists of a "quartet" of a digester, a greenhouse, planting, and breeding that combines anaerobic digestion with planting and breeding industry (Fig. 2.13). In this model, a greenhouse is built in the household courtyard, and a digester is built underground at one end of the greenhouse. Pig house and toilets are built on the digester, and vegetables or fruits are planted in the greenhouse. This model takes solar energy as the power source and biogas as the connection, combining planting and breeding and forming an ecological virtuous cycle.

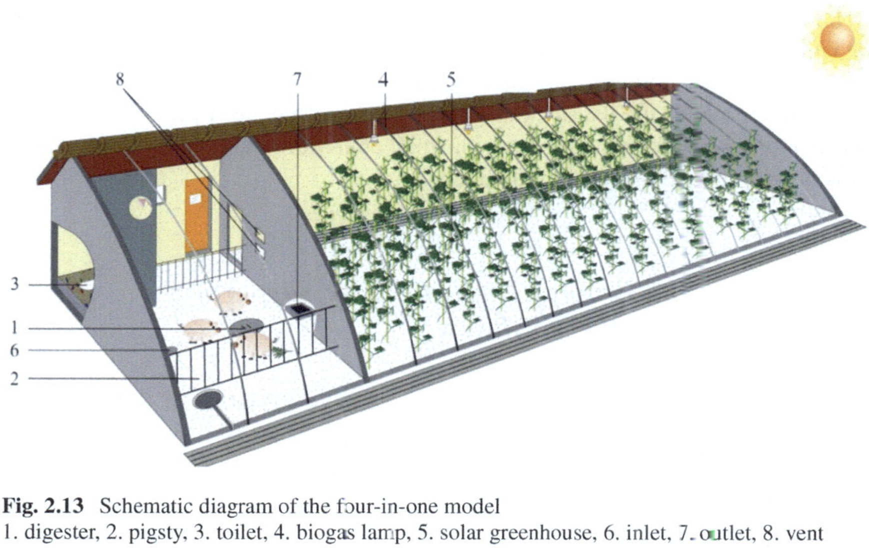

Fig. 2.13 Schematic diagram of the four-in-one model
1. digester, 2. pigsty, 3. toilet, 4. biogas lamp, 5. solar greenhouse, 6. inlet, 7. outlet, 8. vent

Greenhouse is a basic production unit of the model. It utilizes the light transmittance, plastic film, and composite thermal insulation walls to convert sunlight into heat to minimize heat and water loss, thus achieving the goal of temperature rise and heat and moisture preservation. As a result, production of anti-season fruit and vegetable is made possible, while safety of livestock and poultry during winter is ensured. Raising livestock and poultry in a greenhouse and combustion of biogas can both increase the temperature inside the greenhouse and provide carbon dioxide as gas fertilizers for crops. Photosynthesis of crops can also increase the oxygen content in livestock and poultry houses. Biogas and digestate produced by anaerobic digestion can be used for daily life and agricultural production while achieving improvement of the environment and living standards and promotion in production.

Digester is the core of the "four-in-one" pattern and plays the role of connecting breeding and planting. The digester is located at one end of the solar greenhouse, producing biogas while providing energy for living and production. At the same time, the residue of anaerobic digestion provides high quality organic fertilizers for vegetables, fruits and flowers.

Greenhouse is the main body of the "four-in-one" model, since the digester, livestock and poultry house, toilets, and cultivation area are all located in the closed greenhouse. A new type of energy-saving greenhouse is designed with reasonable daylight period theory and composite heat-carrying wall theory, and a reasonable daylight period should be more than 4 h.

Livestock and poultry house is the basis of the "four-in-one" model, which is designed according to the principle of daylight greenhouse, so it achieves not only heat preservation and temperature increase in winter, but also plays a role of cooling and sun blocking in summer. This ensures a whole year of growth of animals and shortens growth time while saving feed and promoting cultivation efficiency. Biogas production is enabled all year round.

The required area of the "four-in-one model" is determined by the size of the site, which usually ranges from 100 to 500 m^2. This pattern consists of a 20–25 m^2 livestock and poultry house built at one end of the greenhouse, a 1 m^2 toilet built at the north corner of the livestock and poultry house, and a 6–10 m^3 of household digester built underground. Farmers with small courtyards can build their livestock and poultry houses to the north of the greenhouse.

2.4.7 Energy and Ecology Model of "Five Supporting Facilities" in the Northwest

The "five supporting facilities" model in the northwest consists of five facilities, including a digester, a toilet, a greenhouse, a water cellar, and orchard irrigation facilities (Fig. 2.14). The digester is the core of this model, dynamically connecting fertilizers for rural production and energy for daily use. A valuable ecological cycle

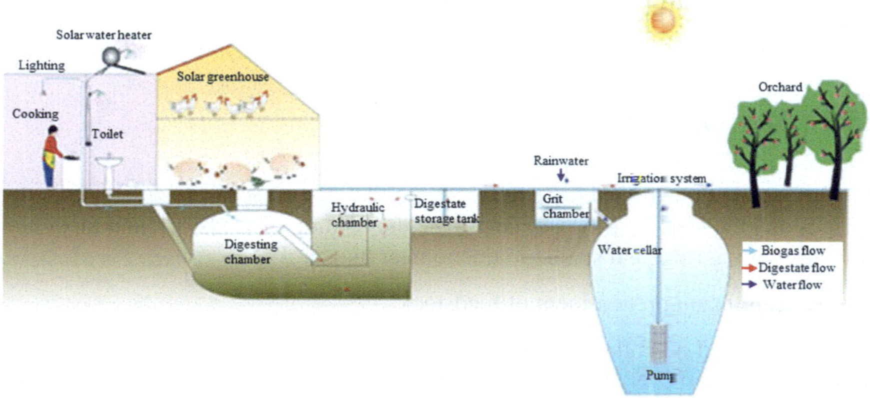

Fig. 2.14 Schematic diagram of northwest the "five supporting facilities" model

of breeding promoting biogas production, biogas production promoting fruiting, and combination of fruiting with breeding.

Digester is the core of this model, which connects breeding and planting, as well as energy and fertilizer production. On one hand, construction of digesters can improve rural ecological environment. On the other hand, the liquid digestate can be used in spraying fertilization and as pesticides on leaf surface of fruit trees. Solid digestate can be used for fertilization in orchards so as to realize the goal of energy utilization, crop and fruit production, and living standards improvement.

Animal house heated by solar energy is the premise of this pattern. By utilizing solar energy, the issue of low temperature in winter is solved, and the growth rate of pigs and biogas production rate are improved.

Water cellars and water collecting tanks are facilities for collecting and storing surface runoff, rain, snow and other water resources. Equipping water collection system for orchards not only provides water for digestion, insecticide spraying, and household use by people and animals, but it also makes up for water use during orchard drip irrigation and hole irrigation during water shortage period, thus preventing negative effect on fruit tree growth by water shortage.

Orchard irrigation facilities carry rainwater stored in the water cellar through water pumps and then transfer water to drip irrigation emitters through water pipes. Water droplets or jets slowly and equably drips to the roots of fruit trees. Combined with irrigation, fertilization can also be achieved by carrying liquid digestate produced by anaerobic digestion to the root of fruit trees, so that the areas around roots of fruit trees are often maintained in a proper condition.

The typical structure of this pattern includes a 0.33 hm^2 mature orchard as the basic production unit, with grass planted in between rows of fruit trees to retain soil moisture and to promote livestock growth. A 12 m^2 solar energy pig house, an 8–10 m^3 rural household digester, a 40–60 m^3 water cellar, and a simple drip irrigation or seepage irrigation system are usually constructed along with this type of orchards.

2.5 Start-up and Operation Management of Household Digesters

2.5.1 Start-up of Household Digesters

Whether newly constructed digesters or those that have discharged a large amount of digestate, the start-up process starts from the feeding of feedstock and inoculum and ends with normal and stable productions of biogas.

(1) Collection and pretreatment of feedstock

Feedstock should be collected in the vicinity of the digester, the ratio should meet the requirements of C/N 20-30:1, and the concentration of TS should be 6–10%. The feedstock from crop waste must be chopped or crushed to under 6 cm in size. Chopping not only destroys the surface of the straw, but also promotes the contact between the feedstock and microorganisms, speeds up the decomposition, and facilitates feed inflow and outflow. In order to avoid a large amount of floating crust after feedstock enters the digester, C/N should be reduced to an appropriate level to facilitate digestion and startup. Before sending feedstocks into the digester, feedstocks that have a higher fiber content can be acidified and composted.

(2) Preparation of inoculants

When preparing feedstock, inoculants should be prepared as well. In general, part of the mixed liquid is taken from the normal operating digester as inoculants. Sludge from the manure pit, sewage ditch, and river mud rich in microorganisms can also be used as inoculants. The inoculation amount is generally 10–30% of the digester volume. The more inoculation is used, the faster biogas production and the shorter the start-up time becomes.

(3) Feeding and water addition

The proportion of ingredient during start-up of the digester follows inoculum: raw material: water = 1:2:5. Taking an 8 m^3 digester as an example, the inoculation is about 1 m^3, and feedstock is about 2 m^3, with about 5 m^3 of water. The collected inoculants and feedstocks are pretreated and put into the digester according to the above ratio. Crop straws can be put into the digester by the movable cover, while inoculants and animal manures can be directly put into the inlet chamber. Water takes a large proportion in the mixed liquid, so water temperature has a great influence on mixed liquid temperature. During startup, it is best to take liquid digestate that is rich in bacteria and is with high temperature from a digester that performs normal biogas production. When starting during warm seasons, water can also be heated by sun inside the water tank in advance, and direct addition of low temperature water should be avoided.

(4) pH determination

After sending in feedstock and water and before sealing the movable cover, pH of the mixed liquid should be tested. If pH is 6.8–7.5, the digester can be sealed.

(5) Digester sealing

Before sealing the digester, the movable cover, its edge, and the water storage ring should be cleansed and sealed with adhesive clay. Water should be then added into the water storage ring to keep the sealing clay moist. At the same time, if there is gas leakage, bubbling at the movable cover can be detected.

(6) Deflation to test gas

At the beginning of digestion, there is some amount of air in the biogas chamber, and a large amount of carbon dioxide is produced at the initial stage of anaerobic digestion. Although a lot of biogas is produced after one or two days, the biogas cannot be ignited due to its low methane content. Generally, when the gas pressure rises to 4000–6000 Pa (40–60 cm), biogas should be deflated to test ignitability. After 2–3 times of deflation, with the increase of methane content in the biogas, the generated biogas can be ignited for use. Note that the fire test should be carried out on the cooker, not near the inlet and outlet of the digester or on the gas-tube in case of explosion caused by backfire.

(7) Normal operation and management after startup.

When biogas generation enters the peak period with stable daily production, the anaerobic microbial activities in the digester have reached optimal state, and the activities of acid-producing bacteria and methanogens have reached an equilibrium. Meanwhile, pH value is relatively suitable. At this moment, the start-up phase of anaerobic digestion is completed, and the biogas production enters the normal operation and daily management phase.

2.5.2 *Operation and Management of Household Digesters*

It is said that smooth operations of a digester depend 30% on construction and 70% on management, which indicates that the daily operation and management is very important for the normal performance of the digester. Special attention should be paid to the following aspects in daily management of the digester.

(1) Feeding and discharging

The feeding and discharging process should be based on the purpose of digester. For the combined three digesters, human and animal wastes are produced on a daily basis and enter the digester automatically. Therefore, feeding and discharging should be conducted in a continuous way. For the purpose of producing biogas and fertilizer, feeding should be intermittent. Attention should be paid to the following three problems. First, discharging should be done prior to feeding, and their rates should be equal. Second, when discharging, the level of mixed liquid should not be lower than the upper edge of the connecting hole between the digestion chamber and the inlet or outlet to prevent biogas from escaping. Third, if it is found that the liquid level is lower than the upper edge of the connecting hole after feeding and discharging, an appropriate amount of water should be added immediately for sealing.

(2) Appropriate stirring

Rural household digesters are generally not installed with stirring devices. There are two ways of stirring. One is using a long stick or other similar tools to insert into the digester from the outlet (or inlet) pipe, and the mixed liquid should be agitated vigorously back and forth for several times to mix the feedstock. The other way is to use submerged pump to pump out a certain amount of mixed liquid into the inlet chamber, and then the mixed liquid automatically flow back to digestion chamber. This can stir the feedstock and break down floating scum.

It should be pointed out that only with sufficient feedstock can frequent stirring improve the biogas output. If the there is no feeding or discharging in the digester, stirring is of little or even no effect.

(3) Maintain the appropriate concentration of feedstock

Because of seasonal utilization of organic fertilizers, digesters need intermittent feeding and discharging. Therefore, water quantity inside a digester should often be adjusted to maintain the appropriate concentration of feedstock. Too much or too little water in the digester is not conducive to normal anaerobic digestion. In general, the liquid content of feedstock in rural household digesters should be 85–95% of the total volume of digester. The concentration of dry matter should be no less than 6% in summer and no more than 12% in winter. Too high or low concentrations of dry matter need to be adjusted.

(4) Maintain a certain temperature

At present, rural household digesters in China are mostly built underground and therefore are prone to be affected by ground temperature. The temperature of feedstock should be above 10 °C so as to ensure normal anaerobic digestion. In winter, atmospheric temperature is low in northern regions, which lowers the temperature inside digester. If temperature of mixed liquid is below 10 °C, normal biogas production stops. In this case, measures should be taken to preserve or introduce heat to ensure the normal activity of microorganisms and facilitate normal biogas production. Common methods of heat preservation include covering greenhouse, digging cold-proof ditches, and covering heat preservation materials. In addition, when temperature is low, biogas pipes are likely blockage by condensed water. When condensed water is found, it should be removed immediately.

(5) Check pH

Excessively high or low pH of mixed liquid in digesters is unfavorable to anaerobic digestion processes. Therefore, pH needs to be tested. pH is usually determined by a pH meter, a pH test paper, or an acid-base indicator. In general, pH can be roughly determined by the color of the mixed liquid. Black color indicates a normal pH.

Yellow color indicates a pH lower than required. In addition, when the pH is low, the mixed liquid gives a sour smell.

During start-up of the digesters, the acidification rate exceeds the methanogenesis rate due to excessive feed or insufficient inoculum, resulting in accumulation of a large amount of volatile acid that reduces the pH to below 6.5. This will affect the microbial activity and the biogas output. Measures should be taken to adjust pH, usually by the following methods:

a. When acidification of mixed liquid is not serious, feeding can be stopped until pH recovers naturally as the volatile acid is consumed by microorganisms gradually.
b. Supplement inoculum to improve the metabolic rate of the volatile acid;
c. Add a proper amount of plant ash and stir evenly;
d. Add lime to adjust pH in an appropriate way. Lime should not be added directly and it should be added with clear limewater. Limewater should be mixed evenly with the feedstock to avoid the influence of strong alkali areas on microbial activities. When adding limewater, check the pH of the liquid with a pH test paper.

After one month of anaerobic digestion, the organic matter is consumed significantly. If no new feedstock is added, pH will gradually increase. When the pH is greater than 8, microbial activities will be affected by the strong alkaline. If this happens, new feedstock should be added, and water should also be added to adjust its concentration.

(6) Reloading

After operating for some while, sediments of a large amount of non-biodegradable solid residues will be found at the digester bottom. The accumulation of residues occupies digester volume, thus reducing the effective volume as well as biogas production. Therefore, based on the type of feedstock used, regular reloading is required. Feedstocks with a high straw content require a short reloading period while those with high manure content require a longer reloading period. In general, reloading takes place 1–2 times per year. It should be carried out when temperature is high. At the same time, the discharged liquid and solid digestates are used as agricultural fertilizers. When reloading, enough feedstocks should be prepared for digestion. The following measures should be taken.

a. Feeding should be stopped before reloading. This can avoid the waste of feedstock and store the saved feedstock for reloading.
b. Prepare enough feedstock. After discharging, put in new feedstock and start up immediately.
c. Reserve enough inoculum. During thorough reloading, sediments should be removed, leaving 10–30% of the original mixed liquid as inoculum.
d. Avoid negative pressure in the digester. When discharging during reloading, the digester should be connected with the atmosphere to prevent vacuum inside the digester. Once there a vacuum forms, it will cause the falling off of paint and poor sealing, sometimes even leakage.

e. Prompt feeding. After discharging, the digester should be checked and repaired immediately, and then feed can be added with water. Since the digester is built underground, the internal and external pressure is in equilibrium when feeding. When discharging, the pressure of mixed liquid exerting on the wall and bottom of digester is zero, indicating loss of equilibrium. Under such circumstances, the groundwater is prone to exert pressure onto the digester and damage its wall and bottom. Risks are high in places where ground water level is high during raining season, so water and feedstock should be added immediately after discharge.

2.5.3 Safety Management of Household Digesters

(1) Safe digestion

In a digester, when anaerobic digestion microorganisms contact with harmful substances, the microorganisms will be poisoned and stop metabolic activities. They may lose activity or even die, leading to the halt of biogas production. Therefore, the following harmful substances should be avoided in digesters.

a. Various highly toxic pesticides, especially organic fungicides, antimicrobials and insect repellents.
b. Heavy metal compounds, toxic industrial wastewater, salts, etc.
c. Sterilized livestock and poultry waste, crop stems, and leaves sprayed with pesticides.
d. Some plant stems and leaves, such as bitter melon vine, paulownia leaf, Chinese parasol leaf, peach leaf, walnut leaf, locust willow leaf, horse fruit and so on.
e. Spicy food, such as onion, garlic, pepper, leek, and radish stem and leaves.
f. Diesel oil, laundry powder, etc.

If it is found that microorganisms are poisoned and the digester stops producing biogas, the mixed liquid should be discharged immediately, only reserving about 20% of it in the digester as inoculum, and then put in new feedstock and adjust to normal biogas production.

(2) Daily safety management

Biogas is flammable and combustible. Hydrogen sulfide in it is a highly toxic substance, so safety management should be reinforced when using digesters.

a. Inlet and outlet chamber must be covered to prevent people and animals from falling in and to prevent accidental casualties.
b. When a digester produces too much biogas, the pressure in it is too high, and the produced biogas should be used immediately to prevent it from breaking out of the movable cover and causing damages or accidents. If the movable cover gets broken, put out the fire nearby immediately to prevent fire disaster.

c. Ignition at the inlet, outlet, or biogas pipes of a digester is strictly prohibited to avoid tempering.
d. Regularly check whether there is leakage of the biogas pipes, switches, and interfaces. If leakage is detected, replace or repair it immediately. When biogas is not used, switches should be turned off immediately;
e. If indoor biogas leakage is detected, one should immediately open doors and windows, cut off biogas sources, forbid the use of open fire, prohibit smoking, and evacuate people to prevent further poisoning. Repair the leakage when there is no odor inside the room.
f. If biogas pipes are blocked by freezing, hot water should be used to melt it. Heating by fire is strictly prohibited.

(3) Safety maintenance

When a digester needs safety maintenance, the following points require attentions

A. If the maintenance personnel need to enter the digester, they should first open the biogas pipe and then open the movable cover, the inlet and outlet cover. Next, they should discharge the mixed liquid in the digester with equipment and tool and rinse them with water for three times before finally discharging the sewage.
B. Digester ventilation. Blow air into the digester until there is enough oxygen for personnel to enter. Let small animals enter first before personnel. If no abnormalities occur within 30 min, the maintenance personnel can enter the digester under the supervision of personnel outside.
C. While staying in the digester, continuous and forced ventilation must be ensured. Flashlight or electric light can be used for lighting. Open fire is prohibited to prevent biogas explosion.
D. Personnel entering the digester must wear a safety belt. If dizziness, suffocation or any other issues are found, they should be evacuated and rescued immediately.
E. Direct ignition and biogas test are strictly prohibited on top of the digester and at the upper part of the biogas pipe so as to prevent explosion caused by tempering and casualties.
F. Biogas lamps, stoves and biogas pipelines should be kept away from combustibles. Biogas lamps should be kept away from the house roof, generally above 50 cm away.

2.5.4 Maintenance and Repair of Household Digesters

To ensure a relatively long service life of rural household digesters under normal biogas production condition, regular repairs and maintenance are necessary.

(1) Inspection of digesters
A. External check

Apply soap water to the biogas pipelines, valves and joints of pipes and wait until biogas pressure increases to see if the soap water bubbles. Leakage is also easy to occur at the connection between the biogas duct and the movable cover plate, as well as at the joint between the movable cover plate and the digester body. In general, the upper part of the movable cover plate has a water storage circle, and leakage can be judged by bubbles in the water.

B. Internal check

After thoroughly discharging the mixed liquid and sediment, and air ventilation to ensure safety, enter the digester and check whether there are cracks, blisters, and small pores on the wall, bottom, cover, inlet, outlet, and the connection parts. Tap the pool with fingers and a small wooden stick, and a hollow sound indicates that the cement has a warped shell or a hole.

(2) Digester maintenance

After checking, the digester should be repaired. Maintenance mainly includes the following aspects.

A. Digesters under normal conditions also requires regular maintenance. Generally, it is carried out at the same time with reloading. After a thorough discharge, use high grade cement to paint the inner wall of digester for 2–3 times.
B. Maintenance of abnormal digester

Repair of wall cracks. If the building material is stone, the cracks should first be dig into "V" shape groove, and embed iron or stone chippings into the groove. Rinse the groove thoroughly with water and plaster it with cement and mortar before painting with high grade cement for 2–3 times. If the digesters are made of concrete or bricks, the area adjacent to cracks can be roughened. Block the leakage with 1:1 cement and mortar, and compact and level it with force before painting with high grade cement for 2–3 times. If the painted mortar on biogas chamber has fallen off or warped, it should be removed entirely followed by cleansing, plastering, and painting with high grade cement for 2–3 times. If gas leakage occurs at the connection between the biogas duct and the movable cover plate, the joint shall be cleaned and re-connected with cement mortar. Next, heighten and enlarge the cement mortar protector base. For places with cracks on the bottom surface of the digester and the joints around the wall, a 2 cm wide and 3 cm deep edge groove should be cut around the cracks first. Next, 4–5 cm thick concrete should be poured onto the bottom of the digester and the trenches around to make them a whole. For biogas chambers with slight biogas leakages, high grade cement and 1:1 cement mortar can be used to alternately paint for 2–3 times. The above painting material includes high grade cement and new-style sealing materials such as water glass or plastic glue with good sealing effect.

(3) Digester maintenance

A. Moisture care for digester
At present, digesters in China are generally made of concrete. Since cement is a porous material, dry weather will open the capillary holes and cause leakage. Therefore, attention should be paid to ensure the moisture of digester for a long time. In

some places, newly built digesters are implemented with a layer of water seal, which has good moisture retaining performance. Some digesters are covered with 25 cm thick soil layers, on which vegetables and flowers be grown in order to maintain moisture.

B. Prevent sun exposures of empty digesters

If the newly built digesters have met quality requirements after inspection, they should be loaded with feedstock immediately to avoid empty digesters. Reloading should not be carried out in dry season. If the agricultural season needs fertilizers, sufficient feedstock should be prepared during reloading, shorten the empty time as much as possible and load feedstock timely. When the feedstock is not enough, the digester cannot be emptied and should be covered to prevent sun exposures.

C. Add moisture layer

In order to prevent evaporation of water from the top of a digester, a layer of asphalt can be coated on top, or a layer of plastic film can also be laid to cut off the capillary pores of the soil and prevent water evaporation. It is required that the covering area of any kind of moisture protective layer should be larger than the cross sectional area of the digester top.

D. Corrosion resistant

The inner layer of a digester is often eroded by feedstock (slightly acidic or alkaline), causing slight corrosions to cement (or other building materials). When a digester has been used for several years, the sealing layer is damaged, and part of the building materials and painted cement will fall off. During annual reloading, the uneven parts of digester wall should be leveled and painted with pure cement for one to two times so as to restore the sealing performance.

Chapter 3
Biogas Digester for Domestic Sewage Treatment

Before the 1980s, manure in small and medium-size cities and towns was collected in cesspit before getting emptied and cleaned by neighboring farmers and then transported to farmland as fertilizer in China. After the 1980s, the township-collective-household system was promoted in rural areas. With this transformation of production system, rural labor forces transferred to cities and industries. In addition, due to easy application and high efficiency of chemical fertilizer, farmers no longer needed to clean manure in cities for fertilizer supplies. As a result, excrements in urban cesspools accumulated and overflowed with wastewater, breeding flies and maggots, spreading diseases, polluting the environment, and seriously threatening health of the residents. Although at the moment the newly built dwelling had septic tanks, their volume was relatively small, therefore the retention time of sewage was short, and treatment effect was poor, which was harmful to health and environmental protection. Septic tanks require short clearance period, otherwise their effective volume would be depleted quickly, losing the function of sedimentation and decontamination. Therefore, it's urgent to find a satisfactory technology to treat sewage. According to the social and economic situation at that time, Chinese researchers engaging in biogas development and promotion were inspired by the latest research advancement in anaerobic digestion and developed biogas digester for domestic sewage treatment. Such digesters are also known as biogas septic tanks, developed on the basis of hydraulic digesters and septic tanks. In the beginning, biogas digester for domestic sewage treatment were mainly used for treatment of domestic sewage from small and medium-sized towns. With the construction of centralized sewage treatment plants and rising discharge standards, biogas digester for domestic sewage treatment gradually shifted to treatment of decentralized domestic sewage in villages and towns, rural public facilities, and tourist attractions.

L. Deng et al., *Biogas Technology*,
https://doi.org/10.1007/978-981-15-4940-3_3

3.1 Characteristics and Amount of Rural Domestic Sewage

Rural domestic sewage refers to sewage produced from daily living activities of farmers, including flushing, washing, and showering, as well as from discharge from kitchen, primary and middle schools, public facilities, tourist spots, hotels, restaurants and so on. Domestic sewage has to exclude industrial wastewater and wastewater from scaled livestock and poultry breeding farm. Production and discharge of rural domestic sewage have their own feature and rules, which differ greatly according to geographic location, climate, and living habit. In general, rural domestic sewage has the following characteristics:

(1) Fluctuation in characteristics and amount of sewage

Unlike urban water sources, rural water sources are more diverse and have included river water, well water, tap water, and so on. Tap water is used as drinking water, while river water and well water are supplementary sources used for laundry, floor cleaning, animal feeding, and so on.

Due to differences in economic development level, living habit, customs, and seasons, characteristics of rural domestic sewage varies. Rural domestic sewage mainly comes from different stages of household water use, among which toilet flush wastewater is the major source of the pollutants. The amount of its COD, total nitrogen, and total phosphorous account for 58.3, 86.3 and 80.5% of the total respective discharge per capita per year. Kitchen sewage is the main source of BOD in rural domestic sewage, accounting for 50.2% of the total BOD discharge per capita per year. Domestic wash wastewater discharge is of the largest quantity, accounting for about 50% of rural domestic sewage discharge (Hou et al., 2012). Rural domestic sewage mainly contains organic pollutants and generally does not contain heavy metals and toxic or harmful substances, although it does contain a decent amount of synthetic detergent, bacteria, viruses, parasitic worm eggs, and so on.

The overall amount of rural domestic sewage is small, and its discharge is uneven with obvious fluctuations. Coefficient of variation (the ratio between maximum daily and hourly sewage discharge and average hourly sewage discharge on that day) ranges from 3.0 to 5.0 and even increase to above 10.0 in some extreme occasions. Due to the similar life patterns among all rural households, there are discharge peaks in the morning, noon and afternoon. At night, discharge is largely reduced to a small amount or to even none, namely discharge is in discontinuity. Moreover, if a village happens to be a tourist spot, the coefficient of variation varies greatly not only from day to night, but also from season to season. (Sun 2010).

The number of villages in China is great, with drastically different population density among various villages. The discharge amount of any specific village or individual household can only be confirmed by on-site investigation. In regions where such data is unavailable, the discharge amount can be calculated by water usage. The calculation method is based on estimated discharge efficiency. Discharge from showering and flushing toilet accounts for 60 to 80% of the corresponding water usage, whereas laundry water discharge is about 70% of laundry water usage. As to kitchen wastewater discharge, the purpose of kitchen water uses needs to be confirmed with local residents, which depends on the livestock and poultry farming situation. If wastewater is all discharged through a drainage system, the discharge amount can be estimated as 60% its water usage. Generally, when designing a wastewater treatment system for a village, the discharge amount is estimated as 60–80% of the total water usage. Different rural areas have different natural environments, economic conditions, and living habits, and their water usages therefore varies greatly. See Table 3.1 for rural domestic water consumption in northeast, north, northwest, southwest, central south, southeast and other six regions of China given in the *Technical Guidelines for Rural Domestic Sewage Treatment* by Ministry of Housing and Rural-urban Development of China.

Rural domestic sewage includes black water and grey water, the former of which mainly refers to flushing wastewater from toilet and livestock barns while the latter of which mainly refers to kitchen drainage, washing drainage shower drainage, and supernatant of black water after treatments in a septic tank or a digester. The concentration of pollutants in black water is significantly higher than that in grey water. A comparison of major characteristics of the two kinds of wastewater is shown in Table 3.2 (Li 2015).

When designing rural sewage treatment plants, characteristics of rural sewage generally needs to be measured on site. When on-site determination is not possible, refer to domestic sewage characteristics listed in Table 3.3.

Table 3.1 Reference values of domestic water consumption of rural residents

Type of rural residents	Water usage (L/person·day)
Good economic condition, with flushing toilets and shower facilities	75–200
Relatively good economic condition, with flushing toilets and shower facilities	40–120
Worse economic condition, without flushing toilets and with simple sanitation facility	30–90
Without flushing toilets and shower facilities; mainly use surface water and well water	20–70
Tourists (who live in rooms with independent shower facilities)	150–250
Tourists (who live in rooms without independent shower facilities)	80–150

Table 3.2 Comparison of the characteristics of the black water and grey water

Parameter	Black water(mg/L)	Grey water (mg/L)	
		Southern region	Northern region
COD	1000–2000	200–350	350–500
TN	200–600	10–30	20–50
TP	20–60	0.5–4.0	2.0–7.0

Table 3.3 Reference range of characteristics of rural domestic sewage unit: mg/L

Parameter	pH	SS	COD	BOD_5	NH_3-N	TP
Value Range	6.5–8.5	100–200	100–450	50–300	20–90	1.0–7.0

(2) Large Number, Wide Distribution, and Small Scale

According to the *Construction Code for Municipal Sewage Treatment Projects (2001)* issued by the Ministry of Housing and Rural-urban Development of China, the scale of sewage treatment plants is divided into five categories, among which the smallest V type scale is 10,000–50,000 m^3/d. In china, a sewage plant with daily treatment capacity lower than 10000 m^3 is defined as a small scale sewage treatment plant. However, the daily treatment amount of rural sewage is far less than 10,000 m^3, which is way below the scale of small sewage treatment plants. Taking Muchuan County of Leshan City, Sichuan Province as an example, 18 village sewage treatment plants were built from 2013 to 2015, with a total treatment capacity of only 6900 m^3/d and an average daily treatment capacity less than 400 m^3/d, which is far less than that of a small scale sewage treatment plant. In the meanwhile, there are a large number of such kind of minimal plants widely distributed. In most villages and towns, the domestic sewage discharge amounts and distributions are similar to those of Muchuan county, featuring in scattered population, a large number of sewage discharge spots, and a small discharge amount at each spot. Therefore, the technological requirement of rural sewage treatment is quite different from that of urban concentrated sewage treatment plants.

(3) High Biodegradability

Although rural sewage differ in characteristics, all sewage shares some characteristics in common overall. The organic matter concentration of rural domestic sewage is higher, with a BOD_5 ranging from 120 to 300 mg/L, a COD of 150–500 mg/L, and a BOD_5/COD ratio of 0.4–0.5. Rural sewage is one with preferable biodegradability, and it is very suitable for biological treatment method.

3.2 Characteristics and Scope of Application of Biogas Digester for Domestic Sewage Treatment

Biogas digester for domestic sewage treatment is a technology and device based on hydraulic digesters and traditional septic tanks. It is suitable for treating decentralized domestic sewage by measures such as utilizing biological packings and filter materials. Germany, India and some other countries call such system DEWATS (Decentralized Wastewater Treatment Systems) (Sasse 1998). This technique is mainly used to treat wastewater of public toilets in small towns and villages, and it can be used in regions where underground water temperature is above 10°, or in sunlight greenhouses where temperature can rise up to 10°. Sewage retention time is normally more than 3 days, which ensures enough time for sewage treatment. Biogas digester for domestic sewage treatment can be constructed in a scattered way in accordance with the reality of rural sewage, which is discharged in a scattered way with obvious amount fluctuation. The effluent quality of biogas digester for domestic sewage treatment is better than that of septic tanks and household digesters, and its effluent quality can meet the national hygiene standard and some lower grade sewage discharge standards. In the process of domestic sewage treatment, there is basically no power consumption, low operation cost, and convenient operation and management. Since the 1980s, rural energy management departments at all levels in China widely promoted this technology and achieved satisfactory results in health and environmental protection.

There are three kinds of scattered anaerobic treatment devices for treating rural domestic sewage, namely household digesters, septic tanks, and biogas digester for domestic sewage treatment. See Table 3.4 for a comparison of these three types. Household digester is mainly used to generate energy, and its feedstock is mainly human and animal manure, organic waste, straw, etc. A septic tank is mainly used for primary treatment of manure, and the purpose of the treatment is mainly to kill pathogenic microorganism, achieve sanitation, and to improve fluidity of manure for easy transportation. The pollutant concentration in the effluent of septic tanks is still high and cannot meet the discharge standard, so the effluent is usually discharged into sewage treatment plant for further treatment. Biogas digester for domestic sewage treatment is mainly used for treating manure and domestic sewage, with the major purpose of decontamination and secondary function of producing renewable energy. Its effluent quality is relatively high and can be directly discharged or further treated.

Table 3.4 Comparison of several traditional devices for anaerobic treatment of domestic sewage

Item	Household Digester	Septic Tank	Biogas digester for domestic sewage treatment
Construction target	Energy, environmental, and sanitation purposes	Environmental and sanitation purposes	Energy, environmental, and sanitation purposes
Treatment process	One stage and simple	Multi-stage and simple	Multi-stage and complex
Treatment feedstock	Human and animal manure and organic waste	Human manure and sewage	Human manure and sewage
Feedstock concentration	High	Middle, low	Middle, low
Retention time	60–90d	0.5–1.0d	3–5d
Effluent treatment	Land use	Further treatment	Land use, discharge, or further treatment

Biogas digester for domestic sewage treatment is a sewage treatment technology that is relatively easy to construct, manage, and maintain. It can be applied in the following scenarios:

1. Residential areas, office areas, living areas of factories, schools, and other public places in small and medium-size towns with no near-future plans to build sewage treatment plants.
2. Treatment of manure and other domestic sewage of decentralized independent residence and public toilets that are not included in urban sewage network.
3. Treatment of domestic sewage of tourist attractions.
4. Treatment of manure and other domestic sewage collected from decentralized residents and public toilets.

3.3 Technical Process and Structure of Biogas Digester for Domestic Sewage Treatment

3.3.1 The Process of Biogas Digester for Domestic Sewage Treatment Technique

Biogas digester for domestic sewage treatment can be divided into separate system treatment and combined system treatment based on whether manure and domestic sewage enter into the same tank.

In a separate system process, manure and other domestic sewage is discharged separately by two independent pipe systems. The process flow can be seen in Fig. 3.1.

In a separate system, back waster enters into a sedimentation digester after the removal of solid impurities (pretreatment area I), so that the digestion time of black water can be increased while other domestic sewage directly enters an anaerobic filtration tank after sedimentation (pretreatment area II).

In a combined system, black water and other domestic sewage share the same pipe, that is, the two kinds of sewage enter the tanks through the same pipe system. The process flow of a combined system is shown in Fig. 3.2.

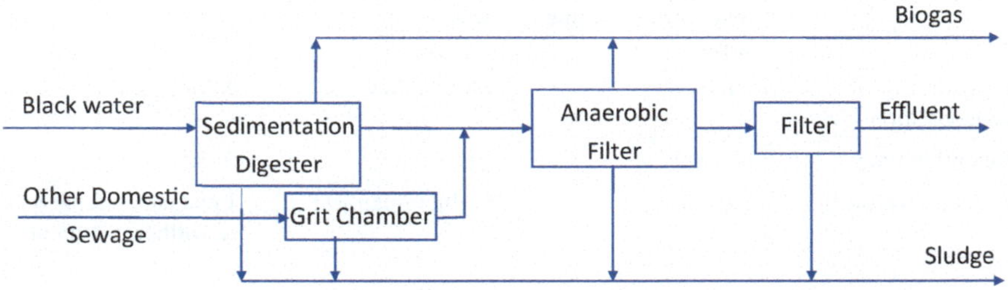

Fig. 3.1 Flow chart of biogas digester for domestic sewage treatment with separate system

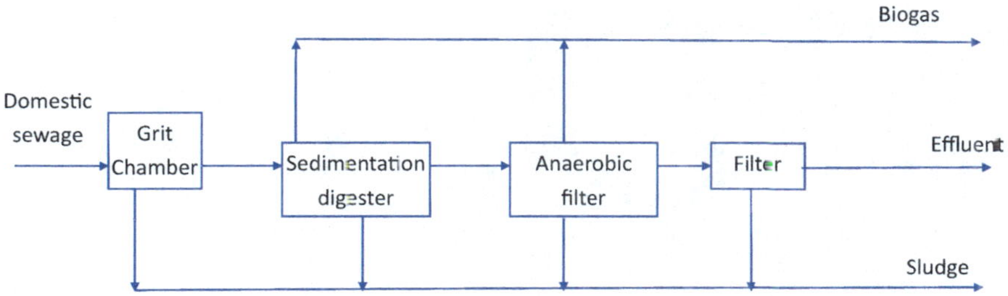

Fig. 3.2 Biogas digester for domestic sewage treatment with combined flow system

3.3.2 The Process Units of Biogas Digester for Domestic Sewage Treatment

A biogas digester for domestic sewage treatment mainly consists of a grit well (tank), a sedimentation digester, an anaerobic filtration tank, and a facultative filtration tank.

(1) Grit well (tank)

A grit well is mainly used for removing relatively large inorganic impurities such as sand, glass, metals, plastics, and dregs in sewage to avoid sedimentation and congestion and to reduce effective space.

(2) Sedimentation digester

In most places, sedimentation areas and anaerobic digestion areas are constructed as a whole, while in some places, the two are separately constructed. A sedimentation digester is also known as the pretreatment area I, with a function similar to that of tradition septic tanks and hydraulic digesters. It sediments and separates inorganic and organic matters, as well as parasite ova in sewage, and it then removes some suspending matters to prevent too much entrance of them into anaerobic filters. The organic matters in sedimentation digesters can be discomposed by anaerobic microbes. In addition, anaerobic digestion can remove some pathogenic microbes.

(3) Anaerobic filter

Equipped with packing, an anaerobic filter is also known as the pretreatment area II, with the function of removing the majority of dissolvable organic pollutants. Following are some common types of packings in anaerobic filters.

1. Soft packing. Its shape imitates a kind of natural algae. It is usually made of vinylon, and the soft filler is inexpensive with anti-corruption and anti-biodegradation characteristics. It has a large specific area and is suitable for up-flow anaerobic filters. Because of an inconsistent influent flow and an overall small flow rate, soft packing can easily cluster and has a short service life (Fig. 3.3a).
2. Semi-soft packing is similar to soft packing, except that a plastic (PVC or PE) ring is added inside soft fillers of semi-soft packing to avoid bonding. It has much smaller specific area compared with soft packing and it is more expensive.

a. Soft packing b. Elastic packing

c. Crushed stone packing d. Plastic hollow packinvg

Fig. 3.3 Four types of packing

3. Elastic packing. Its resilience facilitates even water distribution. It is expensive and its specific area is smaller than that of soft packing (Fig. 3.3b).
4. Hard packing. Hard packing has various types, including inorganic packing such as expanded ceramsite, slags, coke, and gravel, and organic packing such as hollow plastic ball and pall ring (Fig. 3.3d). Hard organic packing is anti-corruptive, small in specific area, expensive, and with a big interspace. Hard inorganic packing is hydrophilic, inexpensive and, with a small interspace, but it is easy to get clogged, and it has a large self-weight and requires intensive material uses for its supporting structures.

Most regions adopt soft packing, some economically advanced regions use hard ball packing, while underdeveloped regions use packing such as gravels and cokes. Efficient semi-soft packing and elastic packing are restricted in application due to their high price (Xiao 2007).

(4) Filter

A filter is also known as the post-treatment area. It is equipped with filter materials and has space for ventilation, which enables a facultative-anaerobic status within this tank. It functions to intercept suspending matters and to remove organic pollutants by facultative anaerobic bacteria. Filter materials are mainly gravel, slag, coke, palm mat, and ceramic, while polyurethane foam is also applied in some regions.

In practical utilization, the above-mentioned filter materials may suffer from some extent of congestion, which is related to factors such as material grading, concentration of suspending matters, and flow rate. Therefore, these factors should be taken into account in design and construction. In practical operation, filter materials should be cleaned regularly to avoid congestion. In addition, after operation for some time, polyurethane foam filter usually suffers from serious congestion that can lead to foam cracking if not treated in time. The consequences are bypass of sewage from filter cracks and failure of treatment.

3.3.3 Shape of Biogas Digester for Domestic Sewage Treatment

Originally, the shape of biogas digester for domestic sewage treatment imitated that of hydraulic digesters or septic tanks. With the accumulation of experience and trials in practical application, biogas digester for domestic sewage treatment becomes more diverse in terms of process combination and tank structure. Many regions have made improvements according to local requirements including effluence quality requirement, topography, and construction level (Xia et al. 2008). At present, the shape of biogas digester for domestic sewage treatment includes rectangle, circular arch, and mixed type. Some narrow places also adopt the concentric ring shape. the inner ring is a sedimentation digestion tank while the outer ring is an anaerobic filter.

(1) Rectangle tank

This is a basic shape of biogas digester for domestic sewage treatment in China. The body of this tank consists of a sedimentation area, anaerobic areas (I and II), and a facultative-anaerobic area, each accounting for 20%, 50% (I: 40% and II 10%), and 30% of the total volume respectively. The main functions of this kind of biogas digester for domestic sewage treatment are sedimentation and anaerobic treatment, which is similar to the function of a septic tank. Its enhanced measures of sewage treatment are reflected on two aspects. First, packing is added to anaerobic area II

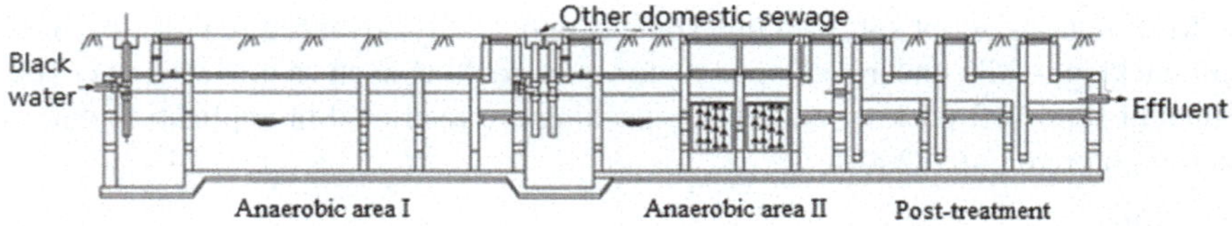

Fig. 3.4 Sectional view of rectangular biogas digester for domestic sewage treatment

to retain more microbes. Second, a facultative-anaerobic area is added downstream of an anaerobic area, which can, to some extent, enhance the performance of low-concentration sewage treatment. However, in application, this kind of biogas digester for domestic sewage treatment has some limitations. One is its weak force-bearing ability due to its linear structure, which is difficult to use in rural areas with complex topography. Another is the limited oxygen content in the wastewater of facultative-anaerobic areas due to natural reoxygenation. This kind of tank is commonly applied in southwest areas such as Sichuan and Chongqing in China (Fig. 3.4).

(2) Arch tank

The structure is a series connection of several hydraulic digesters, in the last two of which packings and filter materials are added. The top of each tank has a dome with a design similar to that of hydraulic digesters. Concentric cylindrical tanks are sometimes used in some places, in which a certain number of guide walls are set. Within the anaerobic area I, a concentric backflow circular wall and a baffle wall are constructed. Sewage enters a small chamber of the concentric circular wall and flow along an S-shaped pathway directed by the baffle wall. The sewage then enters the annulus from the chamber, flowing out after one circulation in the annulus. This design has increased the length of sewage pathway and avoided a short circuit. Anaerobic areaIIis deployed with triangular baffle walls, which can separate sewage and lead separated sewage into different processing units. After removal of organic matters and suspended solids, sewage converges at an outlet pipe and flows out. This is a process of "convergence-separation-convergence" (Fig. 3.5) that allows more thorough mixing and processing. Compared with rectangular tanks, the circular tanks including the arches are more reasonable in terms of mechanical structure and more force-bearing. At the same time, the feed-receiving chamber is located on the arch of the digester within the side wall, which effectively saves the land space. From the perspective of water flow, the circular tank can effectively avoid some flow stagnation points by guide walls and can improve space utilization of the whole device. However, this tank is prone to clogging. Because of specific process requirements, the volume ratio of the anaerobic I area, anaerobic II area and facultative-anaerobic area is about 6:3:1, which means that the volume of the two anaerobic areas accounts for nearly 90% of the total volume, while the facultative-anaerobic area is left with only 10%. The tank depends mostly on the anaerobic areas for pollutant treatment, which is strengthened by the use of partition walls, packing, filter materials, and many other

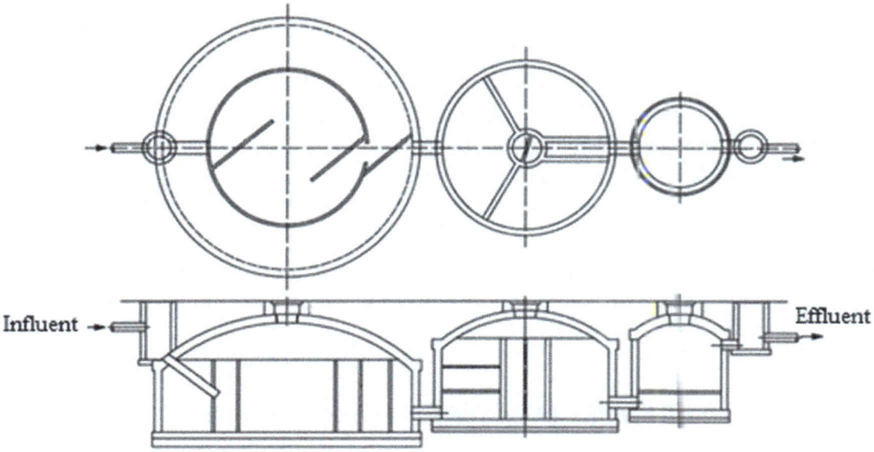

Fig. 3.5 Sectional view of concentric cylinder shape biogas digester for domestic sewage treatment

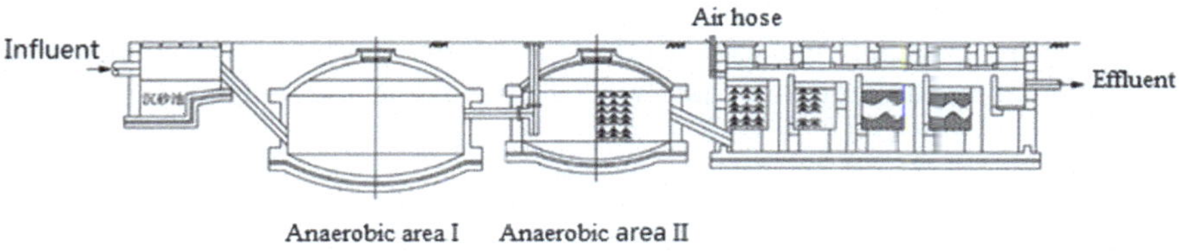

Fig. 3.6 Sectional view of combined biogas digester for domestic sewage treatment

methods. The facultative-anaerobic area mainly plays a role in controlling effluent quality and assisting sewage treatment. This kind of tank is mainly used in southwest regions such as Sichuan, Chongqing in China (Fig. 3.6).

(3) Combined type

A combined biogas digester for domestic sewage treatment is a combination of circular arch tank and rectangle tank, the former functioning as an anaerobic area and the latter as a post-treatment area (facultative-anaerobic filter).

In pretreatment, this type of tank is equipped with a simple sedimentation tank at the front of main tank body. The bottom of the sedimentation tank has a 10% slope.

The shape and structure of the anaerobic treatment areas (anaerobic I and II) are similar to those of a household digester, and each anaerobic area itself is an independent tank body. Anaerobic area I is equipped with no packing for easier dregs clearance whereas anaerobic area II has inlet holes for sewage entrance and is equipped with soft packing. In the post-treatment area, the former two sections adopt soft packing while the latter two sections use filter materials.

The volumetric proportion of anaerobic area I, anaerobic area II and post-treatment area respectively 40%, 25%, and 35% of the total effective volume of the device. The former two areas adopt the structure of hydraulic digesters. Therefore, they can bear large forces exerted and can conveniently store and utilize biogas. In addition, tanks with different functions can be constructed separately, which allows flexible construction plans while catering for the needs of various rural topographies. However, the flexibility brought by independence of each tank of the combined type leads to higher construction cost compared with cost of integrated tanks. Moreover, the combined type has higher requirements for design and construction. This type of tank is commonly used in Zhejiang and Jiangsu provinces in China.

3.3.4 Design of Biogas Digester for Domestic Sewage Treatment

When designing biogas digester for domestic sewage treatment, the volume of the whole system should be determined. The volume of biogas digester for domestic sewage treatment mainly consists of effective volume and protective volume, and the former can be further divided into wastewater volume and sludge volume.

$$V_{effective} = V_1 + V_2 \quad (3.1)$$

where:
$V_{effective}$—effective volume of biogas digester for domestic sewage treatment, m^3;
V_1—wastewater volume, m^3;
V_2—sludge volume, m^3.
Wastewater volume can be calculated in two ways: one is to calculate based on daily wastewater discharge amount and hydraulic retention time, as is shown in 3.2.

$$V_1 = nqt \quad (3.2)$$

where,

n—the actual number of people using biogas digester for domestic sewage treatment. When calculating the volume of a biogas digester for domestic sewage treatment of an entire building, n is calculated as the total number of people multiplies α (the value of α is shown in Table 3.6);

q—the amount of sewage produced per person per day, usually calculated as 0.08–0.13 m^3/d;

t—the hydraulic retention time of sewage, usually taken as 3–5 days. The determination of hydraulic retention time is related to the amount of sewage produced per person per day and the COD concentration of sewage. The COD concentration of sewage increases when the amount is small. Therefore, the hydraulic retention time

Table 3.5 Hydraulic retention time and volumetric loading rate of biogas digester for domestic sewage treatment (Zeng et al. 1994)

COD concentration(mg/L)	The amount of domestic sewage (L/p·d)	Hydraulic retention time(d)	Volumetric organic loading rate (kg COD/m^3·d)
600	140	3.0	0.200
700	120	3.5	0.200
800	100	4.0	0.200
900	90	4.5	0.200
1000	80	5.0	0.200
1200	67	6.0	0.200
1400	57	7.0	0.200
1500	50	8.0	0.200
2000	40	10	0.200
2500	31	13	0.200
3000	25	16	0.200
3500	22	18	0.200
4000	20	20	0.200
4500	17	23	0.200
5000	15	26	0.200
6000	13	30	0.200
7000	11	35	0.200
7500	10	38	0.200

should also increase accordingly. The values of different hydraulic retention time of sewage under various COD can be referred in Table 3.5.

Another way of calculating wastewater volume V_1 is by volumetric loading rate as shown in Eq. 3.3

$$V_1 = \frac{QS}{L_V} \tag{3.3}$$

where,

Q—domestic sewage flow, m^3/d;

S—COD concentration of domestic sewage, kg/m^3;

L_V—Designed volumetric loading rate, usually set as 0.15–0.25 kg COD/m^3·d. When the local ambient temperature is low, the lower L_V is used. The values of volumetric loading rate can be referred to in Table 3.5.

In fact, the calculation of volume by hydraulic retention time is only meaningful when COD concentration in sewage is relatively fixed. The relatively accurate and reasonable method to obtain wastewater volume is to calculate by volumetric loading rate. If sewage COD concentration is taken into account, both calculations (Eqs. 3.2 and 3.3) base the volume of digester on organic loading rate.

Volume of sludge V_2 can be calculated by Eq. 3.4:

$$V_2 = \frac{1.2knaT(1.0 - b)}{1000(1.0 - c)} \tag{3.4}$$

where,
n—already defined as above;
a—the amount of sludge per person per day (L/person·day). When black water and other domestic sewage are discharged together, the value is set as 0.7 L/person·day and when the black water is discharged independently, the value is set as 0.4 L/person·day;
T—clearance period of sludge (day);
b—water content of fresh sewage sludge that enters biogas digester for domestic sewage treatment, usually calculated as 95%;
c—water content of sludge that has been digested and concentrated in biogas digester for domestic sewage treatment, usually calculated as 90%;
k—reduction of volume after digestion, usually calculated as 0.8;
1.2—volumetric coefficient of sludge that should be remained after clearance.
When it is impossible to collect sufficient data to calculate wastewater volume with the equation above, an empirical formula shown as Eq. 3.5 can be referred to.

$$V_{effective} = RN \tag{3.5}$$

where:
R—effective coefficient per person (refer to Table 3.6);
N—total number of persons.

Protective volume is usually 10–50% of effective volume. The smaller the effective volume is, the greater the protective volume should be so as to withstand the impact of sewage fluctuation. For example, protective volume of individual household biogas digester for domestic sewage treatment is usually above 45% of effective volume. Besides, when there is a lack of basic data, we can take a larger safety coefficient to ensure that the quality of effluent can meet design requirements.

Table 3.6 reference of α value and R coefficient

Type of building	α value(%)	R coefficient
Caring houses, boarding nursery, individual household	100	0.30–0.35
Dormitory, hotel	70	0.25–0.30
Office building, teaching building	40	0.06–0.10

Biogas digester for domestic sewage treatment consist of several parts, including anaerobic areas and a post-treatment area which is facultative-anaerobic. When sewage is mainly urine and feces from toilets, the volume of anaerobic areas should be no less than two thirds of the total effective volume. In addition, we should pay attention to the following points when designing biogas digester for domestic sewage treatment:

1. Wastewater flow should be unidirectional to ensure smooth water flow.
2. The inlet and outlet of Biogas digester for domestic sewage treatment should have a level difference, and the bottom of the digesters should have a slope to make up for the head loss. The hydraulic slope of Biogas digester for domestic sewage treatment should be $\geq 5‰$, and the level difference between inlet and outlet should be ≥ 200 mm.
3. The horizontal layout of Biogas digester for domestic sewage treatment should be based on local conditions and should be easy to clean and manage. The distance between Biogas digester for domestic sewage treatment and underground pump house should be no less than 30 m, and the distance from digesters to any building should be greater than 5 m.
4. The height, width, and length of Biogas digester for domestic sewage treatment should be no less than 1.60 m, 1.00 m and 2.00 m respectively. Diameter of a circular well head should be no less than 0.60 m. Width of a rectangular well mouth should be no less than 0.50 m and length should be no less than 0.70 m. Diameter of a feedstock inlet pipe should be greater than 200 mm.

In practical construction, side pipes should be set with biogas digester for domestic sewage treatment to ensure smooth wastewater discharge in case of malfunction. When designing wastewater inlet and outlet, the requirements for water distribution should be taken into account to avoid incidents such as short circuiting and flow stagnation points. In the middle of construction, adjustments should be made according to local condition.

3.4 Other Anaerobic Technology for Domestic Sewage Treatment

As early as in 1860, Jonh Mouras from France built the first septic tank. In the following years, various anaerobic treatment technologies emerged, among which the double layer sedimentation tank developed in 1905 by Imhoff from Germany is best-known. In the next several decades, anaerobic treatment technologies underwent rapid development and gained broad utilization. By 1914, 14 cities in America had built anaerobic digestion tanks. In 1940, continuous stirred anaerobic digesters emerged in Australia (Lens P et al. 2001). And in later years, highly efficient anaerobic reactors such as AF, UASB, and EGSB were developed in succession. Traditional and highly efficient anaerobic reactors both have utilization in different scenarios.

3.4.1 Septic Tank

Originated in Europe, septic tanks are a kind of common equipment of preliminary treatment of domestic sewage. In 1860, Mouras and Moigno built the earliest single-chamber septic tank and named it "MOURAS tank" (Lens et al. 2001). In later years, British researchers modified MOURAS tanks and get them patented, renaming them as "septic tank". Since then septic tanks were widely used across the world (Butler and Payne 1995). Septic tanks were then introduced to China. The discharge system of nearly all domestic sewage treatment equipment is connected to septic tanks whether in urban or rural areas. Figure 3.7 presents a common three-chamber septic tank. As a primary equipment of treating domestic sewage, septic tank mainly works by the principles of anaerobic digestion and gravity separation. Because of specific weights, suspended solids in domestic sewage either settle down (form sediments) or float up (form scum) under gravity. At the same time, microbes decompose some organic matters via anaerobic digestion to achieve preliminary sewage treatment for easier further treatment, or to satisfy relatively lower standards for discharge or recycling.

As shown in Fig. 3.8, sewage separates into three layers in a septic tank: the scum layer, the middle (clear liquid) layer, and the sediment layer. Solids with light weight or flocs float up with bubble to the surface, forming a layer of scum, while heavy solids settle at the bottom. Pollutants in the sewage is degraded by anaerobic microbes, producing gases such as CH_4, CO_2, and H_2S. Although a septic tank has a cover, gases produced will be directly released into atmosphere. Retention time of a septic tank varies with situations. Septic tanks that require stabilization and harmless treatment usually take 30 days (GB19379-2012 *Hygienic Specification for Rural Latrine Appendix A*), whereas septic tanks for sewage pretreatment generally

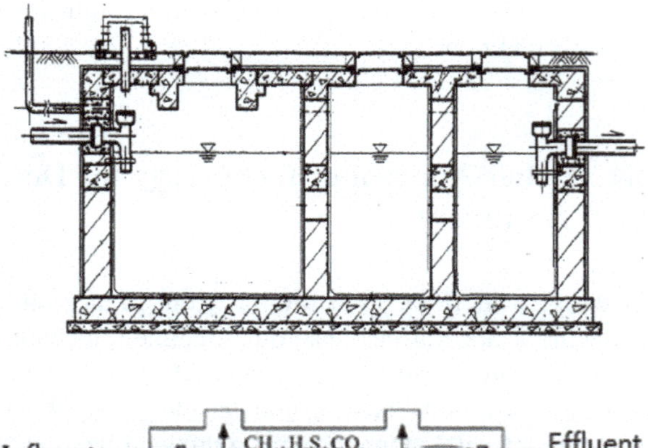

Fig. 3.7 Three-chamber septic tank

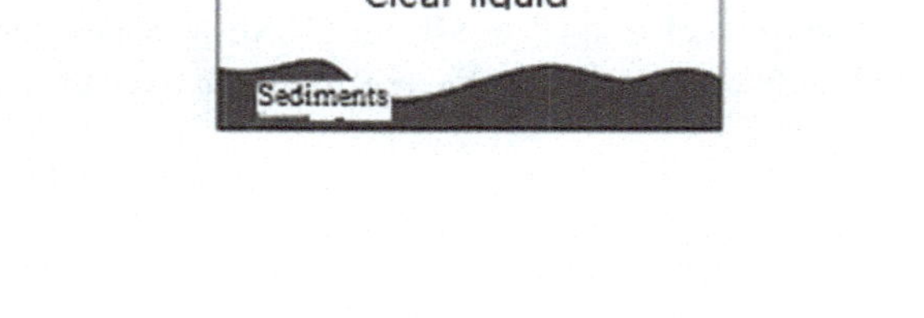

Fig. 3.8 Schematic diagram of a septic tank

takes12–48 h. Scum and bottom sediments need to be regularly cleaned to ensure effect of the treatment. Cleaning frequency is usually 3–12 months once. Solid produced in this process can be used as fertilizers, and the liquid in the middle layer can be directly discharged when standards are meet. Otherwise, the liquid will be processed at post-treatment units. A septic tank is unable to completely mineralize pollutants, and its effluent still contains a relatively high content of pollutants (TN, TP). For this reason, septic tanks are regarded as preliminary and low-efficiency anaerobic treatment technology.

3.4.2 Anaerobic Filter

An anaerobic filter (hereinafter abbreviated as AF) is an effective anaerobic reactor with packings as its microbe carrier. Anaerobic microbes are attached to the packings to form biofilms, and the combination of biofilm and packings again form filter beds. Wastewater after pretreatment flows through filter beds and fully contact with microbes in the beds, and then pollutants in incoming sewage can be decomposed. The features of AF include long sludge age which can be over 100 days, high biological concentration, strong resistance to high organic loading, short start-up time, relatively easy restart after operation halt, and no need sludge backflow in most cases. The disadvantages of AF include easy congestion and short circuiting like other anaerobic processes and low efficiency in nitrogen and phosphor removal (Seggelke et al. 1999).

AF can operate under mesophilic (30–35°C), thermophilic (50–55 °C), and ambient temperature (10–25 °C) conditions. Generally, domestic sewage is treated by ambient temperature. To avoid congestion, effluent can be reflux to dilute influent and enhance surface hydraulic loading. AFs can be categorized into the upflow and the downflow type by water flow direction. Sewage flows upwards in upflow AFs while flowing downwards in downflow AFs (Fig. 3.9). Upflow AFs have two layers, the lower of which (the part below the straight line in Fig. 3.9) is left blank without packing, while the upper layer is equipped with packing. This type is called upflow blanket filter or UBF. This way, clogging can be avoided to some extent. On the other hand, water flows downward in downflow AFs, and agitation from uprising

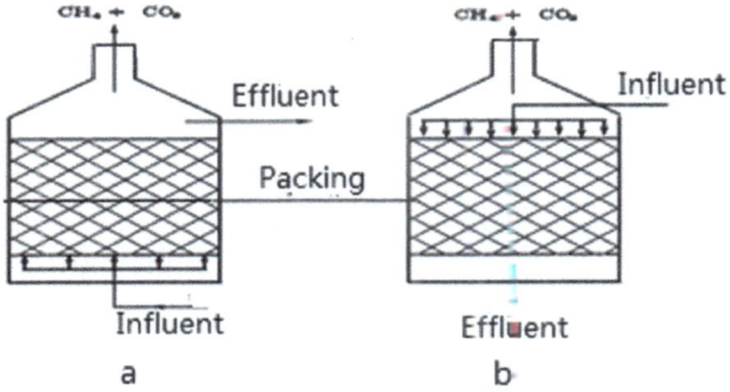

Fig. 3.9 Schematic diagram of an AF **a** upflow anaerobic filter **b** downflow anaerobic filter

biogas together with the existence of sludge in packing slits create a condition that resembles continuous stirred tank reactor (CSTR), whereas the mechanism of upflow AFs is more similar to plug flow reactor (PFR). To balance the loads of all treatment units, two AFs can be connected in series, and the flow sequence of the two AFs can be regularly switched to fully utilize each unit.

A single AF can be applied in small-scale domestic sewage treatment equipment (Kobayashi et al. 1983). At the same time, anaerobic area II of biogas digester for domestic sewage treatment adopts anaerobic filter technique, which is in most cases upflow.

3.4.3 Upflow Anaerobic Sludge Blanket (UASB)

Upflow anaerobic sludge blanket (abbreviated as UASB) technology developed based on upflow AFs. The lower part of an upflow AF get easily clogged. The UASB process, however, replaces packing layers by the granule sludge layers that intercept, absorb, and decompose organic particles. When sewage enters the reactor from the bottom, the force exerted by up-flowing water and rising of gas product create a natural stir. In this process, some sludge forms a thin layer of sludge suspension, and a three-phase separator is set at the top of the reactor to separate effluent, biogas product, and sludge. (Fig. 3.10).

UASB reactor consists of three sections: the inlet section, the reaction section, and the three-phase separation section. The inlet section is at the bottom of the system and mainly functions to enable the entrance and even distribution of sewage to avoid surge or stagnation zone. The reaction section is in the middle of the reactor where

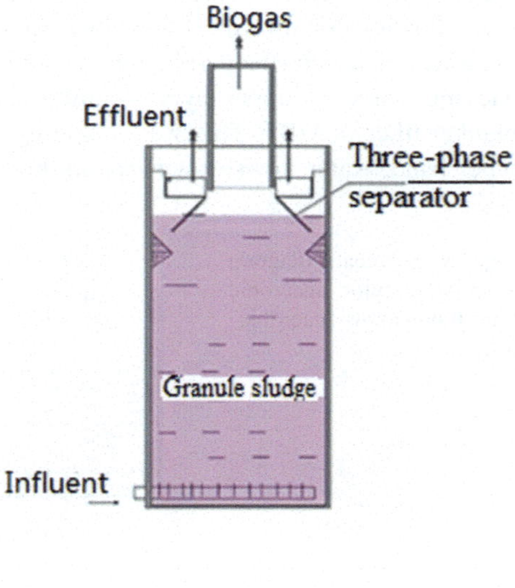

Fig. 3.10 Schematic diagram of a UASB reactor

sewage and anaerobic sludge fully contact, and most pollutants are removed by interception, decomposition and absorption. The three-phase separation section is at the top, separating gas, liquid, and solid. Separators have a wide variety of structures, and the separation performance directly determines the treatment effect.

Up to now, UASB is a widely applied anaerobic biological technology. It has a series of advantages including high volumetric load, good interception performance, size compatibility, and so on. In the 1990s, application of UASB to treat domestic sewage was started. Tropical countries such as Brazil, Mexico, Colombia, and so forth have created many sewage treatment plants using UASB technology and these plants have been operated smoothly. (Barbosa and Jr 1989; Uemura and Harada 2000).

3.4.4 Expanded Granular Sludge Bed (EGSB)

Expanded granular sludge blanket (abbreviated as EGSB) reactors are developed based on UASB. In this technology, high upflow velocity is achieved by backflow of effluent and high H/D (height-diameter) ratio, which enables expansion of bio-beds. Generally, the expansion rate of EGSB is around 10–20%. Fluidized particles come into contact with liquid so that mass transfer is enhanced. In addition, this reactor can adopt small granular bio-packings, which can avoid plugging that usually occurs in fixed bio-film reactors. As shown in Figs. 3.10 and 3.11, EGSB and UASB are similar in structure, but they differ in details. The upflow velocity of EGSB is faster, so the requirements for water distributors are not as strict, unlike the requirements for three phase separators. The high hydraulic loading leads to loss of sludge, thus the design of three phase separators is key to stable and effective operation of EGSBs.

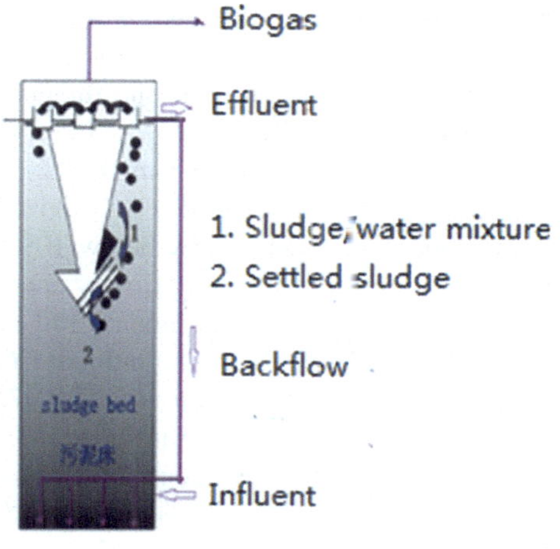

Fig. 3.11 Schematic diagram of an EGSB reactor

Meanwhile, high H/D (3–8) of EGSBs means small land occupation. Like UASBs, an EGSB is mainly used for treating high concentration wastewater, and its application of treating domestic sewage is still at research stage (Seghezzo et al.1998; Mario et al. 1997).

3.5 Treatment Performance of Biogas Digester for Domestic Sewage Treatment

3.5.1 Removal of Organic Pollutants and Generation of Biogas

Biogas digester for domestic sewage treatment removes pollutants in sewage through decomposition by anaerobic microbes. Treatment performance can also be improved by complementary means such as sedimentation, filtration, and treatment by facultative microbes. Under anaerobic conditions, microbes can decompose macromolecules into micro molecules and even carbon dioxide and water. This way, organic pollutants are removed.

Chen et al. (1988) conducted a research on treatment of domestic sewage by combining primary anaerobic digestion and secondary facultative aerobe technology under 25 ± 2°C. The total volume of the equipment was 15 L, and the primary anaerobic digester was divided into three sections, with the latter two sections filled with packings. The secondary facultative anaerobic digester was also divided into three sections and filled with filter materials, with hydraulic retention time of 2 days, 2.5 days, and 3 days respectively and a volumetric load of around 0.18–0.53 kg COD/(m^3·d). Results showed that when the influent had a COD of 753–1058 mg/L, the effluent COD was 70.4–190 mg/L, and the COD removal efficiency was around 73.8–84.8%. When the influent BOD_5 was 350–640 mg/L, the effluent BOD_5 was 60.0–100 mg/L, and the removal efficiency of BOD_5 was 77.1–84.4%. When the influent suspended solids (SS) was 148.3 mg/L, the effluent SS was 4.7 mg/L, and the removal efficiency of SS was 96.8%. The average COD removal efficiency of the primary anaerobic treatment was above 70% while the that of the secondary facultative anaerobic was usually below 10%. This means that the organic pollutants were mainly removed in the primary anaerobic treatment unit. Zeng et al. (1994) conducted a simulation experiment (30 L) and achieved better results. When treating black water from toilet, the influent COD concentration was 3127–5591 mg/L, and the effluent COD concentration was 97.9–216 mg/L, and the COD removal efficiency was 93.1–97.9%. The influent BOD_5 concentration was 1329–2777 mg/L, and the effluent BOD_5 concentration was 86.4–98.3 mg/L, so the BOD_5 removal efficiency was 93.5–98.3%. The influent SS concentration was 289–1476 mg/L, and the effluent SS concentration was 37.6–59.4 mg/L, achieving a removal efficiency of 96.0–98.7%. When treating domestic sewage, the influent COD concentration was

737–1041 mg/L, and the effluent COD concentration was 73.79–90.5 mg/L, achieving a COD removal efficiency of 90.0–90.5%. When the influent BOD_5 concentration was 304–450 mg/L, the effluent BOD_5 concentration was 29.3–35.3 mg/L, and the BOD_5 removal efficiency concentration was 90.4–92.2%. When influent SS concentration was 397–661 mg/L, the effluent SS concentration was 22.8–31.7 mg/L, and the removal efficiency was 92.0–96.6%.

Shen et al. (2005) adopted shorter hydraulic retention time(1 day and 2 days) under ambient temperature, and the treatment effect of tubular anaerobic bioreactor filled with hollow filler was compared with the effect of baffled anaerobic bioreactor (volume is about 1 m^{3}). The removal performance of COD, BOD_5, and SS is shown in Fig. 3.7. The removal efficiency of COD, BOD_5, and SS was 66.1–80%, 26.8–85.7%, and 11.7–90.2% respectively. The concentration of organic pollutants in this pilot test was similar to that of the experiment done by Zeng et al. (1994). This showed that the COD concentration of effluent after treatment in biogas digester for domestic sewage treatment can decrease to as low as 100 mg/L, while BOD_5 and SS concentrations were as low as 40 mg/L.

Study results demonstrated that COD, BOD_5, and SS removal performance of anaerobic baffled reactor (ABR) was better than that of tubular anaerobic reactor. After the 2-day retention time, COD, BOD_5, and SS removal performance of ABR could reach Class 1B required in *Discharge Standard of Pollutants for Municipal Wastewater Treatment Plant* (GB 18918–2002).

Due to factors such as temperature and fluctuation of sewage concentration, the treatment performance in actual plants cannot live up to the effect in experiment. The treatment performance results of 78 biogas digester for domestic sewage treatment in four counties of Sichuan province showed that the average value of effluent COD, BOD_5, and SS was 152, 49.7, 56.1 mg/L respectively (Zeng et al. 1994). The treatment result of 88 biogas digester for domestic sewage treatment in a city in Sichuan showed that from 1998–2000, the yearly average effluent COD concentration was 174, 174, and 240 mg/L respectively, while the yearly average effluent BOD_5 concentration was 81, 84, and 94 mg/L respectively and the yearly average effluent SS concentration was 43, 33, and 46 mg/L respectively (Tian et al. 2002). Bremen Overseas Research and Development Association, abbreviated as BORDA, and Biogas Institute of Ministry of Agriculture jointly constructed biogas digester for domestic sewage treatment to treat wastewater from residential areas, hospitals, schools, and public toilets. Their treatment result shows that, as the influent COD concentration was 1686–11206 mg/L, the effluent COD concentration was 104–256 mg/L, and the COD removal efficiency was 84.5–98.3%. When SS concentration of influent was 708–5633 mg/L, the effluent SS concentration was 32–179 mg/L, and the SS removal efficiency was 37.3–99.3%. Sedimentation and digestion played a major part in that process (Deng et al. 1997). Xie et al. (2005) conducted an investigation that showed similar removal results. When the influent COD concentration was 352–1034 mg/L, the effluent COD concentration was 128–166 mg/L, and the average COD removal efficiency was 84.5%. When the influent BOD_5 concentration was 821–835 mg/L, the effluent BOD_5 concentration was 80.0–84.5 mg/L, and the average BOD_5 removal

Table 3.7 Organic removal effects of pilot scale test tube device and the baffle anaerobic bioreactor (Shen et al. 2005)

Retention time	Item	Influent concentration	Effluent concentration of tube device	Removal efficiency	Influent concentration	Effluent concentration of baffle reactor	Removal efficiency
(d)		mg/L	mg/L	%	mg/L	mg/L	%
1	COD	287.2	97.5	66.1	288.4	91.	68.3
	BOD_5	120	35.0	70.8	139.8	32.5	76.8
	SS	73.3	14.3	80.5	96.8	9.5	90.2
2	COD	288.1	67.0	76.7	291.6	58.3	80.0
	BOD_5	120	26.8	77.7	139.8	20.0	85.7
	SS	73.3	11.7	84.0	96.8	15.0	84.5

efficiency was 90.1%. When SS influent was 350–421 mg/L, the effluent SS concentration was 31.0–45.2 mg/L, and removal efficiency was 90.2%. Although in small and medium scale tests, the effluent COD of biogas digester for domestic sewage treatment can be as low as 100 mg/L while BOD_5 and SS of effluent concentration can be as low as 40 mg/L, COD concentration of the effluent from scaled equipment can only be lowered to 150 mg/L, and a BOD_5 concentration of 80 mg/L and a SS concentration of 50 mg/L.

Small-scale experimental biogas digester for domestic sewage treatment could only reach a a volumetric biogas production rate of 0.06–0.102 L/(L·d), with 60.0–70.2% of methane content (Zeng et al.1994; Chen et al. 1988). Biogas digester for domestic sewage treatment of public toilet wastewater treatment could have a volumetric biogas production rate of 0.05–0.15 L/(L·d) (Deng et al. 1997), which showed that the biogas product efficiency of biogas digester for domestic sewage treatment was low, and their major function was to remove pollutants (Table 3.7).

3.5.2 Nitrogen and Phosphorous Removal

Biogas digester for domestic sewage treatment play a minor role in removal of nutritious matters (nitrogen, phosphorous etc.). This is due to the fact that biogas digester for domestic sewage treatment operate under anaerobic or facultative anaerobic conditions, thus lacking biological treatment process under aerobic conditions. If sewage contains nitrate and nitrite under anaerobic or facultative anaerobic conditions, bacteria can utilize NO_2^- or NO_3^- as electron acceptor to oxidize NH_4^+ into N_2 and gaseous nitrous oxides in anammox or denitrification reaction, during which nitrogen will be removed (Chou et al. 2004). However, since concentration of nitrate and nitrite is limited, nitrogen in wastewater cannot be removed only by anaerobic process. Current major biological denitrification methods remove nitrogen mainly by ammonification, nitrification and denitrification. Nitrification occurs in aerobic conditions, transforming ammonia and other forms of nitrogen into nitrate or nitrite.

There are two ways to remove phosphorus in sewage. One is absorption by microbes for cell synthesis, and one typical such kind of microbe is phosphorus-accumulating organisms. When sewage is in aerobic condition (aeration), this kind of bacteria will absorb phosphorous in sewage, while in anaerobic environment, phosphorus-accumulating organisms will release absorbed phosphorous into water. Therefore, it is hard for biogas digester for domestic sewage treatment with only anaerobic or facultative anaerobic condition to remove phosphorous. The second way of phosphorous removal is particle sedimentation and sludge absorption. The latter requires good hydraulic condition to remove phosphorous by full contact of sewage with sludge. Generally speaking, due to lack of aerobic condition, Biogas digester for domestic sewage treatment is limited in its ability to remove nutritious matters.

The pilot-scale test results of tubular and baffled anaerobic reactors (1 m^3 volume) with hollow-shaped packings was conducted. Under the condition of the hydraulic retention time (HRT) of 2d for both tubular and baffled reactors, the average influent total nitrogen (TN) concentration being 66.10 mg/L and 66.65 mg/L respectively, and the average effluent TN concentration of 53.0 mg/L and 51.5 mg/L respectively, a total removal efficiency of 19.9% and 22.9% were obtained, respectively. The average influent total phosphorus (TP) concentration was 7.27 mg/L and 7.40 mg/L for the two reactors respectively, and the average TP of effluent was 3.93 mg/L and 4.05 mg/L respectively, giving a TP removal efficiency of 45.9% and 45.3% respectively. The removal efficiency of TP was higher than that of TN because the removal of nitrogen relies mainly on microbes, whereas removal of phosphorous was determined by the effect of both absorption and sedimentation. When HRT was increased, the concentration of NH_4^+−N was found to rise, which was due to the conversion of organic nitrogen compounds into ammonium in the process of organic decomposition (Shen et al. 2005).

Nitrogen and phosphorus removal is poor by biogas digester for domestic sewage treatment. The monitoring results of the 88 biogas digester for domestic sewage treatment in a city in Sichuan for three consecutive years showed that the average ammonia nitrogen concentration of effluent from 1998 to 2000 was 20, 32, and 24 mg/L, respectively (Tian et al. 2002). In *Discharge Standard of Pollutant for Municipal Wastewater Treatment Plant GB 18918–2002*, the secondary discharge standard requires that the ammonia nitrogen concentration should be $\leq$ 25 mg/L (when the water temperature is less than 12 °, the concentration should be $\leq$ 30 mg/L). So biogas digester for domestic sewage treatment alone is difficult to achieve the discharge standards.

3.5.3 Sanitation Effect

In terms of sanitation effects, Biogas digester for domestic sewage treatment are similar to household digesters and septic tanks. The anaerobic treatment, sedimentation, and filtration process of biogas digester for domestic sewage treatment can remove pathogenic microorganisms in sewage, achieving sanitation and epidemic prevention. Shen et al. (2005) used underground unpowered anaerobic reactor (UUAR) for sewage treatment, and the sanitation index were tested. The test result is shown in Table 3.8. Tubular anaerobic reactor had high removal efficiency of fecal coliform, total bacteria, and ova of roundworm. When HRT was 1 day, their average removal efficiency from two tests was 95.8%, 82.9% and 100% respectively. When HRT was 2 days, their average removal efficiency from two tests was 95.8%, 86.1%, and 100% respectively. As opposed to tubular anaerobic reactors, anaerobic Baffle reactor had loss of sludge during every second sampling, so the removal efficiency of total coliforms, total bacteria, and ova of roundworms was relatively low. When HRT was 1 day, the average removal efficiency from two tests was 99. 8%, 37. 4% and 78. 7%. When HRT was 2 days, the average removal efficiency from two tests was 99.8%, 59.3%, and 98% respectively.

Monitoring results of the 88 biogas digester for domestic sewage treatment in a city in Sichuan Province for three consecutive years from 1998 to 2000 show that the average number of effluent parasite eggs was 0, 0.55, and 0/L respectively (Tian et al.

Table 3.8 Hygienic indexes of the underground unpowered anaerobic reactor (Shen et al. 2005)

Number	Reactor type	Sample	Total number of acteria	Removal efficiency	Total coliform	Removal efficiency	Ascaris egg	Removal efficiency
			(\ml)	(%)	(/ml)	(%)	(/ml)	(%)
First test	Tubular anaerobic reactor	influent	9.3×0^6	–	3.5×10^5	–	85	–
		1d effluent	2.7×10^6	71.0	2.4×10^4	93.1	0	100
		2d effluent	2.2×10^6	76.3	2.4×10^4	93.1	0	100
	Anaerobic Baffle reactor	influent	7.9×10^6	–	5.4×10^6	–	225	–
		1d effluent	2.0×10^6	74.7	2.4×10^4	99.6	15	93.3
		2d effluent	4.2×10^5	94.7	2.4×10^4	99.6	0	100
Second test	Tubular anaerobic reactor	influent	1.9×10^6	–	2.4×10^6	–	70	–
		1d effluent	1.0×10^5	94.7	3.5×10^4	98.5	0	100
		2d effluent	8.0×10^4	95.8	3.5×10^4	98.5	0	100
	Anaerobic Baffle reactor	influent	2.1×10^6	–	2.4×10^7	–	125	–
		1d effluent	2.1×10^6	0	2.4×10^4	99.9	45	64
		2d effluent	1.6×10^6	23.8	3.5×10^3	99.99	5	96

2002). For biogas digester for domestic sewage treatment processing sewage from homes, hospitals, schools, and public toilets, the influent fecal coliform value was 10^{-5}~4×10^{-7}, and the effluent value was 10^{-3}~4×10^{-5}, giving a removal efficiency of above 99.0–99.7%. The parasitic ova count in the influent was 160–320/L while that in the effluent was 0–28/L, giving a removal efficiency of 91.3–100%. In the meantime, salmonella, shigella, bacillary dysentery, mycobacterium tuberculosis, and other pathogenic bacteria were not detected in effluent (Deng, et al. 1997). These results indicate that biogas digester for domestic sewage treatment can achieve better sanitation effect.

According to *Hygienic Requirements for Harmless Disposal of Night Soil GB 7959–2012*, removal efficiency of ascaris eggs by anaerobic digestion process at ambient temperature shall be higher than 95%, and fecal coliform value shall be greater than 10^{-4}. According to most test results, removal of fecal coliform and ascaris eggs in anaerobic reactors was notable, but some fecal coliform values still cannot meet the relevant standards. In anaerobic reactors, removal mechanisms of ascaris eggs and fecal coliform are different. Most Escherichia coli are aerobic bacteria, which cannot survive in anaerobic conditions for a long time, and some Escherichia coli are facultative anaerobic microorganisms and can survive only when there is nitrate that replaces oxygen (Bertero et al. 2003). The digester internal is anaerobic with no oxygen and low nitrate content, so it is difficult for Escherichia coli to survive in reaction processes. The average size of ascaris eggs is larger than that of Escherichia coli, generally around 45 × 60 μm, so sedimentation and filtration are the most likely mechanisms for the removal of ascaris eggs. It is difficult to completely precipitate ascaris eggs with a short hydraulic residence time, i.e., with relatively high flow rates. This experiment also reflects that the extension of retention time can improve the sanitary effect of effluent to a certain extent. The treatment effect with an HRT of 48 h was better than that with an HRT of 24 h. In China, the minimum hydraulic retention time of septic tanks is required to be 12 h.

3.5.4 Treatment Effects of Other Anaerobic Biological Process

Various types of anaerobic bioreactors basically share the same principles of pollutant removal, that means that, pollutants are removed by means of decomposition of anaerobic or facultative anaerobic microbial, filtration, sedimentation, etc. The only difference is the detailed ways through which each process is achieved.

Table 3.9 shows the treatment effect of traditional three-chambe septic tanks in Jiangsu province monitored by Wang et al. (2008). COD removal efficiency was only 30–50%. TN removal efficiency was 3–10%. TP removal efficiency was 13–49%. In the meanwhile, the monitoring results of 30 standard septic tanks in four counties and cities in Sichuan province shows that the average effluent COD, BOD_5, and SS were 544, 47.9, and 291 mg/L respectively (Zeng et al. 1994). The COD concentration of

Table 3.9 Domestic sewage treatment effect of three-chambe septic tanks (Wang et al. 2008)

Region	Influent concentration(mg/L)			Effluent concentration(mg/L)			Removal efficiency(%)		
	COD	TN	TP	COD	TN	TP	COD	TN	TP
South Jiangsu	1370.5	81.8	15.2	641.7	73.2	7.7	53.18	10.55	49.34
Middle Jiangsu	2470.5	71.5	20.8	1116.8	69.4	17.9	54.80	2.94	13.80
North Jiangsu	1350.0	75.0	16.7	914.3	70.1	14.5	32.28	6.49	13.37
average	1730.3	76.1	17.6	890.9	70.9	13.4	48.51	6.83	23.92

septic tank effluent reported by both studies was relatively high, at above 500 mg/L, which indicates that the removal strength of organic pollutants by septic tanks was far weaker than that of biogas digester for domestic sewage treatment.

UASB, AF, and other highly efficient anaerobic bioreactors have made attempts in treating domestic sewage in tropical and subtropical regions. Table 3.10 lists the basic conditions and treatment effects of some such treatment projects (Li et al. 2004).

The hydraulic retention time of UASB, AF, and other high efficient anaerobic bioreactors for treating of domestic sewage was generally within 12 h, but their COD and BOD_5 removal effect was generally better than that of septic tanks and biogas digester for domestic sewage treatment. However, these high efficient anaerobic bioreactors are energy-intensive, more complex to manage and cost-intensive compared with biogas digester for domestic sewage treatment.

3.6 Post-Treatment Technology of the Effluent from Biogas Digester for Domestic Sewage Treatment

Whether the traditional biogas digester for domestic sewage treatment or the improved ones, both feature in anaerobic treatment as the major treatment method assisted by facultative anaerobic measures to strengthen the treatment effect. Such treatment can efficiently remove organic pollutants and pathogenic microbes, but the removal efficiency of nitrogen, phosphorus, and other nutrients is poor. With the development of social economy and the continuous improvement in living standards, the requirements for good nature and living environment become more demanding. With the gradual improving wastewater discharge standards, it becomes harder for biogas digester for domestic sewage treatment to meet the discharge standards. Therefore, in order to meet higher discharge standards and to overcome their limitations, biogas digester for domestic sewage treatment need to be implemented with post-treatment processes combined with other technologies. The commonly used post-treatment technologies include aerobic biological treatment technology and natural ecological treatment technology.

3.6.1 Aerobic Biological Post-Treatment Technology

At present, aerobic biological treatment technologies of biogas digester for domestic sewage treatment mainly include biofilters, biological contact oxidation, etc.

(1) Biofilter

A biofilter is an artificial biological treatment technology developed based on the principle of soil self-purification. In a biofilter, the sewage is sprayed as droplets onto the surface of filter materials, and the surface through which the sewage flows

Table 3.10 Operation results of high-efficiency anaerobic bioreactor for treating domestic sewage (Li et al. 2004)

Item	Netherlands (Bergambacht)	Netherlands (Bergambacht)	Brazil (Sao Paulo)	Colombia (Bucaramanga)	Colombia (Cali)	India (Bombay)	India (Kampur)
Reactor type	UASB	UASB	UASB	UASB	UASB	AF	UASB
Reactor volume(m^3)	20	6	120	35	6	12	6
HRT(h)	10	8	4.7–9.0	5	6	12	6
temperature(°C)	15–16	15–19	20–25	23–27	25	15–25	20–30
Volumetric load(kg COD/m^3·d)	0.4–0.9	0.5–1.0	–	–	2	0.7	–
COD removal efficiency (%)	49	55	70	66	78	49–78	62–70
BOD removal efficiency (%)	–	–	80	80	66	69–83	65–75
TSS removal efficiency (%)	60	60	79	70	75	68	67–79
Effluent COD(mg/L)	170–303	220	961	145	120	92–198	91–103
Effluent BOD(mg/L)	69	40–110	31	39	25	22–55	50--56
Effluent TSS(mg/L)	51	43–80	35	70	30	117	134

will gradually form a biofilm. When the biofilm is mature, it will intercept suspended solids and absorb organic pollutants in sewage, decomposing and metabolizing them to synthesize microorganism body itself while achieving sewage filtration. The flow condition is similar to that of the filter, and the pores and slits of the filter material is relatively small, so it can be clogged easily. Therefore, the concentration of influent SS should be controlled. At present, there are several types of biofilters commonly used in post-treatment of rural sewage, such as the stair-type, the tower type, and the natural enforced ventilation and water spattering biofilter type.

Biofilters have the following major advantages

1. Easy maintenance and low operation cost;
2. Lower energy consumption;
3. Small amount of residual sludge and easy separation.

Biological filters also have the following disadvantages

1. Easy to clog;
2. Breeding mosquitoes and flies; poor sanitation.

The biofilter is added to biogas digester for domestic sewage treatment for the purpose of strengthening removal of SS and nutrients to reach discharge standards. A rural community in Gongju city, South Chungcheong province, South Korea, used a combined process of "anaerobic baffle reactor (ABR) (similar to biogas digester for domestic sewage treatment)-biofilter-constructed wetland" to treat domestic sewage. The process is shown in Fig. 3.12 (Yi et al. 2016).

The operation results demonstrate that the "ABR-biofilter-constructed wetland" combined process for rural domestic sewage treatment gave good pollutants removal efficiency after the system was stabilized. The effluent COD and NH_4^+-N concentrations during most of the time were lower than 15 mg and 30 mg/L respectively, and the average effluent TP concentration was maintained at about 1.5 mg/L. Among those results, removal efficiency of TSS, COD, and NH_4^+-N was 97.7%, 89.6%, and 63.3% respectively, but the treatment efficiency of TN and TP was relatively low, only 26.7% and 32.4% respectively. In the combined process, the TSS removal proportion in ABR, biofilter, and constructed wetland was 35.7%, 9.8%, and 52.1% respectively, while the COD removal proportion was 22.4%, 33.8%, and 31.3% respectively and

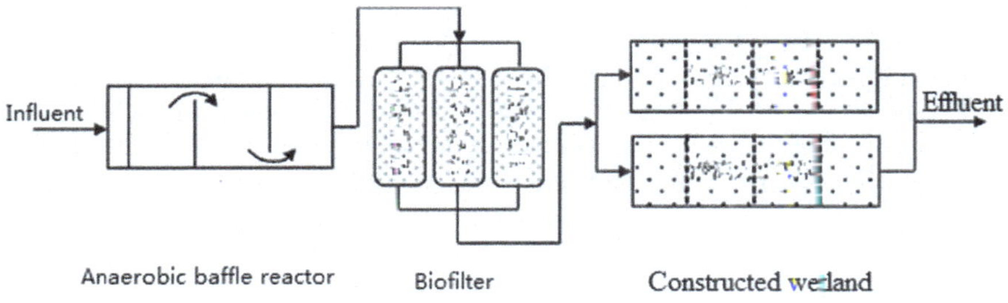

Fig. 3.12 ABR-biofilter-construction wetland process for the treatment of sewage

the TN removal proportion was 5.4%, 0%, and 22.0% respectively. The biofilter is capable of enhancing treatment of COD and NH_4^+-N, while the constructed wetland can effectively remove TSS, TP, and TN.

(2) Biological contact oxidation

Biological contact oxidation technology is a common technology for treating effluents from biogas digester for domestic sewage treatment. The technical essence is to fill packings within an aerobic biological reactor, and the packings are immersed in oxygenated sewage which flows through the packings at a certain velocity. Microbes grow on the packings and form a biofilm. Wastewater and biofilm extensively contact with each other. By functioning of microbial metabolism, organic pollutants are removed, and thus sewage is cleaned.

In general, biological contact oxidation has the following characteristics

1. After adding packings into the aerobic biological reactor, the biomass concentration in the reactor significantly increase (in activated sludge method, the general sludge concentration is 1.5–6.0 g/L, while in the biological contact oxidation the concentration is 10–20 g/L). This can improve treatment capacity, reduce occupation area, and save land and infrastructure investment.
2. High resistance to changes in hydraulic loading and organic loading.
3. Sludge sedimentation performance by contact oxidation method is good without sludge backflow, and this method can effectively overcome the problem of sludge expansion while reducing operation workload and number of maintenance personnel.

However, like biofilters and activated sludge process, if the design or operation is inappropriate, the biological contact oxidation method is susceptible to problems such as uneven water distributions, low aeration efficiency, local flow stagnation points, etc., which requires attention in real application.

There are usually two types of biological contact oxidation tanks, one is the one-stage type and the other is the two-stage type. The effective water depth is 3–5 m. Packings can be installed either partially or completely in the tank in layers, and the height of each layer should be no more than 1.5 m. A 2.0–5.0 kg $BOD_5/(m^3 \cdot d)$ volumetric loading rate is usually suggested when carbon oxidation is desirable, while a 0.2–2.0 kg $BOD_5/(m^3 \cdot d)$ volumetric loading rate is suggested for carbon oxidation and nitrification.

Applying biological contact oxidation process as the post-treatment unit of the biogas digester for domestic sewage treatment can help further remove organic matter. This process also has good conversion efficiency of ammonia, but poor efficiency on removal of total nitrogen. This is due to a lack of electron-donors in the aerobic post-treatment unit during the denitrification process, since most organic matter is removed in the biogas digester for domestic sewage treatment. In a combined anaerobic baffle reactor (ABR)—biological contact oxidation (BCO) system for treatment of rural domestic sewage, the ABR contributes an average of 92% of efficiency to the overall COD removal while the biological contact oxidation tank contributes 6% on average. Removal rate of ammonia in the system was 99%, and that of total nitrogen was 40%. Ammonia was mainly removed by biological contact oxidation, and removal of total nitrogen was mainly done by denitrification and microbial assimilation in local anaerobic area of biological contact oxidation tank (Yuan et al. 2012).

3.6.2 *Ecological Post-Treatment Technology*

Aerobic biological treatment technology, as a supplement to biogas digester for domestic sewage treatment, can further remove organic pollutants and make up for the defects of biogas digester for domestic sewage treatment in nitrogen and phosphorus removal. In general, aerobic biological treatment technology is relatively mature, with high treatment efficiency. However, aerobic biological treatment technology is energy-intensive, demanding for operation and management, and cost-intensive in investment and operation. It requires good economic conditions and professional operation and management. Economically underdeveloped regions with relatively large environmental capacity and lower requirements for effluent quality can adopt ecological treatment technology to further treat effluent from biogas digester for domestic sewage treatment. Ecological treatment technology mainly include stabilization ponds, construction wetlands, and land treatment system.

(1) Stabilization Pond

Stabilization Pond, also known as Oxidation Pond, refers to a sewage pond that is artificially built or appropriately modified with embankment and anti-seepage layers. Such pond cleans sewage naturally through physical, chemical and biological processes of the aquatic ecosystem. There are five major types of oxidation ponds: aerobic ponds, facultative ponds, anaerobic ponds, aeration ponds, and advanced treatment ponds. Stabilization ponds used for treatment of effluent from biogas digester for domestic sewage treatment are mainly facultative and aerobic ones. Generally, the BOD_5 loading rate of aerobic ponds is lower than 3 g/(m^2·d) and the depth of aerobic ponds is shallow, usually between 0.3 and 0.5 m. Sunlight can reach the bottom of the pond. The depth of the facultative pond is usually between 1.0 and 2.5 m and there are abundant creatures residing in the pond. The upper layer of the pond is an aerobic layer, whereas the middle and bottom are facultative and anaerobic layers. The BOD_5 loading rate of the facultative pond is in most cases 5–10 g/(m^2·d). BOD_5 of the effluent from a facultative stabilization pond is generally less than 100 mg/L, and SS is mostly in the range of 30–150 mg/L.

As a simple technology for rural domestic sewage treatment, the stabilization pond has the following advantages.

1. Stabilization pond can be made on the basis of existing river, swamp, valley with modification, resulting in low infrastructure investment.
2. Unlike aeration ponds, other types of stabilization ponds are low in energy consumption and treatment cost, and maintenance is easy.

3. Stabilization pond contains algae, plants, zooplankton, and many other organisms, which can be used for planting and animal breeding for economic values.

Stabilization ponds also have the following disadvantages.

1. Low loading rate resulting in large land occupation.
2. It is a natural open system, greatly affected by natural conditions.
3. It may cause secondary and groundwater pollution, form odour, and breed mosquitoes and flies.
4. Inappropriate management may lead to serious sludge sedimentation.

In rural sewage treatment, stabilization ponds are mainly used in arid and semi-arid regions with ditches, lowlands or ponds, and regions with abundant land areas. Stabilization ponds are often used as the back-end treatment unit of combined treatment processes, which is sequenced after anaerobic treatment units (septic tanks, biogas digester for domestic sewage treatment) or aerobic treatment unit and is mainly used for further removal of pollutants and sewage storage. In many cases, the oxidation pond can be used as a sewage treatment pond in spring and summer and as a sewage storage pond in autumn and winter. Figures 3.13 and 3.14 show stabilization ponds built in rural areas. There are many real-world cases in China that a stabilization pond is used as the post-treatment of biogas digester for domestic sewage treatment. For example, the Chengdu Jianjiang cement plant has been using the technique of biogas digester for domestic sewage treatment with a stabilization pond to treat domestic sewage (Zhao et al. 1995), with a daily treatment capacity of 400 tons. The process flow was as follows: domestic sewage → grit chamber → anaerobic digestion tank → hydraulic pressure tank → baffle filter tank → natural oxidation ditch → stabilization pond → agricultural irrigation canal.

Fig. 3.13 Stabilization pond in rural areas

Fig. 3.14 Stabilization pond with biological floating island

After five years of operation and testing, the average effluent quality is summarized and listed in Table 3.11.

(2) Constructed Wetlands

Based on ecological system that consists of plants, soil, and microbes, constructed wetlands make use of physical, chemical and biological synergy to achieve effective decontamination of wastewater and resource utilization, and resource recycling through filtration, adsorption, coprecipitation, ion exchange, assimilation by plants, and decomposition by microbes. Compared with natural wetlands, constructed wetlands strengthen the physical, chemical, and biological processes in natural ecosystems such as interception, adsorption, conversion and decomposition of organic matters in their design and construction. First, the soil used as packings of constructed wetlands can intercept suspended particles and adsorb partially soluble pollutants. In addition, plants can absorb nutrients such as nitrogen and phosphorus from wastewater, and their root systems can provide oxygen needed for wastewater treatment and microbial adhesion surfaces. Microorganisms in the system can effectively degrade pollutants in wastewater, and the wetland system is rich in microbial species. Under the synergistic effect of these microorganisms, sewage can be treated more thoroughly.

According to the flow pattern of wastewater in a constructed wetland, it can be categorized into surface flow constructed wetland (as shown in Fig. 3.15) and subsurface flow constructed wetland (as shown in Fig. 3.16). Different types of constructed wetlands which have their own advantages and disadvantages have different pollutant removal effects. In terms of process characteristics, constructed wetland load is generally low, and processing capacity is mainly ensured by increasing land area. In the selection of design load for the case with high pollutant concentration and large fluctuation of sewage concentration and quantity, the design load should be as small as possible. For organic pollutants has been removed in biogas digester for domestic sewage treatment, constructed wetland is a post-treatment unit that mainly removes

Table 3.11 Operation effect of biogas digester for domestic sewage treatment with a stabilization pond in treatment of domestic sewage (Zhao et al. 1995)

Sampling Site	Colority	pH	COD (mg/L)	BOD_5 (mg/L)	SS (mg/L)	Ammonia Nitrogen (mg/L)	Coliform	Number of ovum(/L)
Digesters	<50	7–8	85.3	53.3	17.2	16.8	10^{-5}-10^{-4}	0.72
Natural oxidation ditch	<20	7–8	42.1	15.5	13.8	10.8	10^{-5}–10^{-4}	0.5
Stabilization pond	<20	7–8	37.4	12.2	9.7	7.87	10^{-4}	0
Total outlet	<20	7–8	23.6	10.7	6.4	5.8	10^{-4}	0

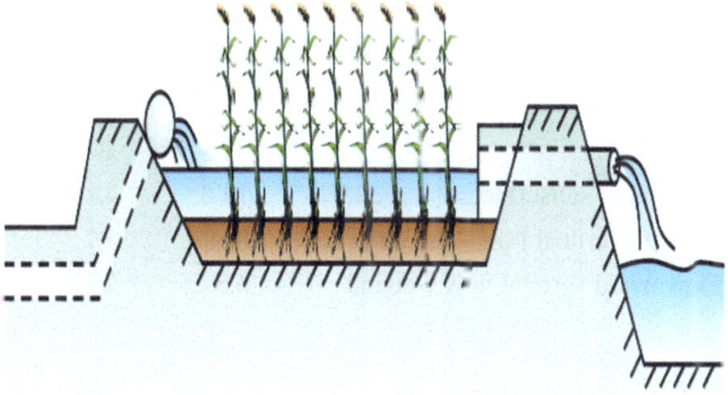

Fig. 3.15 Surface flow constructed wetland

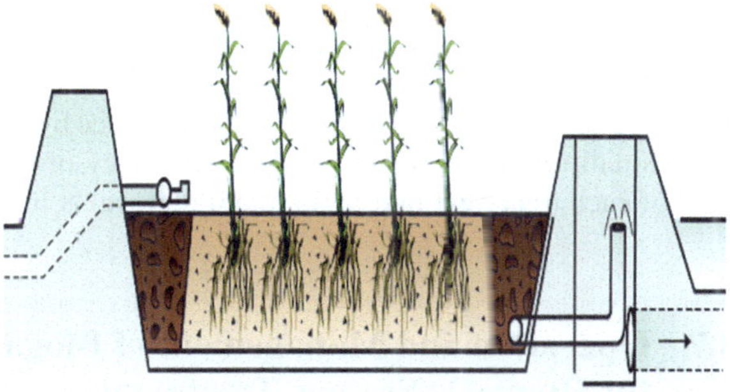

Fig. 3.16 Subsurface flow constructed wetland

nitrogen and phosphorus. Therefore, the design load should be chosen based on the nitrogen and phosphorus removal loading rate.

As a post-treatment technology of biogas digester for domestic sewage treatment, constructed wetland is widely used in domestic sewage treatment in rural areas of China. Its main characteristics are as following.

1. The operation cost is low. In areas with elevation difference, constructed wetland operation almost requires no energy input, and addition of any chemical agent is not necessary.
2. Easy operation and maintenance and low technical requirements. Its daily maintenance and operation is very simple, mainly relying on natural treatment and requiring little manual intervention.
3. Relatively good treatment effect. As long as the constructed wetlands are designed and constructed according to specifications, the wetlands can ensure good effluent quality and are resistant to shock load.
4. Landscape value. Constructed wetlands can be combined with local tourism, planting, etc.

In spite of their merits, constructed wetlands still have problems such as large land occupation and relatively low treatment efficiency per unit area which are not suitable for utilization in areas with land shortage.

Table 3.12 Contribution ratio of each treatment unit in anaerobic—constructed wetland

Treatment unit	Contribution ratio for the removal of pollutants (%)			
	COD	TN	Ammonia nitrogen	TP
ABR	75.4	20.2	12.7	13.2
Horizontal subsurface flow wetlands	14.5	45.8	48.9	41.7
Upward vertical flow wetland	3.8	8.7	12.3	28.7
Downward vertical flow wetland	6.3	25.3	26.1	16.4

Zhu et al. (2014) built ABR-combined constructed wetland to treat sewage from campus, which achieved good results on pollutant removal and effluent quality. During the experiment, COD, $NH_4{}^+$−N, TN, and TP in effluent were 47.1 mg/L, 11.6, 18.2, and 0.6 mg/L respectively. The average removal efficiency of each was 84.6% and 59.3%, 58.0%, and 58.0% respectively. The removal efficiency of $NH4^+$−N and TN of the combined process was greatly influence by temperature, while temperature had a small influence on the removal efficiency of COD and TP. The contribution ratio of each treatment unit to pollutant removal is listed in Table 3.12.

3.7 Operation and Management of Biogas Digester for Domestic Sewage Treatment

The treatment effect of biogas digester for domestic sewage treatment is mainly determined by two aspects. One is engineering construction, which requires reasonable design and careful construction. The other is operation and management, since the purpose of pollution control and environmental improvement is realized with appropriate management and operation after completion of a digester. The key to operation and maintenance is to provide a suitable environment for microorganism growth and metabolism in a digester, improving the decontamination efficiency by the microorganisms and keeping the digester from clogging, water leakage, and air leakage. Specifically, the following points need to be noted:

1. Prevent waste pesticides, calcium carbide, and disinfectants from entering the digester, which will inhibit microorganisms.
2. Sedimentation in the grit well (pond), such as tile, stone, glass, metal, plastic, etc., should be cleaned in time. Pay attention to crusts that might clog the feed tube.
3. The anaerobic zone must be sealed to ensure the gas-tightness of the tank to prevent the tank from cracking and oxygen entering.
4. Regularly clean or replace the packings and filter materials. Soft packings can be rinsed in the tank to prevent agglomeration; hard and semi-hard packings shall be removed from the tank before cleaned and loaded into the tank. Replace

soft packings every 10 years, and replace the polyurethane filter foams every 4–5 years.

5. The sludge should be cleared every 1–2 years. When there is no obvious decline in the quality of effluent nor loss of sludge, clearance period can be prolonged. When cleaning, temperature should be above 10°C. The sewage sludge can be pumped by a manure truck. When pumping, the movable cover shall be opened, and the personnel shall not enter the digester. After proper treatment, the sludge should be used as fertilizer or as inoculum of new biogas digester for domestic sewage treatment.
6. If leakage occurs in the tank, it shall be repaired in time. For example, when the inlet has wastewater while the outlet has no wastewater or when the flow rate of effluent is obviously smaller than that of the influent, it is necessary to check whether there is leakage in the digester. At the same time, check whether there is damage in the connecting pipes between the tanks.
7. The concentration of influent and effluent should be monitored regularly. When the effluent fails to reach the standard, it is necessary to find out the reason s and fix it in time (e.g., clearing sludge in time, replacing packing or filtering material, etc.).
8. People shall not enter the anaerobic digestion tank, anaerobic filter, and facultative filter which produce toxic and hazardous gases. In the occasion of tank maintenance, packing replacement, and filter replacement, approved personnel who enter above-mentioned danger space must comply strictly with relevant specifications to prevent accidents.
9. Effluent from biogas digester for domestic sewage treatment is preferentially used to irrigate vegetables, grain, and fruit trees. This can save irrigation water and providing nitrogen and phosphorus nutrients. After reaching the local discharge standards, it can be directly discharged into the water body.

References

Barbosa, R.A., and G.L.S. Jr. 1989. Treatment of raw domestic sewage in an UASB reactor. *Water Research* 23 (12): 1483–1490.

Bertero, M.G, R.A. Rothery, M. Palak, H.C. Cynthia, D. Lim, F. Blasco, J.H. Weiner, and N.C.J. Strynadka. 2003. Insights into the respiratory electron transfer pathway from the structure of nitrate reductase A. *Nature structural & molecular biology* 10(9):681–687.

Butler, D., and J. Payne. 1995. Septic tanks: problem and practice. *Building and Environment* 30 (3): 419–425.

Chen, Y.G., Z.X. Liu, and X.L. Wan. 1988. Study on application of bio-technology on residential domestic sewage treatment. *Sichuan environment* 7 (2): 1–6. (in Chinese).

Chou, H.S., J.S. Huang, and W.F. Hong. 2004. Temperature dependency of granule characteristics and kinetic behavior in UASB reactors. *Journal of Chemical Technology and Biotechnology* 79: 797–808.

Deng, L.W., B. Wang, and Y. Tang. 1997. Demonstration and study on application of low maintenance wastewater treatment system for the treatment of sewage. *Environmental protection* 6: 15–17. (in Chinese).

GB 19379–2012. Hygienic specification for rural household latrine/National standard of the people's republic of China.

Hou, J.W., B. Fan, Qu Bo, L. Cai, and S.S. Zhu. 2012. Review about characteristics of rural domestic wastewater discharge. *Journal of Anhui Agricultural Sciences* 40 (2): 964–967. (in Chinese).

Kobayashi, H.A, M.K. Stenstrom, and R.A. Mah. 1983. Treatment of low strength domestic sewages using the anaerobic filter. *Water Research* 17: 903–909.

Lens, P., G. Zeeman, and G. Lettinga. 2001. *Decentralised sanitation and reuse-concepts systems and implementation*. London: IWA Publishing.

Leshang City Statistical bureau. 2013. Le Shang Statistics Yearbook 2012. Leshang City Statistical Association. (in Chinese).

Li, L., L. Huang, and Y. Liu. 2004. Present development situation of anaerobic treatment for municipal wastewater. *Techniques and Equipment for Environmental Pollution Control* 5 (12): 1–6. (in Chinese).

Li X, X. 2015. Present situation and suitability of decentralized rural sewage treatment in Chongqing. *Doctoral Dissertation of Southwest University.* (in Chinese).

Mario, T.K., A.F. Jim, and L. Gatze. 1997. The anaerobic treatment of low strength sewages in UASB and EGSB reactors. *Water Science and Technology* 36 (6–7): 375–382.

Ministry of housing and rural-urban development of China. 2010. *Technical Guidelines for Rural Domestic Sewage Treatment*. (in Chinese).

NY/T 1702–2009. Technology specification of biogas digester for domestic sewage treatment. *Agricultural Industry Standards of the People's Republic of China*.

NY/T 2602–2014. Operation maintenance specification of biogas digester for domestic sewage treatment. *Agricultural Industry Standards of the People's Republic of China*.

Sasse, L. 1998. *DEWATS–decentralized sewage treatment in developing countries*. Bremen, Germany: Bremen Overseas Research and Development Association.

Seggelke, K., F. Obenaus, and K.H. Rosenwinkel. 1999. Dynamic simulation of a low loaded trickling filter for nitrification. *Water Science and Technology* 39 (4): 163–168.

Seghezzo, L., G. Zeelnan, and J.B. van Lier. 1998. A review: the anaerobic treatment of sewage in UASB and EGSB reactors. *Bioresource Technology* 65: 175–190.

Shen, D.S., Y.H. He, H.J. Feng, and X.H. Zhou. 2005. Underground unpowered anaerobic reactor for rural domestic sewage treatment. *Transactions of The Chinese Society of Agricultural Engineering* 21 (7): 111–115. (in Chinese).

Sun, R.M. 2010. Analysis on the discharge situation of rural domestic sewage in China. *Energy and Environment* 5: 33–35. (in Chinese).

Tian, H.C., H. Xie, Z.J. Tang, H.L. Zeng, and Q. Wu. 2002. Evaluation on daily life sewage treatment effect by municipal biogas digester. *China Biogas* 20 (4): 33–36. (in Chinese).

Uemura, S., and H. Harada. 2000. Treatment of sewage by a UASB reactor under moderate to low temperature conditions. *Bioresource Technology* 72: 275–282.

Wang, Y.H., Y. Fang, and J. Jiao. 2008. Evaluation of night soil treatment efficiency of " three-grille-mode" septic tanks in the rural area of Jiangsu. *Journal of Ecology and Rural Environment* 24 (2): 80–83. (in Chinese).

Xia, B.S., Q.C. Hu, and L. Song. 2008. Technical analysis of designs of biogas digesters for treating domestic sewage from rural areas. *Transactions of The Chinese Society of Agricultural Engineering* 24 (11): 197–200. (in Chinese).

Xiao, Y.T. 2007. *Biological pretreatment process for polluted raw water by elastic packing for biofilm and fine bubble aeration for oxygen enrichment*. Beijing: China Architecture & Building Press. (in Chinese).

Xie, Y.H., R.J. Dong, and Y.L. Wang. 2005. A survey of anaerobic digestion application in small towns' wastewater treatment. *Renewable Energy* 4: 71–74. (in Chinese).

Yi, Q.T., H. Li, L. Zhang, and Y.Z. Jing. 2016. Evaluation of treatment performance of the combined processes of anaerobic tanks, aerated biofilter and subsurface constructed wetland on rural domestic wastewater. *Chinese Journal of Environmental Engineering* 10 (5): 2395–2400. (in Chinese).

Yuan, H.P., C.S. Peng, and Z.Y. Wang. 2012. Study of domestic sewage treatment of newly developing village by a combined process of ABR and BCO. *Technology of Water Treatment* 38 (4): 91–95. (in Chinese).

Zeng, H.L., X.H. Zhao, H.C. Tian, R.J. He, and J. Zhao. 1994. Study on the design parameters and environmental and hygienic effects of municipal biogas digester for domestic sewage treatment. *China Biogas* 12 (1): 11–17. (in Chinese).

Zhao, X.H., H.L. Zeng, H.C. Tian, T. Liu, and G.N. Chen. 1995. Environmental and hygienic effects of biogas digester for domestic sewage treatment—stabilization pond system on residential domestic sewage treatment. *China Biogas* 13 (2): 29–31. (in Chinese).

Zhu, K., X.P. Liu, J.H. Cheng, and Y.L. Shen. 2014. Pilot-scale study on ABR-hybrid constructed wetland process for rural domestic sewage treatment. *Technology of Water Treatment* 40 (12): 95–99. (in Chinese).

Chapter 4
Biogas Plant

Biogas plants, which refer to the systematic projects with daily biogas production $\geq 5\ m^3$ that are built for treatment of organic wastes (wastewater) through anaerobic digestion, are classified into small, medium, large and mega-scale plants.

Compared with rural household biogas digesters, biogas plants have more complex components: the digesters are generally built above ground or half-buried, and the operation needs power. A plant includes a main body, supporting parts, production management and life service facilities, etc. Professionals of process, structure, equipment, electrical and control are required to do the design according to the feedstock and site conditions.

Biogas plants need not only scientific designs, but also professional construction and elaborate operation management.

4.1 Feedstocks for Biogas Plants and Its Characteristics

There are generally five types of feedstocks for biogas plants: animal waste, crop straws, industrial wastes, municipal wastes, and aquatic plants.

4.1.1 Animal Wastes

Animal wastes, including animal manure, urine, and flushing wastewater, is taken as an ideal feedstock for anaerobic digestion. Organic matters are the substrates for biogas production. Since the organic matters content in urine is quite low, it is a common practice to calculate biogas production based on the quantity of animal manure. The quantity of animal manure usually depends on the animal species, weight, physiological status, feedstuff composition, and feeding modes. Table 4.1

L. Deng et al., *Biogas Technology*,
https://doi.org/10.1007/978-981-15-4940-3_4

Table 4.1 Manure production of livestock and poultry

Animal	Manure production (kg/capita d)
Swine	1.4–1.8
Cow	30–33
Beef cattle	12–15
Goat	1.1–1.2
Rabbit	0.36–0.42
Chicken	0.10–0.15
Duck	0.10–0.12

summarizes manure production of livestock and poultry reported by Guo et al. (2011), Liu et al. (2010), Zhang et al. (2012), and Zhuang et al. (2015).

Animal manure is advantageous as a feedstock for biogas production for the following reasons. ① The C/N (carbon-nitrogen) ratio is usually within 15–30:1, which is favorable for the growth of anaerobic microorganisms. ② There is sufficient buffering capacity against minor acidification. ③ Some animal manure (e.g. cow dung) contains rumen microorganisms that are useful supplements for anaerobic microbes in digesters. However, there are also negative factors when animal manure is used as feedstock for biogas production. ① Animal manure is usually in large volumes and has low content of dry matters (usually <30% in manure and <2% in flushing wastewater), leading to low biogas yield per unit feedstock and high transport cost of feedstock or digestate. ② Some animal manure has high content of heavy metals and antibiotics, which are unfavorable for anaerobic digestion as well as treatment and reutilization of digestate. ③ Ammonia content in poultry manure is high, which can easily cause ammonia inhibition to microbes in anaerobic digestion. To solve aforesaid problems, a common solution is to mix animal manure with straws and municipal organic wastes for co-digestion. A digestion system with mixed feedstock is more stable than that with animal manure only. See Table 4.2 for characteristics and biogas yield of animal manure as feedstock.

Manures of different animals are different in physical and chemical properties that determine the operational performance of a biogas plant, whose design and operation shall be specially noted. Cow manure contains a large quantity of weeds and only a few sediments, so there is more scum than sediments in digesters. Since sand content in cow dung is high, desanding must be included as a necessary step in the pretreatment unit of a biogas plant. Swine manure contains lots of sediments, so the sediments are more rather than the scum. A pig farm produces flushing wastewater in a large quantity. Such flushing wastewater usually has a low total solid concentration and is difficult to heat up. Biogas output is reduced in winter. Chicken manure contains feathers and sands, and sand particles are usually encased in organic matters. Feathers and sands shall be removed in biogas plant. Goat and rabbit manure contains a large amount of grass and a low level of moisture and is mostly granular, so a tank where the manure granular is dissolved with water needs to be built in the pretreatment phase.

Table 4.2 Characteristics and biogas yield of animal manure as feedstock (Al Seadi 2008; Wellinger et al. 2013; Wan 1995; Song et al. 2010; Dong et al. 2011)

Substrate	TS (%)	VS/TS (%)	C:N	Biogas yield			Methane yield	
				m^3/kg FM[a]	m^3/kg TS	m^3/kg VS	m^3/kg FM	m^3/kg VS
Swine manure	20–25	77–84	13–15		0.252–0.352			
Cow dung	16–18	70–75	17–26	0.060–0.120	0.180–0.250		0.033–0.036	0.130–0.330
Beefs manure	17–20	79–83	18–28		0.180–0.250			
Goat manure	30–32	65–70	26–29		0.206–0.273			
Chicken manure	29–31	80–82	9–11		0.323–0.375			
Duck manure	16–18	80–82	9–15		0.359–0.441			
Rabbit manure	30–37	66–70	14–20		0.174–0.210			

[a]*FM* fresh manure

4.1.2 Crop Straws

Crop biomass usually consists of three parts: grain, aerial stem, and root. The proportion of the three parts can be obtained by determining the mass of each part, and the yield of crop straws is estimated based on the grain yield. At present, the grain-straw ratio is widely applied for estimating straw yield. Table 4.3 provides straw yields estimated based on the average grain-straw ratio proposed by Wang et al. (2012) and grain yield issued by the Chinese National Bureau of Statistics.

Table 4.3 Grain-straw ratios and crop straw yields

Type of crops	Grain-straw ratio	Average grain yield (kg/hm^2)	Straw yield (kg/hm^2)	Straw yield(kg/m^2)
Rice	1.04	7779	8090	0.81
Wheat	1.28	4687	5999	0.60
Corn	1.07	5870	6281	0.65
Sorghum	1.60	4101	6561	0.63
Soybean	1.35	1820	2457	0.25
Rape	2.90	1885	5467	0.55
Cotton	2.87	1458	4184	0.42
Tobacco	0.66	2134	1408	0.14
Sugarcane	0.34	68,600	23,324	2.33
Beet	0.37	49,793	18,423	1.84

Table 4.4 Characteristics and biogas yield of crop straws as feedstocks

Substrate	TS (%)	VS/TS (%)	C:N ratio	Biogas yield (m^3/kg TS)
Corn straw	80–95	74–89	51–53	0.33–0.35
Rice straw	82–88	74–83	68–87	0.32–0.33
Wheat straw	83–95	82–84	51–67	0.30–0.31

As a feedstock for biogas production, crop straws contain high levels of dry matters, high biogas yield per unit fresh mass, low feedstock and digestate transportation costs, and reduced production of liquid digestate. However, there are also negative factors when crop straws are used as feedstocks for biogas production. ① Straws contain high levels of celluloses, hemicelluloses and lignin which require a long time to degrade. ② The C/N ratio of straws is high, usually >50, which is unfavorable for normal growth of microorganisms and leads to a long start-up in a straw-only biogas plant. ③ Crop straws are highly seasonal and require long-term storage and large storage facilities. ④ Straw scums generated in digesters can make discharging difficult.

See Table 4.4 for characteristics of straws as feedstocks and their biogas yields. Corn straws have the highest biogas yield, followed by wheat straws and rice straws.

4.1.3 Industrial Organic Wastes

Most industrial organic wastes can be used as feedstocks for biogas production, and these wastes or wastewater are mainly wastes or wastewater from food and beverage processing, feed processing, sugar-making, starch processing, fruit processing. The discharge quantity of industrial wastes is usually estimated based on the ratio of the waste/wastewater to industrial products. Table 4.5 provides the discharge quantity of some industrial organic wastes and wastewater.

The composition, dry matter content and biogas yield of each kind of industrial organic waste differ greatly depending on waste sources (see Tables 4.6 and 4.7). Industrial wastes containing high levels of hydrocarbons, fats and proteins are friendly for biodegradation and suitable for uses as biogas feedstocks. Industrial organic wastewater is usually produced in large quantities. It is difficult to reutilize digested effluent from the anaerobic digestion process, which usually requires a post-treatment unit to further treat digested effluent to meet discharge standards.

Table 4.5 The discharge quantity of waste residues and wastewater in food and fermentation industries (Wang et al. 2000b)

Industry	Description of waste residue/wastewater	Waste-product ratio (W/W)[a]	Wastewater-product ratio (W/W)[b]
Cassava alcohol	Vinasse	15	80
Molasses alcohol	Vinasse	15	60
MSG	Rice dregs	3	400
	Waste original liquid	20	
Citric acid	Dried potato pulp	3	300
	Neutralization solution	10	
	Immersion water	25	
Starch	Bean curd wastewater	10	50
	Peel slags	10	
Starch sugar	Corn slurry slags	0.3	–
Beer	Beer spent grains	0.2	20
	Waste yeast	0.01	
White spirits	Vinasse of white spirits	3	60
Rice wine	Vinasse of rice wine	3	4
Beet processing	Beet pulp	6.7	100
	Mashed beet	1	
Sugarcane sugar	Sugarcane pulp	10	20
	Mashed beet	1	
Dairy products	–	–	8
Canned food	–	–	100
Soft drink	–	–	100

Note [a]Only includes main residues and wastewater
[b]Includes washing water, flushing water, and cooling water in addition to main residues and wastewater

Table 4.6 Characteristics and biogas yield of industrial waste/wastewater as feedstocks (Wang et al. 2000a, b; Hu 1995; Han et al. 1988)

Substrate	COD (mg/L)	BOD_5 (mg/L)	Biogas yield (m^3/kg COD)
Vinasse of molasses alcohol	80,000–110,000	40,000–70,000	0.360–0.495
Vinasse of cassava alcohol	50,000–70,000	20,000–40,000	0.331–0.678
Wastewater of starch processing	3000–9000	1500–5000	0.340–0.440
High-concentration wastewater of beer	4000–6000	2400–3500	0.420
Wastewater of citric acid	10,000–44,000	6000–25,000	0.430–0.530
Wastewater of MSG	30,000–70,000	20,000–42,000	0.340–0.520

Table 4.7 Characteristics and biogas yield of industrial waste residues as feedstock (Dong et al. 2011)

Substrate	TS (%)	VS (%)	C:N	Biogas yield		Methane yield	
				(m^3/kg FM)	(m^3/kg VS)	(m^3/kg FM)	(m^3/kg VS)
Wheat vinasse	20–25	70–80		0.105–0.130		0.062–0.112	0.295–0.443
Grain vinasse	6–8	83–88		0.030–0.050		0.018–0.035	0.258–0.420
Potato vinasse	6–7	85–95		0.026–0.042		0.012–0.024	0.240–0.420
Wastes of fruit processing	15–20	75	35		0.250–0.500		
Crude glycerol				0.240–0.260		0.140–0.155	0.170–0.200
Rape seed cake	92	87		0.660		0.317	0.396
Potato pulp	13	90		0.070–0.090		0.044–0.050	0.358–0.413
Potato slurry	3.7	70–75		0.050–0.056		0.028–0.031	0.825–1.100
Beet pulp	22–26	95		0.060–0.075		0.044–0.054	0.181–0.254
Apple pulp	25–45	85–90		0.145–0.150		0.098–0.101	0.446–0.459
Grape pulp	40–50	80–90		0.250–0.270		0.169–0.182	0.432–0.466

4.1.4 Municipal Wastes

(1) Source-separated organic wastes

Municipal garbage refers to solid wastes produced in daily lives of people and it mainly includes household wastes, commercial wastes and cleaning wastes. Source-separated organic wastes consist of the organic fraction of household waste such as food waste, garden waste and other similar organic wastes, which can be used as feedstocks for anaerobic digestion. The production of municipal garbage is mainly dependent on population density, energy structure, geographical location, season, living habits, economic development and waste recovery practices. For the moment, municipal garbage production in China has reached 1.2 kg/(person day) (Fu et al. 2014). Considering the complexity of household waste and ensuring continuous and stable operation of anaerobic digestion process (including pretreatment, feedstock transport, mixing, digestion as well as utilization of digestate), it is necessary to effectively sort out household garbage from the sources so that each category of organic waste is simply composed and the content of impurities is minimized.

There are two major considerations for utilizing source-separated organic wastes as the feedstock of anaerobic digestion. ① The cost for collecting, sorting and separating garbage. ② The effect of sorting and separating garbage from the sources. Impurities such as metals, glass, plastics and sands/stones in garbage can seriously affect normal operation of a biogas plant and must be removed during pretreatment. Organic garbage usually contains some pathogenic bacteria, which must be sterilized before and after use to ensure a higher safety level.

Compositions of household waste is closely related to living habits and economic development of a region. In a coal-using area, the organic component in household wastes is usually around 30% while in a fuel gas-using area, this ratio can be as high as 75% (Guo and He 2001). The VS/TS ratio of organic garbage is usually around 80% and the C/N ratio is about 12–20, making it a suitable feedstock for anaerobic digestion, with a biogas yield of 0.22–0.27 m^3/kg TS or 0.30–0.40 m^3/kg VS (Xiang et al. 1989; Zhang et al. 2005; Ye 2007).

(2) Food/kitchen waste

Food waste includes restaurants waste and kitchen waste. Restaurants waste refers to food leftovers from restaurants, hotels, or canteens as well as wastes from processing of fruits, vegetables, greases, food made of flour and so forth in a cookhouse of restaurants. Kitchen waste includes fruits, vegetables, food offcuts, leftovers, peels and other perishable organic wastes produced in household daily lives. In China, the treatment of food/kitchen waste mainly refers to the treatment of restaurants waste while kitchen waste is usually categorized as household garbage. The production of restaurants waste depends on eating habits, economic conditions and ways of dining. For the moment, production of restaurants waste is about 0.1 kg/(person day). Different correction coefficient can be multiplied for different cities, which ranges from 1.05–1.15 for cities where economy or tourism is developed or universities are concentrated, whereas a higher coefficient of 1.15–1.30 is recommended for cities where both economy and tourism are highly developed or for coastal cities with highly developed economy.

In comparison with other biogas digestion feedstocks, food/kitchen waste contains high levels of fat and salt and comes from highly varied sources. The composition of food/kitchen waste varies greatly with regions of different eating habits. Table 4.8 provides the composition of food/kitchen waste of a city in south China reported by Liu et al. (2016). Table 4.9 provides some physical and chemical characteristics of food/kitchen waste of a city in north China reported by Ma and Liu (2012), and these characteristics are highly typical for food/kitchen waste in China. Food/kitchen waste usually has a water content of 80–90%, a fat content of 1–5% and a salt (NaCl) content of 1–2%. Such garbage is rich in organic matters, of which volatile organic matters usually account for 90–94% (dry-base) (Jing et al. 2013). Each ton of fresh garbage can produce 70–110 m^3 of biogas. The biogas yield of food/kitchen waste is 0.45–0.50 m^3/kg TS (Ma and Liu 2012; Liu et al. 2016). Food/kitchen waste has the following characteristics when used as a feedstock for anaerobic digestion. ① It contains solid impurities such as bones, knives, forks, chopsticks, bowls, etc. that may damage pumps, pipes and other equipment and must be removed during

Table 4.8 Composition of food/kitchen waste of a southern city (Liu et al. 2016)

Types	Percentage of dry basis (%)	Percentage of wet basis (%)	Water content in wet-basis (%)
Water content	84.82		
Food residue	10.1	87.16	88.41
Paper	0.08	0.29	72.41
Metals	0.45	0.52	13.46
Plastics	0.48	0.75	36.00
Wood/bamboo	0.03	0.05	40.00
Bones	1.59	3.72	57.26
Fat	2.2	7.22	69.53
Glass/ceramics	0.25	0.29	13.79

pretreatment. ② It contains high levels of salt and protein, which easily introduces ammonia and salt inhibition during anaerobic digestion. ③ A high level of hydrogen sulfide in produced biogas is expected. ④ There is a high ratio of degradable organic matters, which can be easily acidized.

(3) Municipal sludge

Municipal sludge includes various sorts of sludge and waste produced by municipal sewage treatment plants during wastewater treatment. Municipal sludge includes residues in screens, grit chambers, primary sedimentation tanks, and secondary sedimentation tanks. Municipal sludge has a high water content and is produced in large volumes with instability. Sludge of primary and secondary sedimentation tanks is rich in organic matters which is easy to degrade and is suitable for anaerobic digestion. During wastewater treatment, production of sludge is usually 0.3–0.5% of treated wastewater (sludge water content 97%) (Zhang 2000). Sludge of primary and secondary sedimentation tanks has similar potential to produce biogas as animal manures, with a biogas yield of 4–14 m^3/m^3 sludge and 7.54 m^3/m^3 sludge on average (Wu et al. 2008). Biogas yield can be improved by taking some pretreatment measures including chemolysis, pyrolysis and enzymolysis, etc.

Sludge contains some hazardous substances, especially one in primary sedimentation tanks which contains pathogens and some heavy-metal compounds, is unfavorable to reutilizing of digestate as a fertilizer. This is the greatest limitation on utilizing municipal sludge as a feedstock for anaerobic digestion. In some countries, digestates are prohibited to be used as farmland fertilizers when municipal sludge is used as the feedstock of a biogas plant.

Table 4.9 Physical and chemical characteristics of food/kitchen waste of a city in north China (Ma and Liu 2012)

Items	TS (%)	VS/TS (%)	Organic matter (%)	N (%)	C/N (%)	P_2O_5 (%)	K_2O (%)	pH
Scope	20.19–26.15	90.71–95.87	78.60–95.48	1.70–2.35	22.28–27.68	0.60–2.84	0.12–0.27	3.87–4.55
Average	23.28	93.75	89.16	2.05	25.38	1.36	0.20	4.15

Note TS and VS/TS are counted as per wet basis while organic matters, N, P_2O_5 and K_2O are counted as per dry basis

Table 4.10 Characteristics and biogas yield of aquatic plants as feedstocks (He J et al. 2008; Liu et al. 2011; Huang et al. 2006, 2013)

Substrate	TS (%)	VS (%)	Biogas yield	
			(m^3/kg TS)	(m^3/kg VS)
Water hyacinth	8	80	0.245–0.302	0.292–0.362
Alligator weed	10	75	0.184–0.334	0.250–0.447
Blue algae	4	90	0.255–0.345	0.285–0.366

4.1.5 Aquatic Plants

Aquatic plants produce renewable energy while treating wastewater at the same time. Some aquatic plants such as water hyacinth, alligator weed, macroalgae and microalgae that grow in aquatic plant ponds, constructed wetland, or eutrophic water can quickly reproduce to large populations. These plants also contain high levels of organic matters and show high digestibility. Therefore, they are ideal feedstocks for anaerobic digestion. Table 4.10 introduces the characteristics and biogas yield of some aquatic plants.

4.1.6 Resource Availability and Supply Chain

Whichever feedstock is used, it shall meet the following conditions of anaerobic digestion:

(1) COD concentration > 3000 mg/L, or TS > 0.4%;
(2) C/N ratio: 20–30:1;
(3) BOD_5/COD ratio: > 0.3;
(4) pH: 5.0–8.0;
(5) Concentrations of anaerobic digestion inhibitors such as heavy metals, salts, pesticides, antibiotics, etc. shall not exceed the inhibition level.

In theory, there are abundant resources for anaerobic digestion. However, when taking perspectives at ecological, technological and economical levels, it is often found that not all resources are available. The potential utilization of feedstock for biogas production depends on the spatial distribution of the feedstocks, their availability for collection, transport distance, and costs. According to experiences of biogas plants in Europe, the economic radius to transport feedstock such as animal slurries and energy crops shall be less than 5 and 15 km respectively. In addition to distances between feedstock origins and biogas plant construction sites, transport distance of digestates is also a critical factor to be considered, which shall be less than 15 km around a biogas plant.

Safety and hygiene in anaerobic digestion shall also be highly valued. Measures shall be taken against pathogen transmission if animal manures come from different farms.

4.2 Composition and Process Flow of a Biogas Plant

4.2.1 Components

A biogas plant is a waste treatment and energy recovery system composed of a series of sophisticated facilities, and it consists of multiple process units. There are usually five units defined by their functions (Fig. 4.1):

(1) Feedstock pretreatment unit: composed of feedstock collection, transport, and storage subunits, as well as various pretreatment facilities.
(2) Biogas production unit: composed of an anaerobic reactor, a feeder, a discharger, a heater and insulator, mixers, protective devices, etc.
(3) Biogas cleaning and storage unit: composed of dewatering, desulfurization and biogas upgrading equipment.
(4) Biogas utilization unit: composed of subunits for centralized gas supply, biogas power generation, connection of bio-methane into natural gas pipe network, utilization of bio-methane as vehicle fuel, etc.
(5) Digestate processing and utilizing unit: composed of subunits for solid-liquid separation, fertilizer production from liquid digestate, concentration of liquid digestate, digestate transmission, distribution and farmland application, treatment for standard-met discharge, etc.

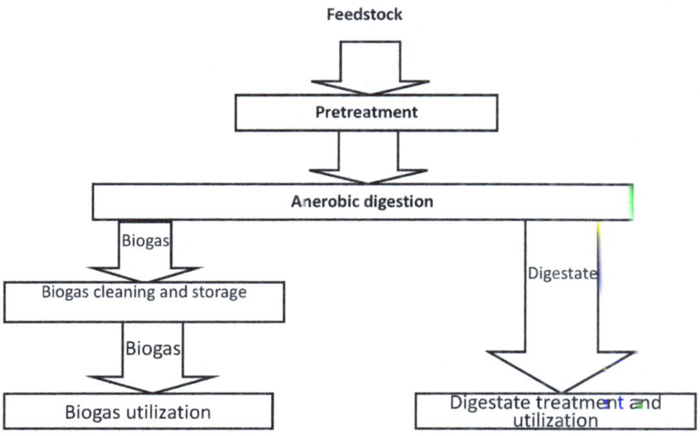

Fig. 4.1 Components of a biogas plant

There are common characteristics between biogas cleaning, storage, utilization, digestate treatment in plants and those in rural biogas digesters and septic tanks, which will be introduced separately in the following chapters. This chapter mainly deals with the pretreatment unit and the biogas production unit.

4.2.2 Process Flow in Biogas Plant

The type and design selection of a biogas plant is mainly related to quantity and type of available feedstock. Feedstock quantity decides the project scale while substrate species and characteristics decide the technical route to be adopted. Feedstock diversity decides the variety of technical route, digestion equipment, and operations. Some typical technical processes for anaerobic digestion are introduced below.

(1) Anaerobic digestion process for soluble organic wastewater

One of the features of soluble organic wastewater is that the organic matters in feedstock mainly exist in soluble forms, in which SS is low (usually <1500 mg/L) or SS is low in separated liquid after solid-liquid separation. Common examples of the former type include wastewater from beer and starch production, and for the latter type wastewater from alcohol and citric acid production is commonly seen. Figure 4.2 shows the technical process of anaerobic digestion for soluble organic wastewater. This process aims at treating wastewater while recovering energy at the same time. Grit chambers are required for sand-enriched wastewater such as wastewater from cassava alcohol or cassava starch production. Up-flow anaerobic sludge blanket (UASB), expanded granular sludge bed (EGSB), internal circulation (IC) anaerobic reactor and other high-efficiency anaerobic reactors are recommended for treating such kind of wastewater. Since biogas plants treating such wastewater are usually located in suburban areas or industrial parks/zones, it is difficult to reutilize digested effluent for agricultural purposes. Therefore, the digested effluent is usually subject to aerobic and physical-chemical post-treatments before discharges complying with relevant requirements. Biogas produced from anaerobic digestion is mainly supplied to the boiler or power generation unit to generate electricity utilized by the factory itself that discharge wastewater.

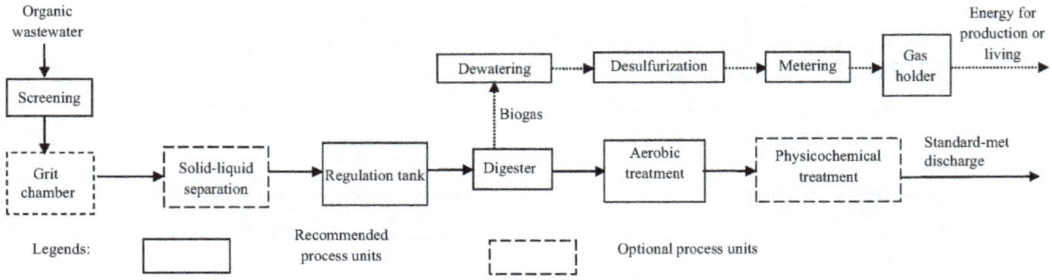

Fig. 4.2 Process flow chart of anaerobic digestion for soluble organic wastewater

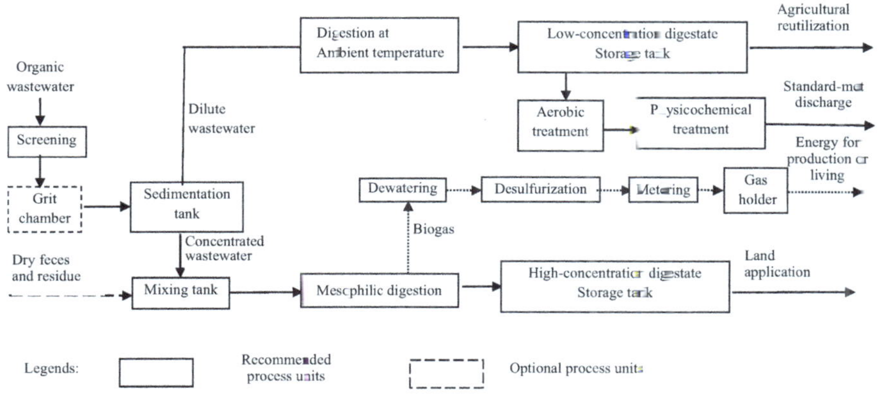

Fig. 4.3 Process flow chart of anaerobic digestion for medium-concentration and moderate-solid organic wastewater

(2) Anaerobic digestion process for medium-concentration and moderate-solid organic wastewater

In medium-concentration and moderate-solid organic wastewater, organic matter and SS contents are moderate, with a TS of 0.5–4%, COD of 3000–35,000 mg/L, SS of 1000–30,000 mg/L. Such wastewater include wastewater from animal manures, slaughter wastewater, and food processing wastewater. Figure 4.3 shows the technical process of anaerobic digestion for medium-concentration and moderate-solid organic wastewater. This process aims at treating wastewater while recovering energy at the same time. High SS level makes it hard to form granular sludge in anaerobic digestion and thus the high-efficiency anaerobic digestion process can be hardly achieved. Besides, heating is difficult due to the small amount of organic matters in such wastewater, thus heating all wastewater to mesophilic digestion or near-mesophilic digestion can hardly be guaranteed. When pretreating such wastewater, it is necessary to separate the concentrated and dilute portions, i.e. to separate the wastewater by gravity sedimentation to supernatant (thin wastewater) and bottom sediments (concentrated wastewater). Manures collected by dry method as well as waste residues may be mixed with concentrated wastewater for anaerobic digestion. After the separation, the volume of concentrated wastewater accounts for about 20% of the volume of original wastewater, but the biogas production of concentrated wastewater takes up about 70% of total biogas production. The available heating source is used to heat up concentrated wastewater so that concentrated wastewater can reach mesophilic digestion or near-mesophilic digestion, whereas the dilute wastewater anaerobically digest at ambient temperature. The digested effluent of concentrated wastewater will be used for agricultural purposes as a priority. The remaining part, if any, is subjected to aerobic and physical-chemical treatments together with dilute wastewater before conforming discharges (Deng et al. 2012, 2014). Continuous stirred-tank reactors (CSTR) and Up-flow solids reactors (USR) are recommended for mesophilic digestion of concentrated wastewater. For anaerobic digestion of dilute wastewater at

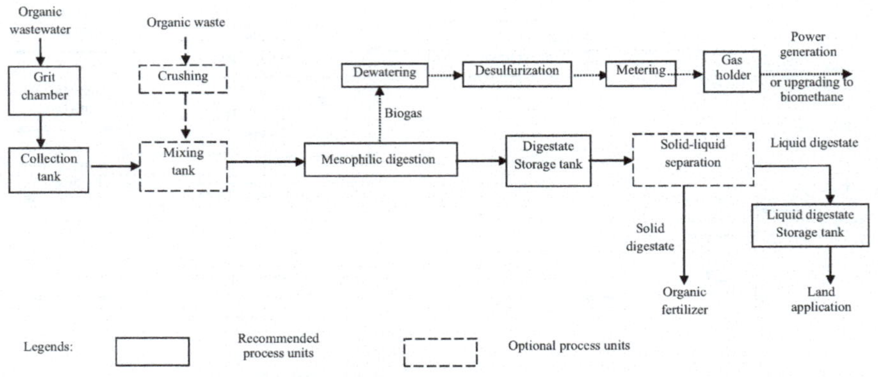

Fig. 4.4 Process flow chart of anaerobic digestion for high-concentration and high-solid organic wastewater/mixed feedstocks

ambient temperature, the anaerobic contact (AC) process, anaerobic baffled reactors (ABR) and up-flow blanket filters (UBF) are recommended. Biogas produced from anaerobic digestion is supplied to livestock farms and factories as living fuels, or to power generation unit to generate electricity utilized by livestock farms/factories.

(3) Anaerobic digestion for high-concentration and high-solid organic wastewater/mixed feedstock

In high-concentration and high-solid organic wastewater, organic matters level and SS are high, with a TS of 4–12%, COD of over 30,000 mg/L, SS of over 35,000 mg/L. Such wastewater include food/kitchen waste, vinasse liquid, high-concentration animal manure, etc. In a biogas plant using mixed feedstock, even if the organic level is not high in wastewater, the feedstock will become high-concentration and high-solid organic wastewater by adding solid organic wastes such as animal manures, organic garbage, industrial organic wastes and crop straws. Figure 4.4 shows the technical process of anaerobic digestion for high-concentration and high-solid organic wastewater/mixed feedstock. This process aims at producing biogas or bio-methane. Considering high concentration of feed solids, CSTR is used for anaerobic digestion at mesophilic or thermophilic conditions. The produced digestate are also highly concentrated, mainly used for agricultural purposes as fertilizers. Biogas produced from this process is for power generation, power grid access, bio-methane production via biogas upgrading, and connection into natural gas pipe network or for vehicle uses.

(4) Anaerobic digestion process of straw

Straws have the following characteristics when used as a feedstock. ① Straws are rich in celluloses, semi-celluloses and lignins, which are intertwined with each other to form a tight spatial structure that is difficult to be directly utilized by anaerobic microorganisms and enzymes. ② Straws can easily produce scums due to its small density and large volume. Discharge is not easy considering the poor flowability. ③

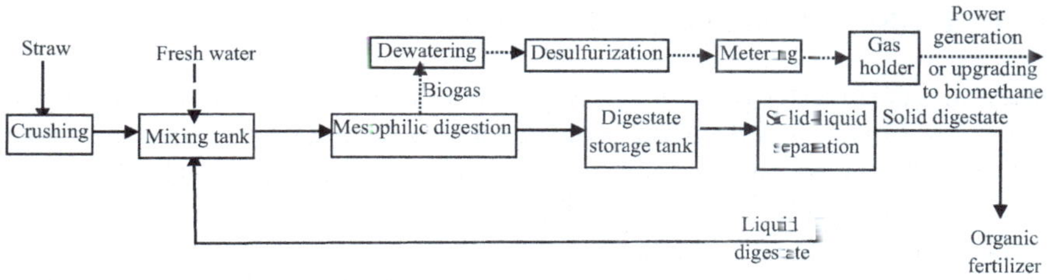

Fig. 4.5 Process flow chart of anaerobic digestion for straw

Straws have a high C/N ratio, leading to imbalanced acid production and consumption at the beginning of digestion, causing acid accumulation and acid poisoning. ④ High concentrations of solids result in uneven mass and heat transfer in the reactor, incomplete contact between substrate and inoculum, and difficulty in controlling digestion conditions. ⑤ High concentrations of solids lead to poor flowability and higher requirements for mixing, resulting in larger power consumption. Figure 4.5 shows the technical flow of anaerobic digestion for straws. This mode aims at producing biogas or bio-methane. CSTR or Vertical Plug Flow (VPF) reactor is recommended for this anaerobic digestion process. Due to the low C/N ratio of straws, the reflux of liquid digestate can be utilized to increase the nitrogen content. Since liquid digestate can reflux for a long time, there is barely any liquid digestate discharged to the external environment. The digestate produced has a high level of humus and a low level of hazardous substance and is therefore a kind of high-quality organic fertilizer. The biogas produced is mainly used for civil gas supply. Bio-methane can be generated by upgrading biogas and then injected to natural gas pipe networks or used as vehicle fuels. Such biogas is also used for power generation.

4.3 Process Units in Biogas Plant

4.3.1 Pretreatment Unit

The pretreatment unit includes collection and transportation of feedstocks, homogeneity of quality and quantity, temperature and pH, as well as removal of impurities and sediments. Pretreatment facilities and equipment should be set up based on the feedstock type. After pretreatment, the temperature and concentration of the feedstock shall be uniform and shall contain no solid materials with diameter or length greater than 40 mm.

4.3.1.1 Feedstock Collection and Transport

Collection is the first step of ensuring feedstock availability. Feedstocks are collected in two steps. First, it is collected in proper containers or transport vehicles from the place of feedstock production, and the collection areas shall be divided based on feedstock characteristics and amount of collection so that appropriate arrangement of collection method and gathering location. Next, feedstocks stored in containers or transported by vehicles are delivered to the target biogas plant site from their place of production or a transfer station. During transfer, proper protection and pollution-prevention measures shall be taken against fire, poisoning, corrosion, leakage, airborne dust particles, rain, etc. The following paragraphs introduce collection and transporting of two common types of feedstocks used for anaerobic digestion: manures and straws.

(1) Collection and transporting of manure

Manure is first collected in a temporary storage facility at the farm. Liquid manure may be stored in a collection tank while solid manure may be temporarily stored under a shed. There shall be special channels for collecting manure to avoid cross infection. Besides, protection against rain, snow and seepage should be performed during manure collection to prevent leakage and loss.

Proper transport equipment shall be employed based on the physical form, characteristics and transporting conditions of manure. For solid manure (water content <85%), automatic-loading trucks are ideal. For semi-solid or liquid manure (water content >85%), channels, pipelines or tank trucks are recommended. Vehicles shall be properly sealed against water entrance, leakage, and loss. Tank and driver's cabin must be separate. The tank shall have smooth and flat inner walls for easy cleaning and disinfection, and it shall be made from waterproof and corrosion resistant materials. There shall be measures taken against liquid leakage from tank bottom, and collection devices for cleaning wastewater shall also be set up.

(2) Collection and transporting of straws

Collection areas shall be split based on physical form and quantity of straws. Straw silage harvesters and straw combined harvesters are usually employed for straw collection. For straw silage, whole-crop corns are harvested and crushed after reaching postulation period. Vehicles are usually loading trucks operated together with a straw silage harvester. There are also self-propelled combined harvesters in addition to the suspension-type and the traction-type, which are employed as a set with tractors. Yellow silages (such as corn and wheat straws) are usually harvested stage by stage. That is, after harvested with a combined harvester, straws are spread over on ground for drying. Once the water content has met the requirements, dried straws are collected with a pick-up baler.

4.3.1.2 Feedstock Storage

Collected feedstock is usually stored in a buffer storage facility. It is stored to make up the insufficient supply or to solve the imbalanced supply due to seasonal changes or fluctuation in production cycle, so that feedstock can be supplied in a steady manner. Feedstocks such as manure and industrial wastes are generated and utilized on a daily basis, which is usually stored for a very short time (from several hours to 2 days). Straws are often collected within a specific period of time, but the use of such feedstocks lasts all year long, thus straws need to be stored for a long time (usually 10–12 months). Storage facilities shall be designed based on substrates. Facility size depends on daily feed rate of anaerobic reactor and the time required for storing feedstocks. When feedstocks with potential risks of epidemic disease are taken, it is necessary to isolate the feedstock acceptance area from the operation area of a farm.

The type of a storage facility depends on characteristics of the feedstock. There are usually storage bins for solid feedstocks (such as energy crops and straws) and storage tanks for liquid feedstocks (such as liquid dung). In general, a storage bin is capable of storing feedstock for over one-year uses while a storage tank can only store feedstock for several-day uses.

(1) Straw storage bin

In a biogas plant using straw as feedstock, a site for short-term stacking of straws is usually built inside the plant while that for long-term stacking is usually located outside, either with proper measures taken against dampness, rain, snow and fire. Straw storage capacity shall be determined based on straw characteristics, harvesting frequency, consumption, and storage cycle to ensure sufficient feeding for continuous operation of a plant. There are ventilated and wet storage bins for storing straws. Dry straws are often stored in a ventilated or dry bin, where straws are subjected to convection drying under hot-air forced circulation or passive air ventilation, so that hot air undertakes heat exchange by flowing through straw bales or gaps to ensure a safe storage water content of straw bales (12–15%) (Wang et al. 2008).

Wet storage bins are divided into two categories by their purposes of service: silage bins and yellow silage bins. Silage bins were initially used to store silage feed of animals and now are also used to store straws for biogas plants.

For silage, corn stalks are harvested and crushed after reaching postulation period when water content is 65–75%. The obtained corn stalks are cellared in a closed and anoxic environment, where soluble hydrocarbons are synthesized into lactic acid with the aid of fermentation of lactobacillus, so that the pH of corn stalks is reduced to 3.8–4.2 to inhibit the reproduction of other hazardous microorganisms. As a result, crop straws are softened for easy digestion. Silage can reduce utilization of carbohydrates by microorganisms while suppressing the lignification of straw itself, and free sugar in green straws is retained to the maximum degree. After silage, most organic matters in straws are transferred to the liquid phase to significantly improve the efficiency of anaerobic digestion and the utilization rate of straws.

The procedure for silage preparation is described as follows. ① Harvesting. whole-crop corns are usually harvested during the initial ripening stage of corn seeds. For corn straws after harvesting cobs, corn cobs are often harvested immediately during waxen maturity of around 70% of complete ripeness when leaves are not withered yet or 1–2 bottom leaves of corn stalk start to wither. Corn stalks are harvested for silage at the same day or at the latest by the second day of harvesting corn cobs. ② Crushing. Before silage, straws must be crushed to 1–2 cm long for proper compaction during silage. ③ Loading. Cellaring is a common practice for storage. A cellar is usually rectangular, and its depth, width and length depend on the quantity of straws and the time required for storage. In general, a total of 500–600 kg of whole-crop corns can be stored in each cubic meter of cellar. Walls of a silage cellar shall be flat and smooth, preferably constructed with bricks or reinforced concrete. Before cellaring, a layer of straws (10–15 cm thick) shall be spread over the floor to absorb liquids. Plastic film is fixed on cellar walls to prevent water and air leakage. ④ Compaction. Corn stalks shall be compacted, possibly with a bulldozer or a tractor to squeeze out as much air as possible till the top reaches about 60 cm over cellar edges before roofing. ⑤ Sealing. A layer of curtal straws shall be spread for roofing, with a layer of plastic film on top before tires or sandbags are placed. A drain ditch that prevents inflow of rainwater shall be dug 1 m away from the cellar.

For yellow silage, straws are crushed and loaded into a yellow silage cellar after cobs are harvested, with an appropriate amount of water and microbial inoculum added before compaction for storage. In a suitable anaerobic environment, microorganisms decompose large quantities of celluloses, semi-celluloses. These substances are transformed into carbohydrates that are further fermented into lactic acid, acetic acid and propionic acid while inhibiting the reproduction of destructive bacteria including butylic acid bacteria and molds. Eventually, the same function as a silage is achieved. The procedure for yellow silage is described as follows. ① Crushing. Straws are shredded to a dimension of 1–2 cm. ② Loading. Crushed straws shall be immediately cellared. Except for the bottom layer, each layer shall also be moisturized evenly to allow a moisture content of 65–70%. ③ Compaction. Straws shall be compacted layer by layer, with air squeezed out as much as possible. A tractor or a loader may be employed for repeated compression, especially along all edges and at four corners. ④ Sealing. The cellar shall be sealed when straws are filled up to 40–50 cm above the cellar edges, and the center should be piled highest while the surrounding is lower (looking like a bun). A plastic film shall be placed on top of the straws and all edges are tightly sealed. Tires or sandbags can be used as the fixing weight for sealing. A drain ditch shall be dug around the cellar to prevent rainwater inflow, and regular inspection is necessary. If sagging, cracking or damages are found, corrective measures must be taken immediately to prevent air leakage.

(2) Solid manure shed

Solid manure can be stacked up under a shed. To reduce the smell released during stacking, both storage and preparation of such feedstock shall be carried out in a confined space. For example, a ceiling shall be added to a storage facility to cover the spaces for feedstock receiving, preparation and storage. Waste gas can be pumped and transferred to an appropriate purifying device (such as a scrubber or a biofilter). A negative pressure system is usually provided in a shed for the storage of solid feedstock to collect waste gas, remove odor, and prevent the diffusion of odor to a maximum possible extent.

(3) Liquid feedstock storage tank

Liquid feedstock is usually stored in a concrete storage tank which normally contains feedstock for uses of 1–2 days and has a cover. Feedstock can be crushed, homogenized and finally mixed into pumpable mixture in a storage tank. A mixing system is required for each storage tank to avoid blockage, sedimentation, floating and other stratification behaviors of mixtures in the tank. A Liquid feedstock storage tank needs regular maintenance to remove bottom sediments and sands.

4.3.1.3 Screen

A screen is composed of one or several sets of parallel metal grid bars, a plastic hook or a metal net, a frame, and an associated device. It is usually installed at the inlet of a wastewater channel or pump house sump well or upstream of a biogas plant in a tilted manner for intercepting coarse and large floating matters and suspended matters in wastewater. Such matters include long fibers, garbage, leather pieces, hairs, vegetable pieces, wood pieces, cloth strips, plastic items, stones and packaging materials, etc., which may block and twining water pump, pipe valves, inlet and outlets as well as other equipment and devices. Screen installation in this manner ensures that downstream facilities can operate normally. According to different cleaning methods, screens are categorized into manual and mechanical cleaning ones. Manual cleaning screens are recommended when the treatment flow is small or the quantity of slags to be intercepted is small. Such a screen is usually composed of straight steel bars. The space between bars is usually 25–40 mm. The screens is usually installed at an angle of 30°–60° to the horizontal plane to facilitate manual cleaning and to prevent wastes on screens from flowing into water. When the quantity of slag interception is greater than 0.2 m^3/d, screens shall be cleaned up mechanically, i.e. raking residues intercepted on a screen by a machine. There are numerous types of mechanical cleaning machines. One of the most commonly used mechanical cleaning screens is rotary mechanical screens machine. The space between bars should be 16–25 mm and the installation angle should be 60°–90°. The flow rate of wastewater is usually 0.6–1.0 m/s. See Fig. 4.6.

4.3.1.4 Grit Chamber

Some feedstocks such as dung of cow kept in sand-bed, chicken manure, cassava waste vinasse, etc. contain high levels of sand which may block pipes, wear pumps, cutters, mixing blades and mixing axle, adversely impact the feed and mixing systems, as well as settling in digesters to hamper feed or reducing the effective capacity of digester and thus biogas yield. Therefore, sands must be removed in a pretreatment unit if feedstock contains large quantities of sand. In biogas plants, sand removal is

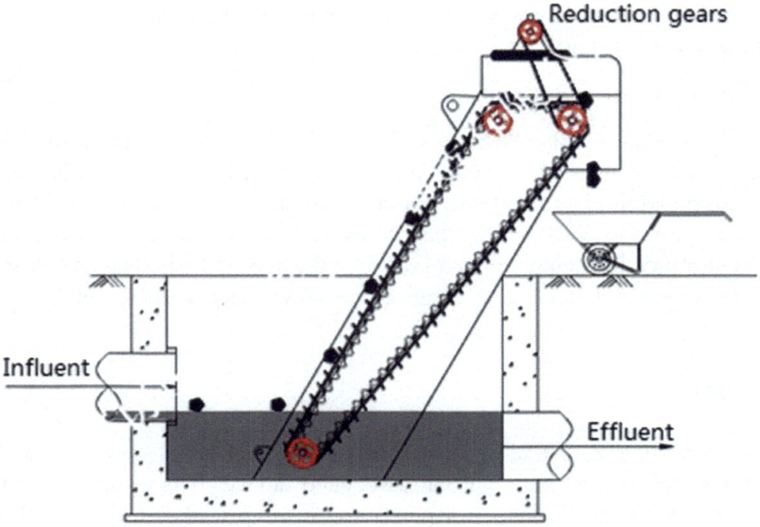

Fig. 4.6 Schematic diagram of a rotary mechanical screen machine

usually accomplished by building a grit chamber or directly using a sand removal device. For low-concentration feedstocks, horizontal-flow, aeration and vortex grit chambers are applicable for sand removal. Settled sands are collected from chamber bottom by a sand scraper or absorber. Then, sand and water are separated by a vortex sand-water separator or hydraulic cyclone with a spiral sand washer. For high-concentration feedstocks, a shallow grit channel, or a special desander is usually adopted alternatively. For biogas plants using manure from a dairy farm, the existing sand removal methods are far from being effective. The most thorough solution is probably the use of digestate or dried cow dung after fermentation as the padding material for cow beds instead of sands.

4.3.1.5 Wastewater Collection Tank

A wastewater collection tank is used to temporarily store wastewater from a livestock farm or a factory. The wastewater collection tank shall have sufficient volume to satisfy the requirements of proper hydraulic layout and installation of a screen and the suction pipe of a pump. However, tank capacity shall be minimized to lower the cost and reduce debris sedimentation and organic decomposition in the tank in the premise that feed wastewater can be pumped away immediately and frequent startup and shutdown of the pump can be avoided. Animal wastewater are not allowed to stay too long in the wastewater collection tank (usually 1 h approximately) prior to solid-liquid separation. A mixing system is required for the wastewater collection tank to avoid blockage, sedimentation, floating and other stratification behaviors of mixtures in the tank. The tank requires regular maintenance to remove bottom sediments and sands.

4.3.1.6 Sedimentation Tank

A sedimentation tank is built to separate dilute and concentrated portions of medium-concentration and moderate-solid organic wastewater so that both dilute and concentrated wastewater can be separately used for anaerobic digestion. There are several types of tanks, i.e., rectangular horizontal flow sedimentation tanks, hopper-bottomed upward flow tanks, and circular sedimentation tanks. The separation of dilute wastewater and concentrated one employ usually the upward flow type. A hopper-bottomed upward flow tank can operate in batch or continuous fashion. In batch operation, the influent enters from the top. After static sedimentation, sediments are discharged from the bottom cone while the supernatant flows out from the middle. In continuous operation, wastewater enters the tank from top to bottom via an inlet pipe located at the center of the tank. Under the pipe is an umbrella baffle that allows wastewater to distribute evenly in the tank and then rises up along the entire cross-section. Suspended matters settle into the sludge cone at the bottom while the supernatant flows out from the overflow weir. There is a baffle and a scum channel installed in front of the weir to ensure high-quality effluent. On one side of the tank is a discharge pipe located near the tank wall, which discharges sediments under hydrostatic pressure. The separation effect is at maximum when a hopper-bottomed upward flow tank operates in batch. The cross sectional plane of a hopper-bottomed upward flow tank may be circular, square or polygonal (Fig. 4.7).

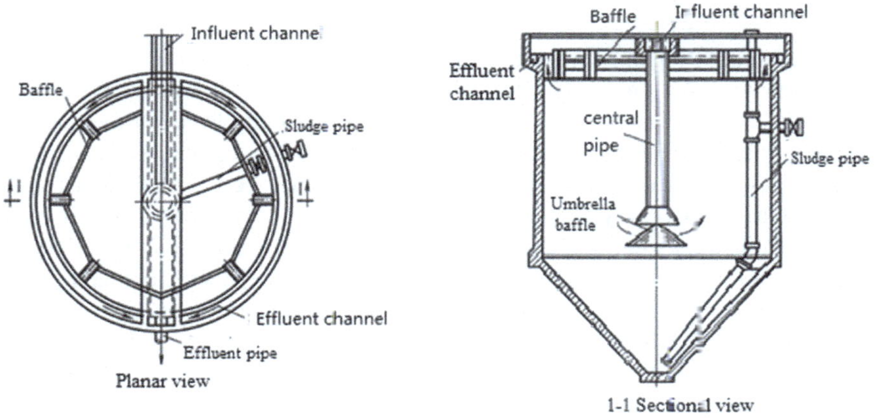

Fig. 4.7 Cross-section of a hopper-bottomed upward flow sedimentation tank

4.3.1.7 Regulation Tank

For industrial enterprises, wastewater from different workshop sections and at different time may differ greatly due to technology and process adopted, especially in cases of abnormal operations or equipment leakage when quality of wastewater changes abruptly and wastewater quantity increases greatly, which usually exceeds the normal treatment capacity of a treatment and utilization facility. In livestock and poultry farming, flushing is intermittent, and there is unsteady water consumption and uneven distribution of pollutants in discharged wastewater. Therefore, the flow rate or concentration of wastewater varies greatly within each 24 h. These problems can add complexity to operations of biogas plants and interfere with normal operation of facilities in biogas plants. Therefore, when feedstock with large quality and quantity fluctuations are used, it is necessary to introduce the feedstock into a regulation tank for quality and quantity uniformization before it is sent into digesters, which will create a stable and optimized operation environment for the anaerobic digestion system in the next stage. A regulation tank mainly homogenizes quality and quantity, and it may also play the roles of sedimentation, mixing, chemical dosing, neutralization, heating and pre-acidizing. Such a tank should be rectangular or square in shape. The bottom of a rectangular tank usually has a slope of 5%, tilting towards the emptying port or the suction port of pump. The bottom of a square tank should be coned. Sloped bottom is also acceptable. The retention time of feedstock in a regulation tank is usually designed as one cycle of time spent in feedstock quantity change. When specialized designs are required, the retention time shall be determined based on actual needs or referring to similar plants.

4.3.1.8 Mixing Tank

A mixing tank is for mixing solid feedstock with liquid feedstock, water, or liquid digestate reflux. It also regulates pH and provides heat. Straws and other solid feedstock shall be crushed before mixing. Feedstock coming from various sources such as garbage and food/kitchen waste shall be carefully sorted out. Feedstock with disease risks shall be disinfected before mixing with other feedstock. Pasteurization (70 °C, 1 h) or high-pressure sterilization (133 °C, 0.3 MPa, 20 min) is usually adopted. Pretreated solid feedstock may be fed into the mixing tank by a forklift or some other mobile systems, where it is also crushed, mixed and homogenized to pumpable mixtures. If sands, stones or other disturbing impurities enter the mixture, the mixing tank can isolate these impurities, allowing the impurities to be further removed by a scrapper or a spiral conveyor. The cross section of the mixing tank should be circular, and the bottom usually has a gradient of 5%, tilting towards the suction port of pipe. Measures shall be taken to prevent mixed feedstock from blockage, sedimentation, scum or other stratification behaviors of mixtures in the tank. Therefore, a mixing device, preferably a mixer capable of tearing and cutting, shall be installed in the mixing tank to crush the feedstock. The mixer should operate with a power of 4–8 W/m^3 and the minimum power should be less than 2.2 kW.

The mixing tank shall be large enough to accommodate feedstock for one feeding cycle. Where there are special requirements, the retention time shall be determined according to actual need or parameters adopted in similar plants. The mixing tank shall be regularly maintained, e.g. removing sand, stone etc. Sediments may be discharged with a crapper, spiral conveyor, pit drain pump, sink hole gathering system, etc. To reduce odor, the mixing tank shall be located in a confined room where ventilation measures are required. Facilities can be well preserved indoors. Mixing and monitoring activities can be conducted indoors regardless of climate or weather.

4.3.1.9 Feeding

In a biogas plant, semi-continual feeding is a common practice i.e. feedstock are fed into the biogas digester in batch. Correspondingly, feed transmission equipment does not operate continuously in general. The equipment and technology for feed transmission shall be determined based on uniformity of feedstock. In practice, there are feed transmissions of liquid feedstock and solid feedstock.

1. Feeding of liquid feedstock

Liquid feedstock is usually fed from a wastewater collection tank or a mixing tank. Centrifugal pumps, submerged sewage pumps, rotor pumps and screw pumps are commonly employed for feeding, which can be controlled by a timer or a computer. The entire process can be realized in a completely automatic or semi-automatic way. There shall be enough space for pump installation and maintenance (i.e. troubleshooting). Even if feedstock is subjected to ideal pretreatment, pumps can still be blocked, in which case rapid clearing is required. To ensure normal operation of the biogas plant, it is necessary to install globe valves on pipes to isolate pumps from pipes and the digesters for easy maintenance. In addition, a cutter or crusher may be installed at pump inlets to protect the pump, or a pump combined with a cutter can be used.

2. Feeding of solid feedstock

Solid feedstocks are directly or indirectly fed into a digester. During indirect feeding, solids shall be first introduced into a mixing tank or a feeding pipe. During direct feeding, solids may be directly introduced into a digester without getting mixed into slurry in a mixing tank or a feeding pipe.

(1) Indirect feeding

Solid feedstocks such as household organic garbage, silage, solid manure, etc. can be fed into a pipe via a suitable metering device (such as a bin pump). In practice, solid feedstocks may be pushed into a liquid feeding pipe, or liquids are directly pressed into the pipe by a bin pump. Liquids in the pipe are mainly liquid manure and digested effluent. The bin pump can work at a pressure as high as 4.8 MPa, allowing a discharge flow rate of 0.5–1.1 m^3/min for liquid feed and 4–12 t/h for solid feed. This way of feeding applies to pre-crushed feedstocks that barely contains any impurities. The bin pump is resistant to wearing and of good suction and discharge performances.

Its substrates are measurable, and it can crush feedstocks using a cutting device. However, in some cases, it will be affected by disturbances from impurities (such as stones, long fibers, metal blocks).

(2) Direct feeding

Feedstocks can be directly fed by a plunger feeder or a spiral conveyor.

The plunger feeder operates on hydraulic power and directly pushes feedstocks to the bottom of the digesters through the opening on one side. Fewer scums are produced as feedstocks are submerged in liquid manure. The system consists of a reversely rotating screw to allow feedstocks to fall into the lower column. Meanwhile, the system can also crush long fibers. The feed system is usually connected to a receiving bin or directly installed under the receiving bin.

When feeding feedstocks with a spiral conveyor, the variable-pitch spiral pushes feedstocks to below the liquid level in digesters to ensure that no produced biogas will escape from the spiral. The simplest way is to install a feeding unit on a digester. Only one vertical spiral conveyor is needed to directly send feedstocks to the digester. The spiral conveyor can take feedstocks from the receiving container at any angle, and the receiving container is equipped with a crushing tool.

4.3.2 Anaerobic Digestion Unit

4.3.2.1 Volume Determination for Anaerobic Reactor

The volume of the anaerobic reactor (digester) may be determined by Eq. 4.1.

$$V = \frac{q}{r_p} \tag{4.1}$$

$$r_{p(T)} = r_{p(20)}\theta^{(T-20)} \tag{4.2}$$

where

- V volume of the anaerobic reactor (digester), m^3;
- q biogas output, m^3/d;
- r_p volumetric biogas production rate, m^3/m^3 d; $r_{p(T)}$ and $r_{p(20)}$ represents the volumetric biogas production rate at T °C and 20 °C respectively;
- θ temperature activity coefficient, which comes from test results or data adopted in similar plants, taking θ = 1.02–1.04 for T = 20–35 °C and 1.20–1.40 for T = 15–20 °C when no reference data are available.

4.3.2.2 Anaerobic Digestion Process

There are two major processes for anaerobic digestion depending on the state of microorganisms in reactor: the anaerobic activated sludge process and the anaerobic biofilm process. The anaerobic activated sludge process consists of a traditional digestion tank such as the hydraulic biogas digester, continuous stirred-tank reactor (CSTR), anaerobic contact (AC) process, anaerobic baffled reactor (ABR) and up-flow anaerobic sludge blanket (UASB). The anaerobic biofilm process consists of an anaerobic filter (AF), an anaerobic fluidized bed reactor (AFBR), and an anaerobic rotating biological reactor. An Up-flow blanket filter (UBF) is a combination of the anaerobic activated sludge process and the anaerobic biofilm process.

1. *Continuous stirred-tank reactor (CSTR)*

Continuous stirred-tank reactor (CSTR), developed in the 1950s, is a technology applying mixing, heating, and thermal insulation to a traditional digester to greatly improve the biodegradation rate in a reactor. A CSTR is also called a high-speed anaerobic digester.

Feedstocks are regularly or continuously fed into anaerobic reactor from bottom. Sludges and digested effluents are respectively discharged from bottom and top, and produced biogas is discharged from the top (Fig. 4.8). To allow even contact between microorganisms and substrate, and to make bubbles generated escape in time, a mixer is installed in CSTR to regularly stir mixed liquid in the reactor. Intermittent stirring was practiced at an interval of 2–4 h for early CSTRs. Stirring is often paused when discharging effluents, and digested effluents are discharged from top after separated from sediments. Nowadays, 24 h continuous stirring is applied by some plants. The feedstocks or mixtures in the reactor need to be heated for mesophilic and thermophilic digestion. Usually, a heating coil is installed inside the reactor, and hot water is circulating inside the coil to provide heat. A CSTR is suitable for treating sludge from municipal sewage plants, degradation-resistant organic wastewater, animal manure, food/kitchen waste and other feedstocks having high levels of SS and organic matters.

There are generally three forms of stirring in an anaerobic reactor.

(1) Hydraulic agitation

For hydraulic agitation, mixed liquid is pumped out from middle section of the reactor by a pump outside the reactor and then pumped in from the reactor bottom or top to

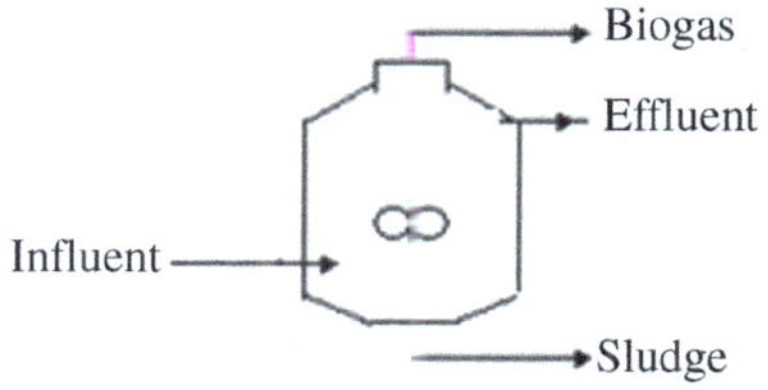

Fig. 4.8 Schematic diagram of continuous stirred-tank reactor (CSTR)

realize circulated agitation. Sometimes there is an ejector installed inside the reactor. Mixed liquid pressed by the pump is ejected from the ejector to create vacuum at the throat, where a part of mixed liquid is sucked from the reactor to form strong agitation. Hydraulic agitation requires simple equipment that is easy to maintain, thus favorable for reactors with low feedstock concentrations. However, hydraulic agitation can easily cause short flows and the mixing effect is poor, usually applicable to small anaerobic reactors. Larger flows are necessary to ensure complete mixing of liquid. Based on current experiences, about 0.005 kW of power is required for the mixing of every cubic meter of mixed liquid.

(2) Mechanical agitation

Mixing is realized by impellers or blades in a reactor with mechanical agitation. A motor drives an impellor to rotate, pressing the mixed liquid towards the blades in motion. In some cases, a draft tube is used. Rotation of blades pushes mixed liquid inside the draft tube to move vertically and drives mixed liquid inside the reactor to circulate. Mechanical agitation is advantageous in its larger agitation radius and better mixing effect. However, it is necessary to install a gastight device for stirring shaft to pass through the tank cover. About 0.0065 kw of power is required for mechanically mixing every cubic meter of liquid.

(3) Biogas agitation

In biogas agitation, biogas is extracted from the reactor or gas holder and then pushed back to the reactor by an explosion-proof blower. When biogas is released from mixed liquid, the mixed liquid inside the reactor is driven by the biogas to circulate during suction and winding when it rises. One of the major advantages of biogas agitation is that changes in liquid level of the reactor has only minor impacts on mixing efficiency. There are no moving parts in the reactor and thus failure is less likely to happen. Moreover, the mixing force is strong and the whole reactor is within the mixing range. Nevertheless, when the concentration of feedstocks is high, it is difficult to achieve ideal mixing effect via biogas agitation. Such technology can be hardly applied to anaerobic digestion with high-solid and high-concentration. Biogas agitation is rarely adopted in farm-based biogas plant as it needs complicated safety facilities such as explosion-proof blowers, flame arrestors, filters, safety valves, etc. About 0.005–0.008 kW of agitation power is required for the mixing of every cubic meter of liquid.

The volumetric biogas production rate of a CSTR shall be determined experimentally or by referring to similar plants. When parameters of volumetric biogas production rate are unavailable by experiments or reference, data provided in Table 4.11 may be referred to, and the volumetric biogas production rate at other temperatures may be calculated according to Eq. 4.2.

CSTRs are usually designed as vertical cylinders with an effective height of 6–12 m. The ratio of total height against diameter should be 0.8–1.0. The top cover should be a spherical cap or a cone-shaped cap. The bottom should be an inverted spherical shell, cone cover, or circular plate.

Table 4.11 Volumetric biogas production rate of CSTR using different feedstock (35 °C)

Feedstock	Pig manure	Chicken manure	Cattle dung	Corn straw	Wheat straw	Rice straw	Alcohol vinasse
Volumetric biogas production rate, $m^3/(m^3\ d)$	1.00–1.20	1.10–1.30	0.80–1.00	0.90–1.10	0.80–1.00	0.70–0.90	1.90–2.10

A CSTR has the following advantages:

(1) A CSTR has a mixing system which allows full mixing of substrate and microorganisms and thus has a higher rate of biochemical reaction; in addition, mixing can also prevent short flows in the reactor.
(2) The anaerobic reactor is vertically installed on ground for an easy discharge of sludge (sediments).
(3) There is a heating and thermal insulation system to increase digestion temperature with the coordinating effect of heating and thermal insulation, consequently improving the efficiency in anaerobic digestion.
(4) The system is highly resistant to impact loads. The mixing system allows fast mixing and diluting influent with the mixed liquid in the reactor.

Disadvantages of CSTR:

(1) It cannot retain microorganisms. In a CSTR, flows are completely mixed, and microorganisms reproduced in the reactor are discharged together with overflow of digested effluent, leading to low sludge concentration in reactor (around 5 g mL VSS/L).
(2) Low loading rate. Severe sludge loss occurs in case of short hydraulic retention time and low-concentration feed. As a result, CSTR requires a longer hydraulic retention time (HRT), usually 15–30 days, to maintain stable operation of the reactor.
(3) Energy consumption is large. A longer retention time leads to greater energy consumption on heating and mixing.

2. *Anaerobic contact (AC) process*

To overcome the disadvantages of CSTRs that anaerobic microorganisms cannot be retained in reactor, a sedimentation tank is built downstream of the digester. Settled sludge is refluxed to the digester so that as many microorganisms as possible are retained in the digester. The separation between hydraulic retention time and solid retention time allows a longer solid retention time without prolonging hydraulic retention time, which results in a higher organic volume loading rate, a higher treatment efficiency, and a lower anaerobic reactor volume. Sedimentation of effluent also leads to lower levels of pollutants in discharged fluids (Fig. 4.9). Anaerobic contact process usually applies to treatment of medium-concentration organic wastewater and animal manure.

AC's volumetric biogas production rate shall be determined via experiments or by referring to similar plants. When the parameters of volumetric biogas production

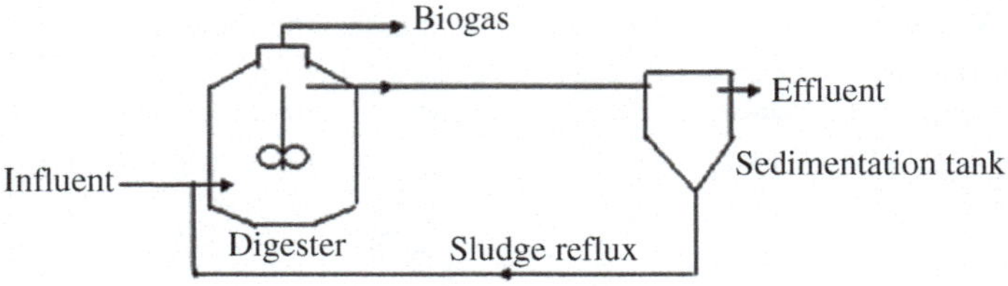

Fig. 4.9 Schematic diagram of an anaerobic contact (AC) process

rate are unavailable via experiments or by referring to similar plants, data provide in Table 4.12 may be referred to, and the volumetric biogas production rate at other temperatures may be calculated according to Eq. 4.2.

Sludge reflux rate shall be determined based on sludge quantity in digester, pH of mixed liquid, and temperature, and a desirable reflux range would be 50–200%. Proper measures such as vacuum degassing, cold impact and so forth shall be taken to accelerate sludge-water separation in the sedimentation tank as sludge cannot settle easily. The geometric dimensions of a digester used in AC are similar to those of a CSTR. Hydraulic agitation is often adopted in AC. A hopper-bottomed upward flow tank is recommended for the AC process.

Advantages of AC process:

(1) Sludge reflux increases sludge concentration in the digester. VSS (Volatile Suspended Solids) may be as high as 5–10 g/L. The digester is highly resistant to impacts.
(2) The volumetric loading rate is higher than that of a CSTR. The COD volumetric loading rate is usually 1–5 kg COD/m^3 d and HRT is about 8–15 d, with a COD removal rate of 70–80%.

Disadvantages of AC process:

(1) The process flow is quite complex due to the installation of a sedimentation tank and a sludge reflux system.
(2) Anaerobic sludge is poorly settled, and a large quantity of anaerobic sludge is contained in the mixed liquid discharged from the digester. On one hand, micro biogas bubbles are absorbed by sludge flocs. On the other hand, anaerobic sludge produces biogas in a sedimentation tank, leading to poor sedimentation

Table 4.12 Volumetric biogas production rate of an anaerobic contact (AC) process using different feedstock (20 °C)

Feedstock	Swine wastewater	Chicken farm wastewater	Dairy wastewater	Slaughtering wastewater
Volumetric biogas production rate, m^3/(m^3 d)	0.50–0.70	0.60–0.80	0.40–0.60	0.40–0.60

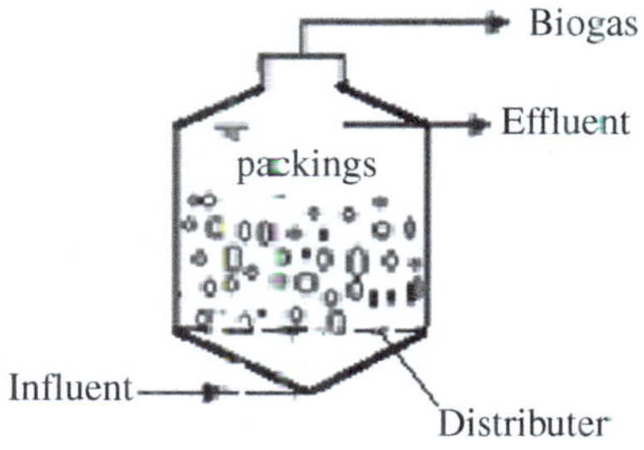

Fig. 4.10 Schematic diagram of anaerobic filter (AF)

of sludge. Quite a part of sludge rises up toward water surface and flows out with effluent. For the moment, agitation, vacuum degassing, addition of coagulant, or ultrafiltration membrane is used in lieu of sedimentation tanks to achieve higher sludge-water separation efficiency.

3. *Anaerobic Filter (AF)*

Anaerobic filter (AF) is a kind of anaerobic reactor with microorganism carriers as packings (Fig. 4.10). Packings of an anaerobic filter usually consist of crushed stones, pebbles, coke, and plastic pieces in various shapes. Some anaerobic microorganisms attach to the packings and form an anaerobic biofilm, while some microorganisms are suspended in the gaps of packings. A water distribution system is located at the bottom of an AF, through which wastewater flows into an infilled reactor at the bottom. Under the action of microorganisms attaching or interception by packings, organic matters in wastewater are degraded and transformed to biogas, which is released from reactor top and then collected for use. Digested effluent are discharged from the rector via a drain pipe. The biofilm in the reactor constantly metabolizes. The exfoliated biofilm is discharged with effluent, thus a facility for sedimentation and separation is required downstream of AF.

There are up-flow and down-flow AFs differentiated by their patterns of feeding. In an up-flow AF system, wastewater flows in from the bottom and then flows upwards through the packings. In a down-flow AF system, the wastewater distribution system is located at the top. Wastewater evenly flows from top to bottom through the packings. Produced biogas flows upwards, causing agitation. A down-flow AF system needs no complicated wastewater distribution mechanism. The reactor can hardly be blocked, but the sedimentation of sludge or solids at the bottom may bring complexities to operations. The up-flow pattern is a popular feeding pattern of traditional AFs.

In the past, the packings usually accounts for 2/3 of filter height. To prevent blockage, one packing layer at the bottom and in the middle of the filter was maintained separately, thus the UASB-AF reactor (UBF) with partial packings was developed. Since then, researches in China have focused on the UASB-AF reactor (UBF) with partial packings instead of AF.

Advantages of AF:

(1) Solids retention times of microorganisms are long, usually for more than 100 d, thus the concentration of anaerobic sludge can be as high as 10–20 g VSS/L.
(2) The system is highly resistant to impact loads.
(3) The system starts within a short time and easily restarts after shutdown.
(4) Organic loading rate is high. When block packings are applied at a temperature of 25–35 °C, the volumetric loading rate can be as high as 3–6 kg COD/m^3 d, 2–3 times as that obtained in an ordinary digester. If plastic packings are applied, the loading rate can be increased to 5–10 kg COD/m^3 d. In general, the COD removal efficiency can be over 80%.

Disadvantages of AF:

(1) AF gets clogged easily, especially at the bottom. When block packings are applied, SS in influent should be usually ≤200 mg/L.
(2) Short flows can be frequent when sludge concentration is high in AF.
(3) The use of large quantities of packings leads to higher cost.

4. *Up-flow anaerobic sludge blanket (UASB)*

A up-flow anaerobic sludge blanket (UASB) is a kind of anaerobic bioreactor with incubated granular sludges and a gas-solid-liquid separator installed at the top. The structure is shown in Fig. 4.11. Granular sludges that are highly concentrated and show high sedimentation performance are found at the bottom of the reactor, which is also called a sludge blanket. Wastewater to be treated flows into the sludge blanket from reactor bottom, where organic matters in wastewater are decomposed by anaerobic microorganisms and converted into biogas. Biogas is released continuously in the form of micro bubbles, which continue to combine when rising up, gradually forming larger bubbles. Stirred by influent and biogas produced, the sludges in the middle and upper parts of the reactor remain suspended, forming a rather thin sludge suspension layer. The gas-solid-liquid mixture flows into the sedimentation zone of the separator, where sludge in wastewater flocculates, with particles increasing gradually and settling under the action of gravity. Sludges settled to skew wall of the

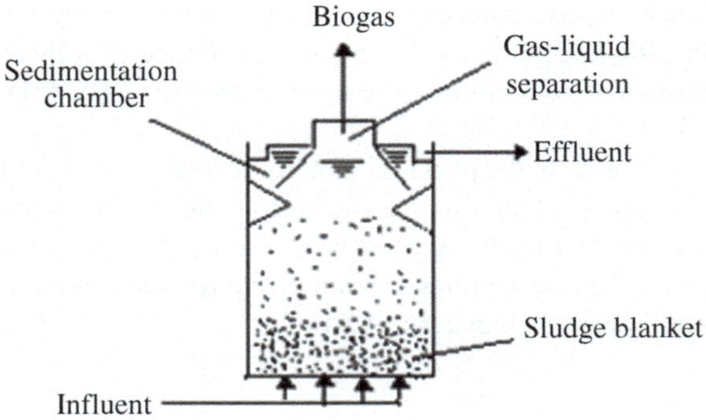

Fig. 4.11 Schematic diagram of an up-flow anaerobic sludge blanket (UASB)

Table 4.13 Volumetric biogas production rate of up-flow anaerobic sludge blanket (UASB) for flocculent sludge and granular sludges (35 °C)

Sludge type	Flocculent sludge	Granular sludges
Volumetric biogas production rate, $m^3/(m^3\ d)$	1.5–3.0	2.0–3.5

Note take the higher value when COD > 9000 mg/L and the low value when COD is 3000–9000 mg/L

separator slip back into the anaerobic reaction zone along the wall, so that a large quantity of sludge is accumulated in the anaerobic reaction zone. With sludge separated, liquid overflows from the sedimentation zone and then gets discharged. Biogas bubbles generated in the reaction zone rise up and deflect around the reflector when colliding with the reflector of the separator. Next, bubbles pass through the water layer and enter the gas chamber where biogas is piped out. When feeding wastewater to a UASB, two measures are recommended to distribute wastewater evenly: installing a water distribution system or conducting pulse feeding by increasing the instantaneous flow so that water flows through holes evenly.

A UASB is favorable for treating feedstock with SS $\leq$ 1.5 g/L. The volumetric biogas production rate of a UASB shall be determined by experiments or by referring to similar plants. When parameters of volumetric production rates are unavailable by experiments or reference, data provided in Table 4.13 may be referred to, and volumetric biogas production rates at various temperatures may be calculated according to Eq. 4.2.

The height of a UASB shall depend on sludge condition and wastewater characteristics, but in most cases it is preferably 4–10 m.

The hydraulic loading rate in sedimentation zone of the three-phase separator shall be maintained below 0.8 m^3/m^2 h. The velocity of water flowing through a gas chamber gap shall be maintained below 2 m/h on average. The total water depth in the sedimentation zone shall be $\geq$1.0 m. The retention time of wastewater in the sedimentation zone should be 1.0–1.5 h.

The gap area of gas-collecting hood of the three-phase separator should be 10–20% of the cross-sectional area of the reactor.

The skew wall of the gas collecting hood of the three-phase separator should have an angle of 55°–60°.

The overlap between the separator reflector and the gap should be 100–200 mm.

The gas outlet pipe shall have a proper diameter to ensure that biogas is introduced from the gas collecting chamber and enters the water seal.

When designing a UASB water distribution system, even distribution of water shall be considered, for which multiple influent ports are required. There are multiple ways of introducing influent, but the following principles shall be abided by to guarantee proper water distribution and hydraulic agitation:

(a) Ensure that the quantity of influent is almost the same per unit area, i.e. to set up one influent port at every 2–4 m^2.
(b) Required conditions for hydraulic agitation should be satisfied to the most.
(c) Blockage of the influent pipe can be easily observed.

(d) Blockage can be easily eliminated once detected.

The effluent outlet is located on top of the UASB reactor. Preferably, a multi-channel effluent weir is built for such purpose, with a scum baffle installed in front of the weir. The maximum surface loading rate of the weir should not exceed 1.7 L/s m.

Water head above the weir shall be greater than 25 mm.

The water distribution pipe may be used as sludge outlet pipe as well. A separate sludge outlet pipe may be installed separately at the bottom and in the middle of the reactor (at 1/2 of the reactor height) for discharging sludge.

Advantages of UASB:

(1) High concentration sludge can be obtained in the reactor, with an average value of 20–40 g VSS/L.
(2) Organic loading rates are high and hydraulic retention time is short. The volumetric loading rate for mesophilic digestion can be about 10 kg COD/m^3 d.
(3) No agitation equipment is required as agitation effect is realized via feeding and biogas rising in the reactor.
(4) There is a gas-liquid-solid separation system located at the top of the reactor. Settled sludge automatically returns to the anaerobic reaction zone, thus no sludge reflux equipment is required.
(5) No packing is required in the reactor. Therefore, cost is saved and blockage of reactor due to packings is avoided.

Disadvantages of UASB:

(1) SS in influent shall be limited, usually below 1000 mg/L.
(2) Short flows in the sludge blanket can affect effective treatment.
(3) The reactor is sensitive to abrupt changes in wastewater characteristics and loading rate, and it is poorly resistant to impacts.

The core capability for high-efficiency treatment of a UASB is that highly settleable granular sludges are formed in the reactor. Density of granular sludges is restrained by high suspended solids. As a result, UASBs are only applicable to treatment of animal wastewater after solid-liquid separation. Since a high level of ammonia nitrogen is unfavorable for forming granular sludges, a UASB used to treat animal wastewaters can hardly achieve high treatment loads.

5. *Anaerobic Baffled Reactor (ABR)*

An anaerobic baffled reactor (ABR) is isolated by baffles into several cascaded react chambers perpendicular to the direction of water flow so as to maintain high sludge concentration in the reactor. Water flowing into an ABR move forward in a baffling manner guided by guide plates, passing through sludge blankets in the up-flow and the down-flow chambers one by one, where substrates in the influent sufficiently contacts with microorganisms for degradation and removal. The up-flow chamber is spacious enough for sludge gathering while the down-flow chamber is narrower. A

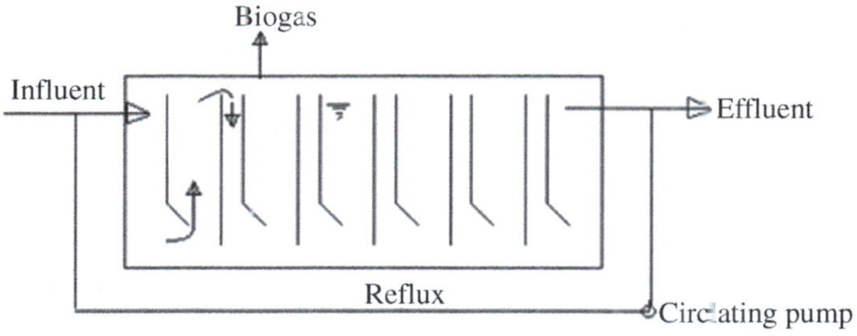

Fig. 4.12 Schematic diagram of an anaerobic baffled reactor (ABR)

50° guide plate is installed at the lower edge of the baffle to the up-flow chamber to send wastewater to the center of the up-flow chamber for sludge-wastewater mixing. The up-flow chamber is a relatively independent up-flow sludge blanket (USB) system, in which sludge exists as granules or flocs. Effluent shall be refluxed if the influent concentration is high. Structurally, an ABR can be regarded as a simple combination of several UASB reactors. However, it is significantly different from individual UASBs in technology. UASBs can be generally regarded as a kind of CSTR. As for ABRs, the hydraulic characteristics of individual reaction chambers are almost completely mixed, which, taking as a whole, is more similar to a push-flow process (Fig. 4.12). Chinese scientist and engineers have improved the ABR technology, mainly including installing packings in the up-flow chamber to retain sludge and filter materials in the last one or several cascades of up-flow chamber to prevent microorganism loss and improve effluent quality. Such a system is usually called biogas digester for domestic sewage treatment (see Chap. 3).

Advantages of ABR:

(1) ABR is simply structured. A large quantity of sludge can be intercepted by the baffle and by sedimentation of sludge. There is no need for a three-phase separator.
(2) Treatment efficiency is quite high. The COD volumetric loading rate can reach 4–8 kg COD/m^3 d at mesophilic temperature.
(3) It has ideal buffering and adapting capability towards impact loads and hazardous or poisonous substances in influent.
(4) Repeated up-and down-baffling creates agitation effect. Hydraulic conditions are ideal, and no mixing system is needed.
(5) The system is usually built underground, which discharges effluent by height difference and needs no power.
(6) The investment and operational costs of the system are low.
(7) The system is easy and convenient to operate and manage.

Disadvantages of ABR:

(1) It is difficult to ensure even feeding, which can easily cause local (first chamber) overloading.
(2) Sludge discharge is not easy from underground.

6. *Up-flow blanket filter (UBF)*

A up-flow blanket filter (UBF) is a kind of reactor which combines AF and UASB, and it is also called a UBF reactor. A UBF reactor consists of a water distributor, a sludge blanket and a packing blanket. Packings are located at the upper 1/3 of the reactor. The three-phase separator is removed from the design, and the thickness of packings is reduced. Some space is reserved between the water distributor at the bottom and the packing blanket so that suspended sludge flocs and granular sludges can grow and accumulate in it. Wastewater flows into the reactor from the bottom, sequentially passing through the granular sludge and floc sludge blankets for anaerobic reaction and then leaving the sludge blankets for further treatment, where gas-liquid-solid separation process is completed. Treated effluent discharges from the overflow weir (pipe) while biogas flows out from top of the reactor (Fig. 4.13).

A UBF is suitable for treating medium-concentration feedstock with SS $\leq$ 1.5 g/L. The volumetric biogas production rate shall be determined by experiments or by referring to similar plants. When the parameters of volumetric biogas production rate are unavailable by experiments or reference, data provided in Table 4.12 may be referred to.

Packings should be loaded at upper 1/3 of the reactor, with a desirable thickness of 0.5–2 m. Above the packing blanket shall be a stable water layer (0.4–0.5 m high).

The feeding, effluent discharging, and sludge discharging systems of UBFs are similar to those of UASBs.

Fig. 4.13 Schematic diagram of an up-flow blanket filter (UBF)

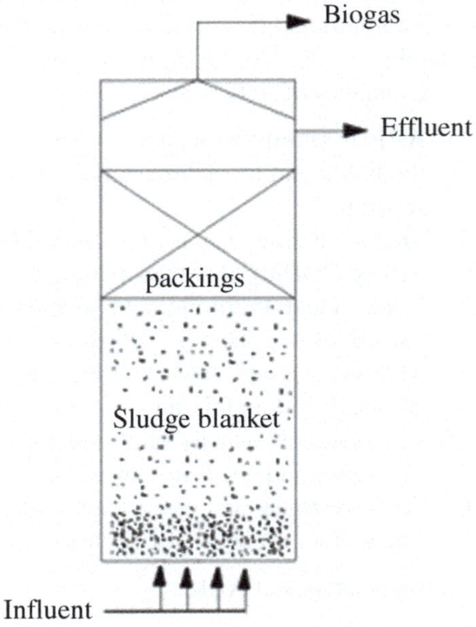

The packings of UBFs shall be selected based on specific surface area, porosity, surface roughness, mechanical strength, weight, and price. Perforated plates or supports are preferred to support packings.

UBFs have the following advantages:

(1) UBFs are more competitive in terms of treating wastewater that can hardly or even not domesticate granular sludges such as high-nitrogen animal wastewater and wastewater containing high salt level or exhibiting bio-toxicity.
(2) In comparison with UASBs, the addition of a packing blanket greatly improves the reactor's capability to accumulate microorganisms. During start up, the system can effectively intercept sludge and reduce sludge loss in a faster manner.
(3) In comparison with AFs, the packing blanket is thinner, which reduces the possibility for blockage.
(4) The system operates stably and is highly adaptable to changes in loading rate, temperature and pH.

7. *Up-flow solids reactor (USR)*

A up-flow solids reactor (USR) is a kind of reactor developed based on working mechanisms of a UASB, which is simply structured and suitable for treating organic wastewater with high SS level, as shown in Fig. 4.14. Wastewater flows into the reactor from the reactor bottom and gets evenly distributed at the bottom via a distribution plate. Then, it flows upwards through a solid bed containing high-concentration anaerobic microorganisms, where organic matters in wastewater fully contacts and reacts with these anaerobic microorganisms before getting degraded and transformed. Rising biogas and water flow produces mixing to promote contact between solids and microorganisms. Undegraded organic solid particles and microorganisms settle down and accumulate in the lower solid bed by gravity, so that a high level of biomass is maintained in the reactor and the degradation of solids is prolonged. Flows passing through the solid bed overflows and are discharged from the effluent outlet channel

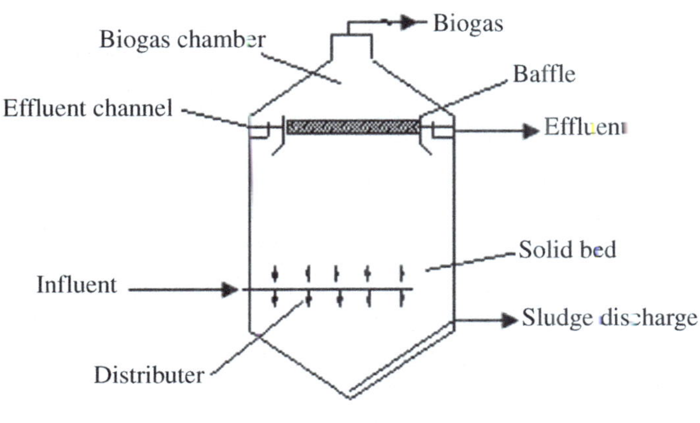

Fig. 4.14 Schematic scheme of an up-flow solids reactor (USR)

at the top of the reactor. A baffle is installed at the outlet channel to reduce the loss of suspended solids. USRs are widely applied in China in treatment of animal wastewater.

USRs propose no strict conditions for pretreatment of feedstock and require no solid-liquid separation pretreatment, three-phase separators, or sludge reflux. However, USRs are only applicable to animal wastewater with TS $\leq 6\%$. Higher SS in wastewater can cause blockage of the water distribution pipes. The lack of agitation can lead to scums, crusting and sedimentation.

8. *Vertical plug flow (VPF) reactor*

Vertical plug flow (VPF) reactor applies to anaerobic digestion of straws. Feedstocks are fed from the top and discharged from the bottom or reversely in a vertical digester. On top of the digester is a liquid sprinkler, ensuring the total TS of mixed liquid exceeds 8%. Before fed, crushed straws and refluxed liquid digestate are mixed. Liquid digestate is refluxed for agitation. For overhead feeding operation, digested straw residues are discharged at the bottom along with water flow. For bottom feeding operation, digested straw residues are discharged via a scrapper or a funnel. Discharged digestate undertakes a solid-liquid separation before refluxed and mixed with straws for another stage of feeding. Continuous layered feeding as well as batched discharges have realized continuous and stable anaerobic digestion of straws. This has solved the problems of mixing, feeding and discharging, overcome unbalanced biogas production of batch digestion of straws, allowed more complete contact between substrate and anaerobic micrograms in an anaerobic reactor, and achieved higher biogas production rate (Fig. 4.15). Liquid digestate reflux can take a long time. With straws used as feedstock, mesophilic- and thermophilic anaerobic digestion processes can be adopted, with the volumetric biogas production of 0.8–1.5 $m^3/(m^3\ d)$.

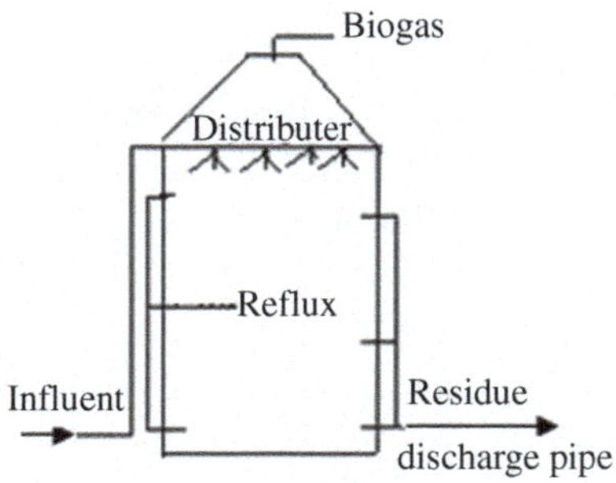

Fig. 4.15 Schematic diagram of a vertical plug flow (VPF)

4.3.2.3 Heating and Thermal Insulation of the Digester

It is a critical to maintain the optimal temperature (>20 °C) of mixed liquid in the digester in order to efficiently produce biogas. Except for some industrial wastewater (e.g. alcohol wastewater, starch wastewater, etc.), most feedstocks for anaerobic digestion are at ambient temperature. As a result, the efficiency of biogas production is quite low in winter or cold regions. To ensure a suitable digestion temperature and to guarantee stable and efficient biogas production, it is necessary to heat up feedstocks (or mixed liquid in digester) for anaerobic digestion when the ambient temperature is lower than 20 °C. Heating sources may be waste heat from power generation, solar energy, heat from biogas combustion (boiler), or the combination of the heat sources. Besides, the temperature of mixed liquid in digesters shall be steadily and uniformly distributed in terms of time and space. Temperature fluctuation should not exceed 2–3 °C within 1 h to guarantee the optimal digestion process. Sharp temperature fluctuations or large deviation from the optimal value will affect the anaerobic digestion process and even cause termination of biogas production. Temperature fluctuations may be caused by the following factors: fresh feed, poor heating and thermal insulation, temperature unevenness due to poor agitation, inappropriate layout of heating system, extreme temperatures in winter and summer, equipment failure, etc. (Dong and Raninger 2000).

There are two ways of heating, internal heating and external heating, available during anaerobic digestion.

For internal heating, heat source is directly introduced into a digester to heat up the mixed liquid, which can be realized by either hot water or steam. Two or more sets of circulating hot water heat-up coils are needed based on the size of the digester. If PVC or PEOC pipes, heat-up coils shall be arranged in a more compact manner as plastics have a lower thermal conductivity. A circulating hot water pipe may be located on the wall or bottom plate of the digester, or inside or outside the wall. When installed inside the digester, the pipes shall be located at about 50 cm from the inner wall of digester. Nevertheless, the heat-up pipes shall not affect the operation of the scrapper, mixer and other equipment. In a horizontal digester, a heat exchanger is usually located on a mixer. The disadvantages of hot water circulation process include low thermal efficiency and frequent fouling of pipe outer surface that reduces heat transfer efficiency. Steam heating can achieve a higher thermal efficiency, but it also increases water content of feedstock, increasing feedstock volume. Both internal heating methods require proper mixing by agitation. For internal heating, thermal loss is minor but there might be complexities in maintenance. Heating efficiency of the base plate can decrease significantly once a deposition layer forms. Heating inside the wall of a digester can lead to fouling and scaling while heating inside a concrete wall can lead to thermal strain.

For external heating, feedstock or mixed liquid is heated outside a digester in either of the following ways: feedstock heating and mixed liquid circulation heating. For feedstock heating, feedstocks are first heated up to the required temperature in the wastewater collection tank or mixing tank before sent to the digester. For mixed liquid circulation heating, mixed liquid is first extracted from the digester, and then heated in an external heat exchanger to the required temperature before returning to the digester. When an external heat exchanger is used, mixed liquid is subjected to continuous heat exchange to maintain a stable digestion temperature, otherwise an additional heater is needed inside the digester. Heat transfer media and heated liquid shall be countercurrent if possible. In general, a coiler or jacketed heat exchanger is preferred. By external heating, temperature change due to fresh feed can be avoided, Maintenance is easier, and temperature can be easily controlled. However, there is a significant heat loss in external heating. In some specific cases, an internal heating system is necessary, and the cost for an external heat exchanger is quite high.

Heating coils for internal and external heating must be equipped with a gas discharger.

To reduce thermal dissipation, a layer of thermal insulation coating shall be applied to the external surface of the digester, heat exchanger and thermal pipes. All materials featured in small thermal conductivity, low density, sufficient mechanical strength, sufficient thermal resistance and low absorption can be used for thermal insulation. Insulation materials shall be selected locally, taking into account the cost performance and complexity in material fabrication and installation. Insulation materials for low-temperature uses are recommended for biogas plants. Insulation materials for digesters or for underground uses can take hard polyurethane foam, foam glass and other closed-pore materials to prevent moisture penetration. Materials shall be strong enough to bear the load when the digester is filled up. External insulation materials above ground could be rock wool/mineral fiber mat, hard foam pad, extruded foam, styrofoam, synthetic foam and polystyrene. Foam concrete, expanded perlite, rock wool products, and so on were often used in the past. For the moment, polystyrene foam plastics, rubber-plastic sponge and polyester foam plastics are popular. Thickness of insulation layers shall be calculated based on designed anaerobic digestion temperature, feedstock temperature, average temperature of the coldest month and thermal conductivity coefficient of the digester materials. Thermal insulation layers usually have a thermal transfer coefficient of 0.02–0.05 W/(m K). A thickness of 5–10 cm is widely adopted for the moment. 20 cm is adopted in cold areas.

Most thermal insulation layers are constructed outside the main structure of a digester, i.e. external thermal insulation. A thermal insulation layer together with a hard cover on top form a complete thermal insulation structure. The cover protects insulation materials from external water vapor, rainwater, and moisture of damp soil that lead to higher thermal conductivity and lower thermal insulation. The cover also protects insulation materials from mechanical damage and ensures the a neat and smooth surface for easy painting. Commonly adopted protective covers include plastering with asbestos mortar, brick wall, iron cover, aluminum cover, A-alloy plate and profiled colored steel. In some cases, the thermal insulation layer is constructed inside the digester, i.e. internal thermal insulation (Zhang 2004).

4.3.2.4 Auxiliary Facilities of the Digester

1. *Effluent overflow system*

A UASB is sealed with a three-phase separator. Discharged effluent overflows from an overflow weir that is similar to a sedimentation tank. Other types of anaerobic reactor are usually sealed off with a fixed top cover. No biogas shall be emitted when digestate are discharged from the discharge pipe. Therefore, the discharge pipe is usually fabricated in an inverted U shape to be used as a water seal. The effective water column head of the water seal must be greater than the designed biogas pressure of the digester. Figure 4.16a, b show a discharge pipe without (CSTR, USR) and with (UBF, AF) an overflow weir in a digester. In both diagrams, water column head A represents biogas pressure in the digester. The outlet of the overflow pipe shall not be located indoors, and the minimum diameter of the overflow pipe shall be no less than 150 mm.

2. *Positive/negative pressure protector*

While a digester is operating, the biogas pressure may be excessively high (positive pressure) inside when discharge is less than feed due to high inflows from the feed pump or blockage of the overflow pipe or the biogas discharge pipe. Negative pressure occurs in the digester when mixed liquid leaks due to breakage of feeding pipe valves (mostly the check valve) or during residue/fluid discharge at the bottom. Positive and negative pressures can cause deformation and even damages of the digester, to prevent which a proper positive/negative pressure protector is required. There shall be a water (liquid) seal installed at a proper depth inside the positive pressure protector to play two roles: relieving pressure by releasing biogas from the water (liquid) seal when excessively high pressure is found inside the digester, and preventing biogas leak by isolating biogas inside the digester from external air with the water(liquid) seal when biogas pressure inside the digester drops to the design pressure. A negative

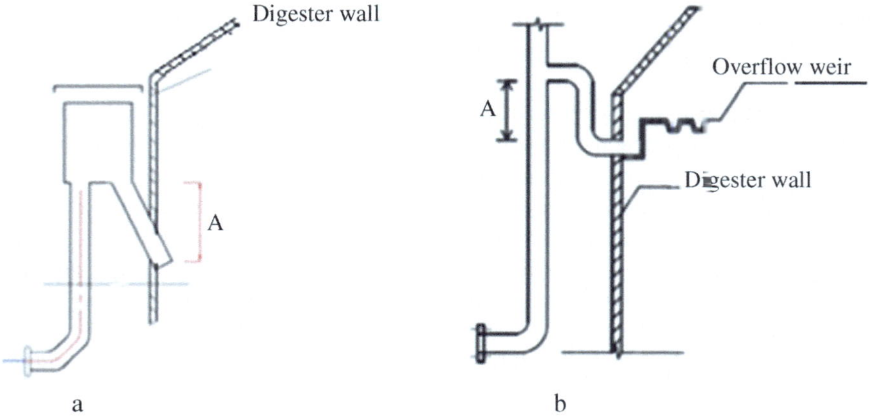

Fig. 4.16 Schematic diagram of an overflow system in a digester

Fig. 4.17 Schematic diagram of a positive/negative pressure protector

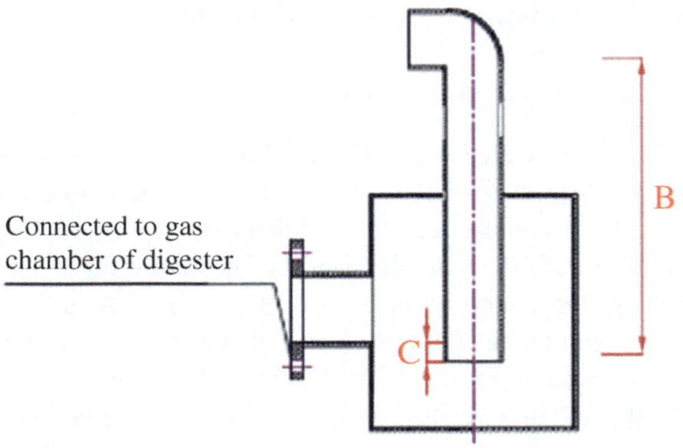

pressure protector is used to prevent excessive negative pressure in a digester. When the internal pressure of the digester is lower than the atmospheric pressure, external air will penetrate through the water (liquid) seal and enter the digester. When external and internal pressures are in equilibrium, the water seal will isolate biogas produced inside the digester from external air. A positive or negative pressure protectors can be connected to a pipe that connects to the gas chamber at the top of a digester. Both could be installed together to prevent both positive and negative pressures, as shown in Fig. 4.17. Water (liquid) column height B represents the height for positive pressure protection while water (liquid) column height C represents the height for negative pressure protection.

In north China where winter is cold, water freezes in a positive/negative pressure protector and the protection fails. As a result, anti-freezing agent shall be added to water as well.

4.3.2.5 Pipe Holes in Digester

Holes for auxiliary pipes include the feeding pipe, discharge pipe, sludge outlet pipe, overflow pipe, biogas collecting pipe, sampling pipe and inspection manhole, and so on.

The feeding pipe is usually located at digester bottom, with some distance from the floor for easy maintenance (at least 300 mm). The feed point and feed port shall be properly located for even agitation and breaking of sediments or scums. The feeding pipe should be at least 500 mm from the bottom of the digester. A small digester (having a volume less than 2500 m^3) usually has 1 feeding pipe. A medium digester (having a volume around 5000 m^3) should have 2 feeding pipes. A large digester (having a volume over 10,000 m^3) should have more than 2 feed pipes, each not less than 100 mm in diameter.

The residue discharge pipe shall be located at the lowest point of digester bottom or a scattered position at the bottom. For a large digester, 1–2 additional discharge pipes shall be provided at different heights above the bottom. There may be a common pipe used as both the emptying pipe and the residue discharge pipe. Both discharge pipes may also be provided separately, with a pipe diameter not less than 150 mm and a cleaning port provided downstream of pipe valve.

When hydraulic agitation is adopted or external heating is applied, the feeding pipe, discharge pipe and sludge outlet pipe shall be properly located for mixing. Double-gate valves shall be used on the feeding pipe and the sludge outlet pipe.

The biogas collecting pipe should be at least 1000 mm from liquid level. Pipe diameter shall be determined by calculation and should be no less than 100 mm.

Sampling pipes are usually located at the top, in the middle, and at the bottom of a digester, at least protruding into mixed liquid for 0.5 m with a minimum diameter of 50 mm.

The manhole at the center of the digester top should be 1.2 m in diameter while that on the side wall and at the bottom boundary should be 0.6–1.0 m in diameter. If necessary, two manholes can be provided to remove sediments and sand. Manhole cover shall be made of cast iron or steel and fixed with anticorrosion bolts.

4.3.3 Biogas Cleaning and Storage Unit

4.3.3.1 Biogas Cleaning

Biogas produced in a digester mainly contains methane and carbon dioxide as well as saturated water vapor and hydrogen sulfide. Hydrogen sulfide is a kind of toxic and hazardous gas with a smell of rotten eggs. Oxide of hydrogen sulfide combines with water vapor can form corrosive sulfuric acid, which corrodes biogas pipes and biogas utilizing equipment such as stoves, generator sets, etc. Biogas must therefore be dewatered and desulfurized to prevent hydrogen sulfide corrosion. Bio-methane production from biogas requires removal of carbon dioxide. The process of removing carbon dioxide is usually referred to as upgrading.

1. *Dewatering*

Biogas contains saturated water vapor which may condense and affect biogas transporting and utilization. Therefore, dewatering is an essential step. A gas-water separator is usually used for biogas dewatering, which is loaded with packings and designed to be a cyclone separator so that water vapor collides with infill, baffle board, or vessel wall to condensate into large water drops. A water condenser is often used for dewatering of biogas pipe, which is located at the lowest point of a biogas pipe to collect and discharge condensate from the pipe.

Another way of biogas dewatering is to use an electric condenser to cool biogas down to below 10 °C to remove water vapor. To prevent condensation of water vapor in downstream pipes, cooled biogas needs to be re-heated.

2. *Desulfurization*

There are chemical and biological methods for biogas desulfurization (hydrogen sulfide removal).

Chemical desulfurization is common for large, medium, and small-scale biogas plants while biological desulfurization is often adopted for mega biogas plants.

Chemical desulfurization can be realized either inside or outside a digester. For the former method, substances such as iron salts are added into feedstocks. This method is also called in situ desulfurization. Chemical desulfurization outside the digester can be realized in a wet or dry manner. In wet desulfurization, alkali (such as sodium hydroxide) or iron-bearing solution is used as the absorbent. In a dry process, iron oxide is used as the desulfurizer.

Biological desulfurization is commonly adopted for mega biogas plants, also applicable inside and outside a digester. For in situ biological desulfurization, air is injected into the digester from the top, of which the quantity is usually 2–8% of biogas output. In the digester, hydrogen sulfide is oxidized into elemental sulfur or sulfurous acid by microorganisms. The biological desulfurization reaction happens in the scum bed at the top or the wall of the digester. When desulfurization happens on the digester wall, the desulfurized product is corrosive. To avoid blockage of any biogas discharge pipe, the air pipe shall be located opposite to the biogas discharge pipe. The in situ biological desulfurization process is simpler and less costly. For ex situ biological desulfurization, a biological desulfurizing tower is installed to precisely control the desulfurization process and accurately regulate oxygen feed. An ex situ biological desulfurizing tower is similar to a scrubber, in which porous packings are loaded to grow microorganisms. Solution is regularly sprinkled onto the infill, which is highly alkaline and is rich in microorganism nutrients. Liquid digestate may be used for such purpose. At 35 °C, the loading rate per cubic meter of packings for biogas treating in each hour is about 10 m^3. In engineering practices, collected desulfurized products are often injected into a digestate storage tank for mixing with digestate and improving manorial effect.

4.3.3.2 Biogas Storage

Biogas is produced in a relatively continuous and stable manner, but its utilization is intermittent and unbalanced. For example, the need for biogas is concentrated when biogas is applied to power generation or supplied to households. Temporary storage of biogas is necessary to ensure the balance between production and consumption. For the moment, varied types of biogas storage facilities are in use. A gas holder on top of a digester is the simplest storage system, which uses a gastight polymeric membrane as the top cover of the digester. Hydraulic storage is an available method for underground digesters. In most biogas plants, an independent gas holder is used. Biogas can be stored at low (gage pressure: $P < 10$ kPa), medium ($10 < P \leq 400$ kPa) and high ($400 < P \leq 1600$ kPa) pressures. A wet gas holder or dual-membrane gas holder is suitable for low-pressure biogas storage. Medium- and high-pressure biogas must be stored in a steel tank or cylinder, of which the energy consumption and operational cost are high. When pressure reaches 1000 kPa, the energy consumption is about 0.22 kWh/m^3. When pressure reaches between 20,000 and 30,000 kPa, the energy consumption is 0.31 kWh/m^3. Due to high operational cost and strict safety requirements, medium- and high-pressure biogas storage is seldom conducted in farm biogas plant.

4.3.4 Biogas Utilization and Emergency Flare

Biogas can be used as a daily fuel, a fuel for power generation, a fuel for injection into the natural gas pipe network after upgraded into bio-methane, or for vehicle fuel uses. The utilization of biogas will be further introduced in the following chapters of this book.

In some special circumstances when the production rate of a digester is too high or the biogas utilization system is in maintenance/failure, produced biogas may not be used up, and emergency solutions such as setting up a temporary biogas storage facility or an additional biogas utilization system are required. The temporary storage of non-compressed biogas is only a short-term solution and is ineffective for long time of maintenance/failure. The addition of a new biogas utilization system such as a second CHP system is usually costly. Therefore, an emergency flare is needed for a large-scale biogas plant. Excessive biogas that cannot be stored or utilized can be combusted with an emergency flare to prevent safety incidents and environmental pollution.

The main purpose of emergency combustion is to transform methane to the maximum degree and reduce the emission of methane and other intermediate products of incomplete burning (such as carbon monoxide). Some hazardous byproducts may be produced during biogas combustion, depending on air content, temperature, and the dynamics of combustion reaction. Emergency combustion depends on two parameters: temperature and retention time. Combustion temperature must be 850–1200 °C and the minimum residence time shall be 0.3 s. To ensure safe and reliable operation of an emergency combustion system, some other factors in addition to burner and barrel shall also be considered, including flame arrestor, failure-safety valve, ignition system, and flame detector. A blower is required to pressurize biogas up to 3–15 kPa, which is needed for the burner.

There are two types of biogas emergency flares: open flares and enclosed flares.

An open flare is essentially a burner, with a small windshield to protect the flame. Biogas control is rudimentary—in many cases, a coarse manual valve. The rich gas mixture, lacks insulation and poor mixing lead to an incomplete combustion and a luminous flame, which is often seen above the windshield. Radiant heat loss is considerable and introduces low-temperature areas at the edge of the flame and quenching of combustion reactions that gives rise to many undesirable by-products. Open flares was popular in the past because of their simplicity and low cost, as well as absence of legislation and control over emissions standards. Henceforward, strict regulation and emission control is likely to limit their uses.

Enclosed flares are usually constructed and fixed on ground, with a protective shell housing a single or several burners that are enclosed within a cylindrical enclosure.

The cylindrical enclosure can prevent quenching while promoting uniform combustions with a lower pollutant emission. Emission monitoring is relatively easy, and process control is achievable by continuous monitoring of temperature, hydrocarbons, and carbon monoxide. Increased engineering and process control provide greater turn down flexibility (the ratio of maximum biogas flow to minimum biogas flow under which satisfactory operating conditions are maintained). Manufacturers typically quote a turndown ratio of 4–5:1 for biogas quality of 20–60% methane (by volume). Higher turndown of up to 10:1 is achievable, but on the expense of combustion quality, since the heat release does not enable temperatures sufficient for complete combustion.

4.3.5 Digestate Treatment and Utilization Unit

Digestate utilization should be prioritized when it comes to digestate handling. Those that cannot be utilized shall be further treated. When used as fertilizer, digestates can be temporarily stored in a biogas plant for days. The in-plant storage tank shall have a volume of at least that of the largest individual digester. Storage tanks are also required in the fields. The total volume of digestate storage tanks shall guarantee a storage duration of several months. According to agricultural stipulations in European countries, animal manures and digestates shall be stored for 6–9 months to ensure the optimal and most effective utilization. Uses in winter shall be avoided. Digestates may be stored in reinforced concrete tanks or the seepage-proofing ponds that are covered by naturally formed or artificial scum beds or polymeric films to prevent the emission of odorous gas.

When liquid digestates will be used for foliage dressing or subjected to further treatment, or when solid digestate residues will be used as solid organic fertilizers, digestates shall undertake a solid-liquid separation. The equipment for separating digestates shall be selected based on feedstock characteristics and the desired extent of separation. The separation equipment employed shall have sufficient capacity to carry the digestates to be treated. When TS level of digestates is >5%, a screw press separator is recommended; when TS level of digestate is <5%, an inclined screen solid-liquid separator can be used.

Solid digestates may be used as fertilizers for crops, raw materials for composting, nutrition pot (soil) of crops, and ingredients for earthworm farming. Liquid digestates may be used for seed soaking, root top dressing, or foliage spraying.

4.3.6 The Control Unit

A biogas plant is a complex system with close interrelations among all parts. For this reason, computer monitoring and controlling is an essential part of plant operation, aiming to track plant behavior, avoid failures, and make contingency planning. Monitoring and documentation is also necessary for process stability, in order to be able to recognize deviations from standard values and to make possible early intervention and corresponding corrective measures. Monitoring includes collection and analysis of chemical and physical parameters. Regular laboratory tests are needed to optimize the digestion process and to avoid inhibition or collapse of biogas production. Monitored parameters include feedstock quantity, feedstock concentration, temperature, pH, biogas quantity and composition, biogas pressure, content of volatile fatty acids (VFA) in feedstock, liquid (filling) level, etc. Site monitoring instruments shall be resistant to corrosion, explosion, leakage, and scaling and shall be capable of self-cleaning.

A biogas plant can be automatically controlled by computers or wireless remote controllers. Most units in a biogas plant can be automatically controlled: feedstock feeding, sanitation (disinfection), heating of digester, agitation (intensity and frequency), sediment removal, desulfurization, solids-liquids separation, biogas power generation and waste heat recovery, etc. Automatic control can be realized via a simple timer or the visualization of computer-supported control with a remote alarm system. Monitoring objectives and automation degree shall be determined based on project scale, investment, operational, and management requirements, the advancement of equipment and instruments used as well as maintenance and management capability. In practice, the measurement and technical control equipment of agricultural biogas plants are often very simple due to economic reasons. Centralized control is recommended for medium- and small-scale biogas plants. In case of a large-scale plant or when lots of reactors are employed, a distributed automatic control system is a better option. An independent control box shall be provided near each key system, which shall be capable of automatically switching between manual control and automatic modes.

References

Al Seadi, T. 2008. *Biogas handbook*. Syddansk Universitet.

Bi, Y.N. 2010. *Study on straw resources evaluation and utilization in China*. Doctoral dissertation of Chinese Academy of Agricultural Sciences (in Chinese).

Banks, C.J., M. Chesshire, S. Heaven, and R. Arnold. 2011. Anaerobic digestion of source-segregated domestic food waste: Performance assessment by mass and energy balance. *Bioresource Technology* 102 (2): 612–620.

Chen, F.Q., and B.Y. Niu. 1995. Preparation of silage feed in China and abroad. *Consultancy for Animal Raising* 1: 50–51 (in Chinese).

Cheng, P. 2008. *Experimental study of secretion coefficient on typical dairy farm in Beijing*. Doctoral dissertation of Chinese Academy of Agricultural Sciences (in Chinese).

China Agricultural University, Shanghai Agricultural Radio and Television Science, South China Agricultural University. 1997. *Animal faecology*. Shanghai: Shanghai Jiaotong University Press (in Chinese).

Cho, J.K., S.C. Park, and H.N. Chang. 1995. Biochemical methane potential and solid state anaerobic digestion of Korean food wastes. *Bioresource Technology* 52 (3): 245–253.

Davidsson, Å., C. Gruvberger, T.H. Christensen, T.L. Hansen, and JIC Jansen. 2007. Methane yield in source-sorted organic fraction of municipal solid waste. *Waste Management* 27 (3): 406—414.

Dębowski, M., M. Zieliński, A. Grala, and M. Dudek. 2013. Algae biomass as an alternative substrate in biogas production technologies—Review. *Renewable and Sustainable Energy Reviews* 27: 596–604.

Deng, L. W., Z. A. Chen, H. Yang, J. Y. Zhu, Y. Liu, Y. Long and D. Zheng. 2012. Biogas fermentation of swine slurry based on the separation of concentrated liquid and low content liquid. Biomass and Bioenergy 45: 187–194

Deng, L. W., H. N., Yang, G. J. Liu, D. Zheng, Z. A. Chen, Y. Liu, X. D. Pu, L. Song, Z. Y. Wang and Y. H. Lei. 2014. Kinetics of temperature effects and its significance to the heating strategy for anaerobic digestion of swine wastewater. Applied Energy 134: 349–355

Department of Natural and Ecological Conservation, National Environmental Protection Administration of China (SEPA). 2002. *Investigation on pollution of large-scale livestock and poultry breeding industry in China and countermeasures for its prevention and control*. Beijing: China Environmental Science Publishing House (in Chinese).

Dong, H.M., Z.P. Zhu, H.K. Huang, Y.X. Chen, B. Shang, X.P. Tao, and Z.K. Zhou. 2011. Pollutant generation coefficient and discharge coefficient in animal production. *Transactions of the CSAE* 27 (1): 303–308 (in Chinese).

Dong, R.J., and B. Raninger. 2000. *Biogas engineering and application*, vol. 4. Beijing: China Agricultural University Press (in Chinese).

Fang, R.X. 1995. *Biomass plant*. Beijing: China Agriculture Press (in Chinese).

Fu, J., X.K. Han, and G. Li. 2014. Exploration on municipal solid waste classification and organic wastes disposal methods. *China Resources Comprehensive Utilization* 33 (2): 30–33 (in Chinese).

Geng, W., L. Hu, J.Y. Cui, M.D. Bu, and B.B. Zhang. 2013. Biogas energy potential for livestock manure and gross control of animal feeding in region level of China. *Transactions of the Chinese Society of Agricultural Engineering (Transactions of the CSAE)* 29 (1): 171–179 (in Chinese).

Ghanimeh, S., M. El Fadel, and P. Saikaly. 2012. Mixing effect on thermophilic anaerobic digestion of source-sorted organic fraction of municipal solid waste. *Bioresource Technology* 117: 63–71.

Gissén, C., T. Prade, E. Kreuger, I.A. Nges, H. Rosenqvist, S.E. Svensson, M. Lantz, J.E. Mattsson, P. Börjesson, and L. Björnsson. 2014. Comparing energy crops for biogas production—Yields, energy input and costs in cultivation using digestate and mineral fertilisation. *Biomass and Bioenergy* 64: 199–210.

Guo, D.J., H.S. Wu, Y. Ma, and Z.Z. Chang. 2011. Study on the amount of manure and urine excreted by sheep and rabbits in intensive pasture. *Ecology and Rural Environment Journal* 27 (1): 44–48 (in Chinese).

Guo, Y.L., and H.J. He. 2001. Study on parameters and characteristics of domestic organic refuse use as anaerobic digestion raw materia. *Shanghai Environmental Science* 20 (12): 611–628 (in Chinese).

Han, S.J., X.L. Tong, and X.T. Zhou. 1988. MSG wastewater treatment by anaerobic fermentation. *Zhejiang Industrial University Journal* 3: 89–101 (in Chinese).

He, J.J., S.H. Yan, X.M. Ye, Z.Z. Chang, W.X. Huang, L. Ge, and H.J. Xu. 2008. Research progress of biogas production by anaerobic fermentation of Water Hyacinth. *Jiangsu Agricultural Journal* 24 (3): 359–362 (in Chinese).

Hu, P. 1995. Anaerobic treatment of sulfate-rich glutamate wastewater. Environmental Pollution & Control 17 (5): 10–13 (in Chinese).

Huang, W.X., L. Ge, and H.J. Xu. 2006. application of foreign invasive biological water peanut in biogas. *Agricultural Environment and Development* 23 (4): 60–61 (in Chinese).

Huang, H., Y.F. Deng, Z.P. Cheng, N. Xu, and J.M. Xu. 2013. Characteristics of mixed fermentation by alternanthera philoxeroides and wheat straw for biogas production. *Research of Agricultural Modernization* 34 (005): 636–640 (in Chinese).

Jing, J. P., S. L. Liu, W. G. Qu and Z. P. Han. 2013. Key Technology of food waste anaerobic treatment. Environmental Sanitation Engineering 21 (5): 5–7 (in Chinese).

Kastner, V., W. Somitsch, and W. Schnitzhofer. 2012. The anaerobic fermentation of food waste: A comparison of two bioreactor systems. *Journal of Cleaner Production* 34: 82–90.

Liu, Z., and Z. Q. Duan. 2010. Distribution of Manure Resources and Environmental Loads of Agro-ecological Regions in China. Resources Science 32 (005): 946–950 (in Chinese).

Liu, S. G., W. D. huang, Z. M. Xu and W. D. Xia. 2011. Pilot-scale study on production of biogas from water hyacinth and alternanthera philoxeroides by anaerobic digestion. Techniques and Equipment for Environmental Pollution Control 5 (6): 1310–1314 (in Chinese).

Liu J. Y., S. Zhao, W. L. Jiang, H. N. Jing, F. Q. Wang, X. G. L. and H. Y. Zhang. 2016. Application of anaerobic digestion in kitchen wastes treatment project. Environmental Science and Technology 29 (5): 43–46 (in Chinese).

Ma L. and X. Liu. 2012. Anaerobic digestion treatment of food waste. Environmental Sanitation Engineering 20 (4): 12–14 (in Chinese).

Regassa, T.H., and C.S. Wortmann. 2014. Sweet sorghum as a bioenergy crop: Literature review. *Biomass and Bioenergy* 64: 348–355.

Rural Socio-economic Survey Department, National Bureau of Statistics. 2013. *China rural statistical yearbook*. Beijing: China Statistical Publishing House (in Chinese).

Searle, S.Y., and C.J. Malins. 2014. Will energy crop yields meet expectations? *Biomass and Bioenergy* 65: 3–12.

Song L., L. W. Deng, Y. Yin, X. D. Pu and Z. Y. Wang. 2010. Biogas production potential and characteristics of manure of sheep,duck and rabbit under anaerobic digestion. Transactions of the Chinese Society of Agricultural Engineering 10: 277–282 (in Chinese).

Wan R. X. 1995. Biomass energy engineering. Beijing: China Agriculture Press (in Chinese).

Wang K. J, J. E. Zuo, H. N. Gan, L. M. Jia. 2000a. The theory and implementation of UASB reactor concept in China. Beijing: China Environmental Science Press (in Chinese).

Wang K. J., and R. W. Qin. 2000b. Treatment of wastewater from fermentation industrial. Beijing: Chemical Industry Press (in Chinese).

Wang J. Y. 2008 Development of harvesting technology for major crops straw. Agricultural machinery 16: 23–26 (in Chinese).

Wang X. Y., S. Xue and G. H. Xie. 2012. Value-taking for residue factor as a parameter to assess the field residue of field crops. Journal of China Agricultural University 17(1): 1–8 (in Chinese).

Wellinger, A., J.P. Murphy, and D. Baxter. 2013. *The biogas handbook: Science, production and applications*. Elsevier.

Wu J., H. Jiang, H. M. Zhou, and L. Bi. 2008. Current operation status of sludge anaerobic digestion system in municipal wastewater treatment plants in China. China Water & Wastewater 24 (22): 21–24 (in Chinese).

Xiang, C. J., S. W. Zhang, D. S. Yu, L. J. Xu, J. P. Chen, and W. R. Mo. 1989. Studies on biogas production from urban domestic waste by anaerobic digestion and utilization of fermentation residue. China Biogas 7 (3):11–14 (in Chinese).

Xu, R., T.R. Gao, E. Pay, and C.F. Xia. 2007. Biochemical methane potential of blue-green algae in anaerobic digestion progress. *Yunnan Normal University Journal, (Natural Science)* 27 (5): 35–38.

Ye, S. Y. 2007. Study on anaerobic digestion condition of municipal organic waste. Master thesis of Hefei university of technology (in Chinese).

Zegada-Lizarazu, W., and A. Monti. 2011. Energy crops in rotation. A review *Biomass and Bioenergy* 35 (1): 12–25.

Zhang, H.J., H. Yang, D.F. Guo, L.C. Liu, Y.L. Zhang, F. Yin, X.L. Zhao, J. Liu and J. Liang. 2013. Experimental study on anaerobic digestion at different temperatures with blue algae from Dianchi Lake. *Yunnan Normal University Journal (Natural Science)* 33 (3): 17–21 (in Chinese).

Zhang, J.S., L. Zhang, and H. Wang. 2005. Anaerobic digestion of organic fraction of municipal solid waste. *Ecology and Environment* 14 (3): 321–324 (in Chinese).

Zhang, T., M.D. Bu, and W. Geng. 2012. Pollution status and biogas-producing potential of livestock and poultry excrements in China. *Chinese Journal of Ecology* 31 (5): 1241–1249 (in Chinese).

Zhang, Z.J. 2000. *Drainage works, Part II*, 4th ed. Beijing: China Construction Industry Publishing House (in Chinese).

Zhang, Z.H. 2004. *Water supply and drainage design manual*, Vol.5. Sewerage. Beijing: China building industry press (in Chinese).

Zhou, M.J., Y.L. Zhang, and J.Y. Lin. 2009. *Biogas practical technology*, 2nd ed. Beijing: Chemical Industry Press (in Chinese).

Zhuang, L., H.P. Zhou, and L.J. Zhang. 2015. Advancement in research on pollutants producing and discharging coefficients of livestock and poultry breeding industry in China. *Journal of Ecology and Rural Environment* 31 (5): 633–639 (in Chinese).

Chapter 5
Construction Materials and Structures of Digesters

Digesters is the core of the whole anaerobic digestion system, which has a decisive impact on waste disposal, biogas production efficiency, and project profit. A good understanding of performances, usage, structural, and features of various construction materials is essential for constructing high-quality digesters. Construction materials for digester include ordinary steel, special steel, concrete, brick, stone masonry, environmental-friendly EPDM (ethylene propylene diene monomer) membrane, environmental-friendly HDPE (high density polyethylene) membrane, environmental-friendly polypropylene random (PPR) membrane, glass fiber reinforced plastic, etc. Selection and use of construction materials are directly related to construction quality, service life, and construction investment. According to the types of construction materials, digesters can be classified into the brick-concrete structure, the reinforced concrete structure, the steel structure (including the welded steel sheet structure, the spiral double fold hem structure, and the assembled enameled steel sheet structure), and the membrane structure. Based on construction position, digesters are classified into underground, semi-underground, and aboveground types.

5.1 Digesters with Brick-Concrete Structure

A digester with brick-concrete structure is a digester with its body constructed of mortared common standard bricks and its internal surface treated by anti-seepage materials. In China, brick-concrete structure is a conventional structure for the body of rural household digesters and biogas digester for domestic sewage treatment system. Usually the bottom is constructed by concrete, while the wall, the vault, the inlet chambers, and hydraulic chambers are from bricks. Advantages of such structure are that the materials can be obtained easily, and the construction technology is relatively mature.

L. Deng et al., *Biogas Technology*,
https://doi.org/10.1007/978-981-15-4940-3_5

(1) **Construction materials**

Ordinary clay brick. A clay brick, also known as a sintered brick, is a small-size man-made unit used in construction. Clays (including shale, coal gangue and other powder materials) as the major raw materials and are subjected to mud treatment, shaping, drying and calcinating in brick manufacturing. Clay bricks are classified into solid and hollow bricks.

Solid brick: no hole or hole proportion <25%;

Porous bricks: hole proportion in between 25% and 40%;

Hollow brick: hole proportion ≥40%.

Bricks with a large number of small-size holes are often used for load-bearing purposes due to their high strength levels. Bricks with a small number of large-size holes are often used in non-load-bearing position due to their low strength levels.

An ordinary clay brick is 240 mm × 115 mm × 53 mm in size, and it can be classified into 6 strength levels: MU30, MU25, MU20, MU15, MU10 and MU7.5 depending on the compressive strength (N/mm^2). Clay bricks are made of materials available locally and are cheap and durable, with functions of fire prevention, heat insulation, sound insulation and moisture absorption, etc.

Ordinary clay bricks are often used for construction of household digesters, biogas digester for domestic sewage treatment, digester in small biogas plants, wastewater collection tanks, regulation tanks, digestate storage tanks, and protective layers of the thermal insulation layer of digesters, as well as buildings and other auxiliary facilities. Such bricks have strength levels of no less than MU10, a unit weight of 1700 kg/m^3, an orderly size, smooth surfaces, and no excessive warpage. Under-burned bricks, crispy bricks and threaded bricks should not be used. The quantity of standard bricks in 1 m^3 of brick masonry shall be 512 bricks (including mortar joints).

Mortar is a paste made of cement and sand mixed with water. In masonry works, it is used to bind bricks, stones, blocks into a whole wall, and it is used to fill voids of masonry and to cement masonry into a whole to achieve a certain strength and compactness. Mortar is classified into cement mortar and mixed mortar according to compositions, or it can be classified into masonry mortar and plastering mortar according to purposes of use.

Brick-concrete digesters shall be constructed with cement mortar, which is applicable for construction of masonry in humid environment and masonry with high strength requirements. Mortar grade is indicated by an average compressive strength (MPa) measured by the standard test method. M7.5 or M10.0 cement mortar shall be used for the construction of digesters.

Plastering mortar is used for leveling structure surfaces and structure protection. It also has the functions of sealing, waterproofing, and anti-seepage. Its mixing ratios are generally 1:2, 1:2.5 and 1:3, and the water-cement ratio is 0.5–0.55. Plastering mortar for digesters can be transformed to waterproof mortar by adding sodium silicate and ferric chloride as the water-proofing agent (3%).

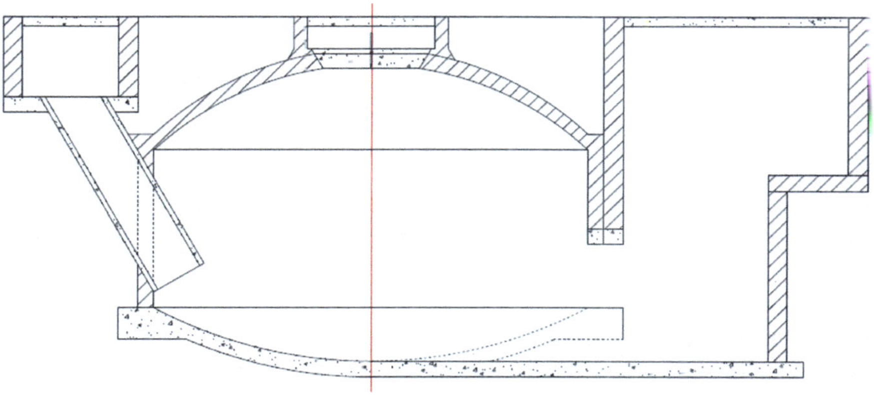

Fig. 5.1 Household digester with brick structure

(2) Structural characteristics

Brick-concrete digesters are usually composed of a spherical bottom, cylindrical walls, a spherical top, and necessary auxiliary facilities. The digester bottom is usually made of the compacted layer of original soil and of the pebble (crushed) stone cushion. The bearing capacity coefficient f_{aK} of the compacted layer of original soil shall be no less than 50 kPa. The pebble (crushed) stone cushion is poured with 1:5.5 cement mortar. Then concrete of above C25 level is poured on cushion, and compacted and plastered into a curved surface. Generally the digester bottom shall be more than 50 mm thick. The clay brick (≥Mu20) and cement mortar (M10 grade) is used for construction of cylindrical wall with 120 mm thickness, and sealing treatment is required. Ring beams should be designed at the top of digester walls to withstand loads above the apex of vault. Brick structure, cast-in-place concrete structure, prefabricated concrete slab installation structure, and FRP dome can be used as the top of an undercut spherical digester. A household digester with brick structure is shown in Fig. 5.1.

The types of biogas digester for domestic sewage treatment include rectangular digesters and their derived types, circular digesters and their derived types, and combined circular and rectangular type digesters. A rectangular digester and its derived types (such as arch digesters and long circular arch digesters) generally have a length-width ratio of above 2 and a larger surface area with poor mechanical properties. However, the process equipment layout is simple, and construction is convenient. A circular digester and its derived types (such as circular digesters with upper and lower spherical shells, circular digesters with upper and lower conical shells) have a diameter-height ratio of about 1 and have a smaller surface area with good mechanical properties, but the process layout is more complicated than that of rectangular digesters, and construction is also more complicated The structure of rectangular digesters is similar to that of septic tanks. The structure of circular digesters is similar to that of hydraulic digesters. The bottom of brick-concrete biogas

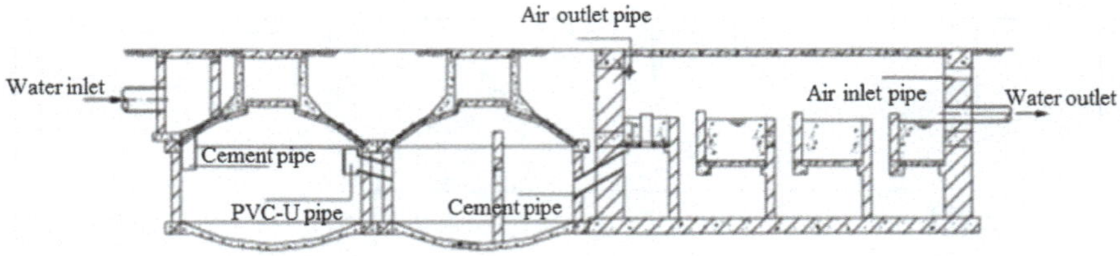

Fig. 5.2 Structural schematic diagram of brick-concrete biogas digester for domestic sewage treatment

digester for domestic sewage treatment is made of concrete. Walls, vaults, inlets, and outlets are all constructed of bricks. Ring beams are designed at the middle and top of the digester wall and on top of tank opening to fix the structure (Fig. 5.2). The characteristic coefficient f_{aK} of bearing capacity of the foundation shall be no less 100 kPa. C15 concrete shall be used for bottom cushion, and C25 concrete shall be used for bottom slab, ring beam, and top. Wall brick shall be of strength levels ≥Mu20, while M10 cement mortar shall be used.

(3) **Jack arch**

Through years of practices, Chinese biogas technicians have summarized a method applicable for construction of digester domes, namely, constructing a Jack arch (a single brick arch), which is to build a dome for a brick digester with a slight camber without formwork. Piles are driven along the edge of the digester top opening. The piles are tied with long ropes, the other end of which is tied with a brick that can be lifted and loaded in the digester so as to get the bricks fixed for dome construction. Alternatively, the camber can be controlled with a pole. Cement mortar (1:2) shall be used in such construction; the outer wall of the dome is compact and is polished with coarse sand mortar at 1:3, and then the outer wall of the dome should be reinforced by wire mesh and sand mortar on top, as shown in Fig. 5.3.

5.2 Digester with Reinforced Concrete Structure

In the past, the brick-concrete structure is mainly used in household digesters, biogas digester for domestic sewage treatment, and underground digesters of small-scale biogas plants. With the strengthening of environmental protection requirements and strict seepage prevention requirements, household digesters, biogas digester for domestic sewage treatment, underground digesters of small-scale biogas plants and some aboveground digesters of large/medium-size biogas plants are constructed using concrete structure. Reinforced concrete digesters usually consists of cast-in-place concrete, prefabricated parts, or prefabricated concrete parts. If underground soil conditions permit, digesters with concrete structure can be partially or wholly located underground. When the concrete digester that is completely buried underground is strong enough, other structures can stand on top of it as long as relevant

Fig. 5.3 Construction of Jack arch

regulations are followed. Such digester should be specially designed and constructed to avoid concrete cracking, leakage, and corrosion.

(1) **Materials for digester body concrete**

The concrete grade for household digester is C15, which no longer satisfies the seepage prevention requirements of hydraulic structures. At the moment, the load-bearing concrete parts of a digester should have a strength level no less than C30, and the cushion concrete should have a grade of at least C15. The minimum concrete strength level of the pre-stressed concrete structure should be C40. For load-bearing reinforced concrete, the strength level of concrete should be at least C30 as well. The impermeability of reinforced concrete digesters should meet the impermeability requirement of concrete itself, and the impermeability grade should be no lower than P_6 (the maximum hydrostatic pressure that can be endured in an impermeability test according to the standard test method is 0.6 MPa for standard concrete specimen of a 28-day age). The aggregate of concrete should be of higher grades possible and the water-cement ratio should be no more than 0.55. Generally, concrete of water treatment structures and underground structures that have already met the impermeability requirements can skip other impermeability treatments. It should be noted that chloride-containing waterproofing agents should not be used in water treatment structures made of reinforced concrete and pre-stressed concrete.

The frost resistance grade F_i of concrete (refers to the strength decrease of a 28-day age concrete specimen after required i times of freeze-thaw cycles) is no more than 25%, and the weight loss is no more than 5%. Exposed reinforced concrete structure in places where the coldest monthly average temperature is below −3 °C should have good frost resistance. In general, frost resistance grade is F_{200}. When antifreeze agent is added to concrete, water-soluble organic compounds (i.e. some alcohol-based organic compounds) should be used as antifreeze additives. Chlorine salt should not be used as admixture for anti-freezing and early strength purposes. Portland Pozzolanic cement and fly ash Portland cement shall not be used when the freeze-thaw effect is considered.

Since the pH of mixed liquid in reinforced concrete digesters will decrease at the initial stage of feeding and start-up, and the pH of some feedstock will even fall below 6.0, anticorrosion treatment should be performed to the digester body. In addition to corrosion-resistant cement for concrete anticorrosion, the following measures can also be taken, including:

(1) Reducing the water-cement ratio in concrete mixing to reduce the voidage of concrete and water absorption rate of concrete;
(2) Strengthening vibration during concrete pouring while reducing concrete voids to reduce pathways for corrosive substances entrance. Meanwhile, it is necessary to vibrate evenly to obtain concrete homogeneity and to prevent local corrosion at the steel bars in heterogeneous medium.
(3) Coating on the surface of steel bars to prevent interstitial solution from contacting with steel bars.
(4) Retaining a concrete layer thick enough to be protective on steel bars while coating on concrete surface to block the invasion of various harmful substances via capillary holes on concrete surface.

Isolating protective coatings can be added to obstruct any erosion and penetration of corrosive medium through concrete surface. That means that, hydrophobic organic coatings are applied on concrete surfaces. The protective coating materials applicable for inner surface of reinforced concrete digesters include epoxy resin, polyurethane, polyurethane asphalt, asphalt, etc.

Steel bar. Steel bars for reinforced concrete refer to straight bars or coiled steel bars. Their shapes can be classified into plain round bars and deformed bars, and the steel bar products are delivered in the form of straight bars and coiled bars.

Plain round bars are in fact small round bars and coiled bars made of ordinary low-carbon steel. Deformed bars are surface-ribbed steel bars, usually with two longitudinal ribs and transverse ribs evenly distributed along the length direction. The shapes of transverse ribs include spiral, herringbone, and crescent. The size of steel bars is expressed in millimeter of nominal diameter. The nominal diameter of deformed steel bars is equivalent to the nominal diameter of plain round bars with equal cross section areas. The nominal diameter of steel bars is usually 8–50 mm, and the recommended diameters are 8, 12, 16, 20, and 25mm. Steel bars in concrete mainly bear tensile stresses, and deformed bars are better at bearing external forces because of the strong bonding force with concrete and the rib geometry. Steel bars in concrete should be free of oil and rust and shall be straightened before use.

(2) **General requirements for concrete digester structures**

Digesters with a reinforced concrete structure can be classified into underground, semi-underground and aboveground types according to the construction level, as shown in Figs. 5.4, 5.5 and 5.6. The underground type is mainly used for rural

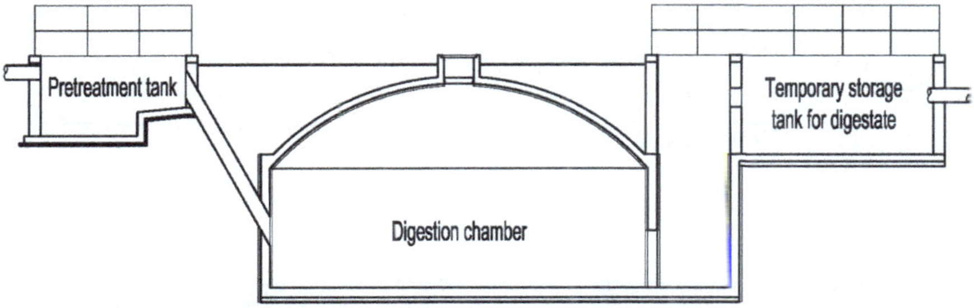

Fig. 5.4 Underground reinforced concrete digester (household digesters and digester of small biogas plants)

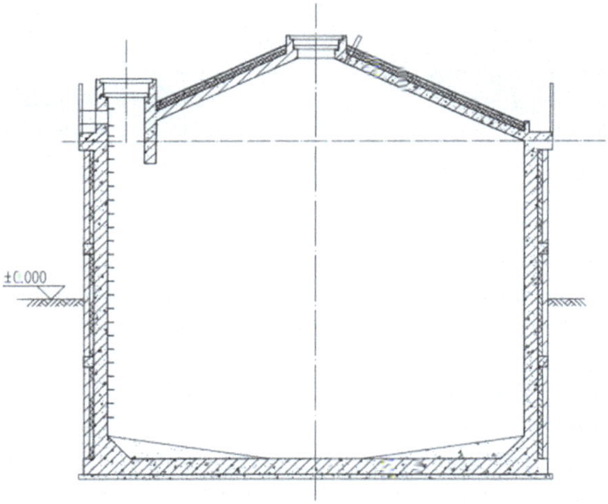

Fig. 5.5 Semi-aboveground reinforced concrete digester

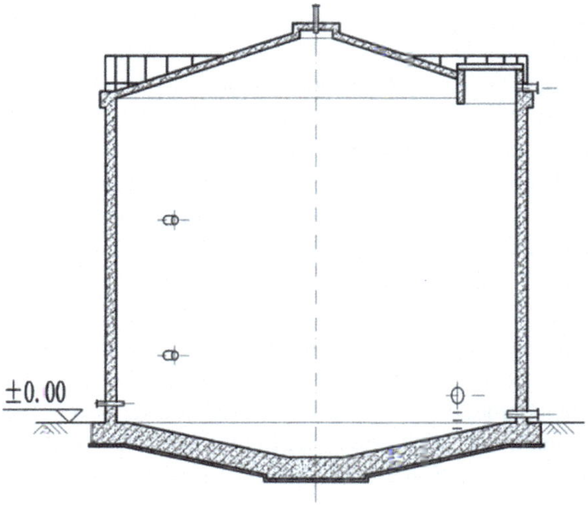

Fig. 5.6 Aboveground reinforced concrete digester

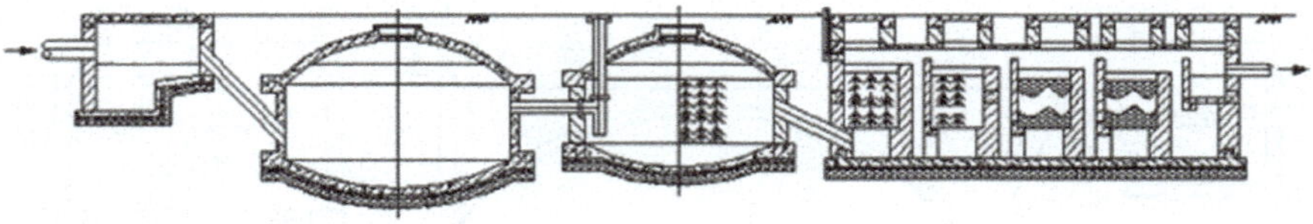

Fig. 5.7 Concrete biogas digester for domestic sewage treatment

Fig. 5.8 Household digesters with prefabricated concrete structure

household digesters, biogas digester for domestic sewage treatment (Fig. 5.7), and digester of small biogas plants. Semi-underground and aboveground types are mainly used as digesters of large/medium biogas plant.

Reinforced concrete rural household digesters can be further classified into cast-in situ concrete digesters and assembled precast concrete digesters as shown in Fig. 5.8. Some large biogas plants in Germany also adopt precast concrete digesters. Compared with brick-concrete digesters, reinforced concrete digesters have better structural stability and a longer service life. Cast-in situ concrete is used for walls, vaults, bottoms and upper/lower ring beams of household digesters. Cast-in situ concrete is also used for hydraulic chambers with a circular internal structure. Brick masonry is used for hydraulic chambers with a square structure. Cast-in situ concrete is used for round feed pipes, which can be also made from concrete prefabricated pipes. Prefabricated reinforced concrete slabs are used for cover plates, central pipes, slurry distribution plates, etc.

Digester with a volume of more than 50 m^3 is recommended to adopt the reinforced concrete structure, with a cushion thickness no less than 100 mm and a stressed wallboard and bottom slab thickness of at least 200 mm. The prefabricated wallboard can be 150 mm thick, and the top should be at least 150 mm thick. A waterproof concrete structure should be no less than 250 mm thick. When a wallboard is more than 200 mm thick, reinforcement should be installed on both sides of the wall. When the wall of a digester is connected to the bottom slab (foundation) and the bottom slab (foundation) is regarded as the fixed support of the digester wall, the bottom slab (foundation) must be thicker than the digester wall. Depending on the soil condition of the foundation, the thickness of foundation may be 1.2–1.5 times of the digester wall, and the bottom slab (foundation) should be stretched out. At the corners of

Table 5.1 Minimum thickness of reinforced concrete protective layer

Component name	Working conditions	Minimum thickness of protective layer (mm)
Plate and shell	Contact with water and soil	30
	Contact with sewage	40
Beam and column	Contact with water and soil	35
	Contact with sewage	50
Bottom slab	Bottom bar with cushion	40
	Bottom bar without cushion	70

Note Thickness of the protective layer herein can be reduced by 10 mm when surface is plastered with cement mortar, or when the quality of other coating materials can be guaranteed

the walls in the digester and the junctions with the top and bottom slab, an axillary angle with width of not less than 150 mm should be set. Diameter of the inclined bar at the axillary angle should equate that of the stressed bar of the digester wall, and the spacing should be twice of spacing between the stressed bars of the digester wall. The minimum overlap length and minimum anchorage length of steel bars are determined by the regulations in GB 50010, *Code for Design of Concrete Structures*. The minimum thickness of the concrete protective layer of stressed bars in concrete digester should comply with Table 5.1.

The design life of a reinforced concrete digester should be longer than 30 years The safety grade of the structure is generally taken as grade two, and the structural importance coefficient shall be 1.0. Concrete should be compacted by vibration and shall be free of honeycomb bubbles, uneven surface voids, and cracks. The formwork and supports should be installed firmly and meet the requirements of the bearing capacity, rigidity and stability of concrete pouring, and can be easily disassembled and assembled. Formwork should be erected at the right position, and the joints should be tight while there is no leak in grouting. The control bolts and heel block should be installed steadily. The reserved holes and embedded pipes in the formwork should not be dislocated or missed, and the installation should be firm. The formwork should also be clean, with demolding agent applied uniformly. No stain is allowed on any joint between steel bars and concrete.

The basic design of digesters must follow the keys of taking advantage of local conditions, taking local materials, ensuring safety and applicability, cost-effectiveness, and comprehensive considerations of construction conditions, and meticulous design. Foundation design should meet the following requirements: (1) Pressure exerted on the foundation should be less than the allowable design capacity; (2) The deformation of groundwork and foundation is less than the settlement required by the digester design; (3) Sufficient overall stability is required for groundwork and foundation; (4) Strength of the foundation itself should meet the construction requirements (Tang and Wang 2013).

The setting and form of construction joints should satisfy the following requirements: (1) Bottom slab and top of the main structure of a reinforced concrete digester should be continuously casted without reserved joints. (2) Horizontal construction joints of digester walls should neither be reserved at which there is maximum shear force and bending moment, nor at the junction of the bottom slab and the digester wall, and the joints shall be located on the digester wall 500 mm above the upper bottom slab surface. (3) When there are manholes and pipe holes on the wall, the distance between the construction joints and the hole edge should be more than 300 mm. (4) Before concrete is casted on a construction joint, the concrete surface at the construction joint should be chiseled, where floating particles and debris shall be removed. The surface shall be washed to stay wet, and be covered with a 20–25 mm thick layer of cement mortar with a mixing proportion higher by one grade than that of concrete. (5) The construction joints should be concave joints, convex joints, stepped joints or flat and straight joints with water stop films.

(3) **General requirements for embedded parts and reserved holes**

Embedded parts of aboveground reinforced concrete digesters include embedded parts for fixing railings, heat insulation materials, and mixing equipment. Before welding the embedded parts, the thickness of steel bars and steel sheets must be checked. Thickness of steel sheets should be no less than 0.6 times of diameter of steel bars, whereas the diameter of steel bars should be no less than 6 mm, and the diameter of the stressed anchor bars should be no less than 8 mm. It is absolutely forbidden to replace ordinary steel bars with cold-processed steel bars. The area and structure requirements of anchor bars for embedded parts for each member of reinforced concrete structures should not only comply with the regulations from GB 50010, *Code for Design of Concrete Structures*, but also comply with the following requirements: (1) Thickness of anchor plates of embedded parts should consider corrosion allowance (usually 1 mm is taken as the corrosion allowance of embedded parts of digester); (2) The exposed portion of embedded parts must be reliably protected against corrosion. Inspections of embedded parts include confirming that anchor bars do not collide with the main bars and are set inside the main bars. Anchor bars should not be terminated within the protective layer. The embedded parts should not protrude onto concrete surfaces and should not be larger than external dimensions of components. Displacement deviation from the central line of embedded parts should be no more than 5 mm, and deviation of horizontal elevation should be no more than ± 3 mm.

The embedded pipes of reinforced concrete digesters of large/medium-sized biogas plants include feed pipes, overflow pipes, sludge discharge pipes, gas exhausting pipes, heating pipes, temperature measurement pipes, mixing pipes, etc. Other than the gas exhausting pipes that are subjected to higher requirements, other embedded pipe fittings may be made, installed, and embedded by waterproof sleeves. The size of waterproof sleeves shall be 2–3 grades larger than the actual diameter of a pipe, and the length of the sleeves shall be shorter by 5 mm than the actual thickness of the wallboard. Waterproof sleeves shall be made of common steel pipes. Water sealing ring with a thickness of 5–8 mm and width of no less than 50 mm shall be welded on both sides tightly in the middle of a sleeve. No offset is allowed during installation. Flexible waterproof sleeves should be used when embedded pipe fittings are connected with mechanical equipment prone to vibrate such as pumps and agitators. Flexible waterproof sleeves should be used for feed pipes and mixing pipes. Rigid waterproof sleeves may be used when the embedded pipe fittings are connected with static equipment.

Reinforcing measures should be taken at the opening of a reinforced concrete digester: (1) When the diameter or width of the opening is larger than 300 mm but smaller than 1,000 mm, stiffeners should be provided along the direction of stressed steel bar on each side of the opening, and the cross-sectional area of steel bars should be no less than 75% of the cross-sectional area of stressed steel bars that is cut off because of opening. Inclined bars should be installed around the rectangular openings. Ring bars should be installed at the circular openings. (2) When the diameter or width of an opening is more than 1,000 mm, rib beams should be added around the opening. When the diameter or width of the opening is more than ¼ of the calculated span of the wall, side beams should be set at the opening, and the reinforcement in the beam should be determined by calculation. (3) When the protective structure of the external heat insulation layer is brick masonry, brick vault should be used to strengthen the opening. Thickness of brick vault should be no less than 120 mm for openings of less than 1,000 mm in diameter and should be no less than 240 mm for openings of more than 1,000 mm in diameter. Openings may also be strengthened partially by concrete.

(4) **Waterproof layer and sealing layer**

A waterproof sealing layer must be thoroughly grouted and compressed by plastering, and the layer shall be free of sand, cracks, hollows, peel-off while having glossy surfaces. Joints shall be tight, and the bonding between layers shall be firm. Cement mortar waterproof layers shall be constructed as follows: (1) The foundation surface should be flat, clean, hard, rough, and fully humidified without water accumulation; (2) Each cement mortar waterproof layer should be constructed in a continuous manner. When construction joints must be reserved, they should be remained at step joints and overlapped sequentially. The overlap length should be more than 40 mm. The distance between the joints and concave/convex angles should be more than 200 mm; (3) Viscosity thickness of cement mortar should be controlled at 7–8 cm and cement mortar shall be used during mixing; (4) Surface temperature of foundations should be kept at above 0 °C and the temperature of the operation environment should be above 5 °C when cement mortar waterproof layers are constructed; (5) Concave/convex angles of waterproof layer should be in an arc shape. The construction order for an impermeable layer of reinforced concrete digesters is as follows:

(1) **2 mm thick plain cement layer**: once the base is treated and saturated with water, 1 mm thick mixed plain cement should be plastered, followed by strong back-and forth scraping for 5–6 times with iron trowel so that the plain cement can fill the pores on the base. The plain cement should be scraped evenly, and then another 1 mm thick layer of plain cement should be plastered. Leveling should be performed to ensure even thickness. The trowel shall not touch the base.

(2) **4 mm thick cement mortar layer (1:3)**: before plastering the second layer, the first layer should be sequentially and uniformly brushed with water. Construction of the second layer of mortar is carried out during the initial setting of the plain cement layer, so plastering should be generally performed to avoid any damages to the plain cement layer. It must be ensured that the thin mortar layer can be plastered onto the plain cement layer, so that the first and second layers can be firmly integrated. After the second layer is done, horizontal stripes shall be carved with tools such as a broom. When sweeping, water should not be added to ensure that the second and third layers can also be combined firmly.

(3) **2 mm thick plain cement layer**: after the second layer of mortar completely stiffens, a suitable amount of water can be sprayed in a way similar to that done to the first layer. When constructing the third layer, if calcium hydroxide is found separated out from the second layer of mortar during hardening and a white membrane forms on the surface, the surface must be brushed by the root of Chinese iris and cleaned with water before the third layer is constructed.

(4) **4 mm thick cement mortar layer (1:2)**: plaster the mortar the same way as what has been done on the second layer. During water evaporation before the mortar is set, use iron trowel to plaster 5–6 times to increase the compactness, followed by a surface plastering with an iron trowel. The time interval for each plastering should be determined according to humidity and temperature of the construction site. Generally, the time interval is shorter when plastering for the first three time. The whole plastering process is usually within 12 h in summer and no more than 14 h in winter in order to avoid the sand precipitation from stiffened mortar due to repeated plastering.

(5) **Waterproof layer**: after the fourth layer of mortar, the fifth layer of plain grout should be plastered the same way as the fourth-layer plastering.

The top (and biogas chamber) of reinforced concrete digesters should also be applied with sealcoat, and coatings that are corrosion-resistant, non-toxic, of good sealing properties, and less irritant should be chosen. Moreover, the coatings should be able to resist high temperatures of at least 80 °C. The interarea of the sealcoat must be clean, dry, and free of scum, water droplets. Proportions of coating ingredients and coating processes should strictly follow the technical requirements. Painting trials are required, and the actual coating is carried out only if the trials have met all requirements. Painting must be even, and at least twice should be done. Coating of a layer should not be started unless a membrane has formed on the previous coating layer; brushing shall be done in the direction perpendicular to that of the previous layer.

5.3 Digester with Steel Structure

Digesters with steel-structure are widely used in mega and large-scale biogas plants in China. The major structures include the welded steel sheet structure, the steel sheet assembly structure, and the spiral double fold hem structure.

5.3.1 *Digester with Welded Steel Sheet Structure*

A digester with welded steel sheet structure is one processed with steel sheets commonly available in market through molding, welding and anti-corrosion processes. Its advantages include mature technology, on-site production no need of special equipment and tools, and diverse choices of manufacturing materials. The disadvantage is that the anti-corrosion technology is relatively complicated. According to the shape of digesters, digesters with welded steel sheet structure can be classified into vertical digesters and horizontal digesters. According to the design requirements of anaerobic digestion process, the horizontal digesters can be used for anaerobic reactor with a volume smaller than 200 m^3, while the vertical digesters can be used for anaerobic reactor with a volume larger than 200 m^3 and is usually built on the ground.

According to the structure of digester top, vertical cylindrical steel digesters can be classified into arch-top digesters and cone-top digesters, as shown in Fig. 5.9. The mechanical property of arch-top digesters is better than cone-top digester, but cone-top digesters are more widely used in biogas plants due to its convenient installation of external insulation layers and protective layers.

A digester with welded steel sheet structure is composed of a bottom, wall and top.

Digester bottom: it is assembled by several pieces of steel sheets, and the way of arrangement usually depends on design. The steel sheets in the middle of the digester bottom is called the middle plate, which adopts the form of overlap weld. The steel sheets along the bottom rim is called the edge plate (side plate), where butt welding shall be adopted. Edge plates can be strip plates or bow plates, depending on digester diameter, volume, and material of the first segment of wallboard welded to the bottom slab (Figs. 5.10 and 5.11).

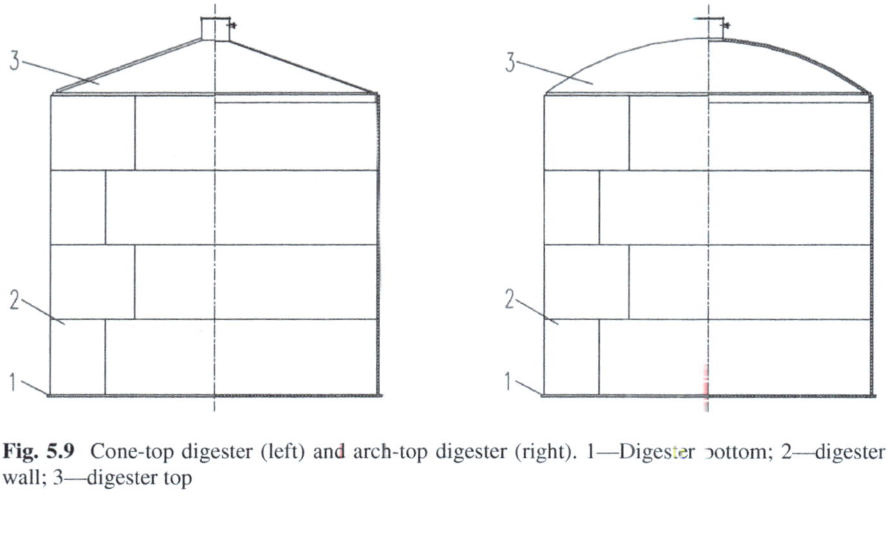

Fig. 5.9 Cone-top digester (left) and arch-top digester (right). 1—Digester bottom; 2—digester wall; 3—digester top

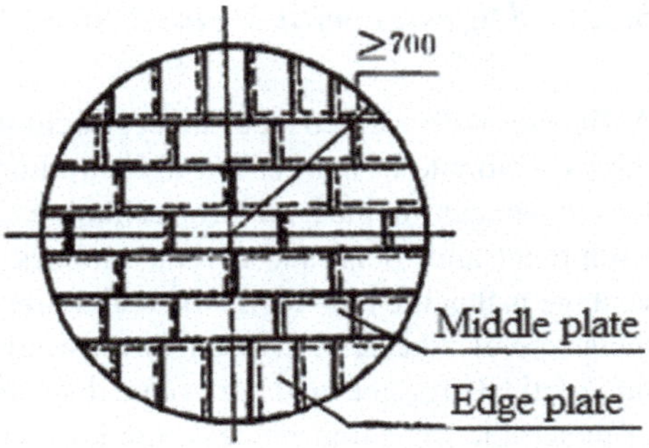

Fig. 5.10 Bottom with strip plates

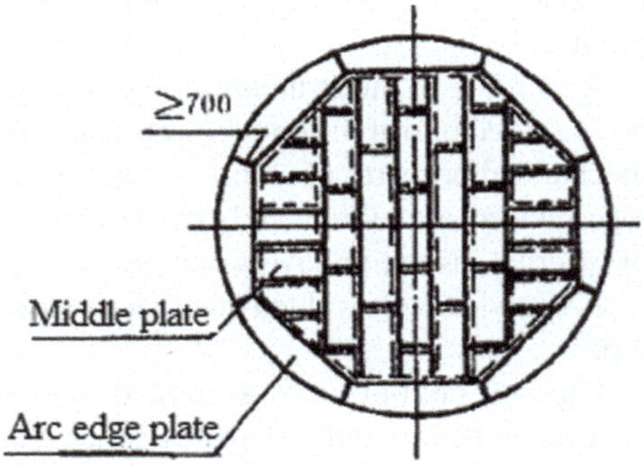

Fig. 5.11 Bottom with arc edge plates

Digester wall: it is formed by butt welding of multi-ring steel sheets, and the thickness of the steel sheets gradually decreases from bottom to top along the wall. The minimum thickness is 4–6 mm. At present, due to the development of installation technology, butt welding is mainly used for digester wallboards whereas lap welding is seldom used.

Digester top: It has a fan-shaped digester top consisting of a number of compressed thin steel sheets with a thickness of 4–6 mm and stiffeners (usually angle steel or flat steel) or a combination of frame and thin steel sheets. Overlap welding is used for tops of various fan-shaped tanks.

There are two installation methods for the digester with welded steel sheet structure: the forward installation method and the backward installation method. The forward installation method is to take the digester bottom as the datum plane, and the digester wall plates are installed upward from the first segment of the bottom, block by block and step by step. The backward installation method is to take the digester bottom as the datum plane, but top ring wallboards and the digester top shall be installed first. Wallboards are then assembled, welded, and jacked-up in turn alternately from top to bottom and ring by ring, until the wallboard at bottom ring is installed completely. Installation companies overseas mostly adopt the forward installation method, while Chinese installation companies mostly adopt the backward method. The backward methods involve center pillar lifting, gas lifting, manual chain block lifting, electric chain block group lifting, hydraulic lifting methods, etc. At present, manual chain block lifting method and electric chain block group lifting method are mostly used.

1. *Digester materials*

Steel sheets for digester welding shall be free of bubbles, cracks, scars, tears, and impurities on the surface, and the steel sheets shall not be stratified. Wallboard materials of digesters should be Q235 and Q345, and materials of angle steels and accessories should be Q235. Materials of seamless steel tube should be 20#. In special occasions, 304 or 316 stainless steel could be used. The thickness of steel sheets shall meet the requirements in Table 5.2.

Local defects on surfaces of welded steel sheets should be removed by grinding, and the thickness of steel sheets should be no less than the minimum allowable thickness.

The use conditions of the digester, the welding properties of materials, processing and manufacturing, as well as economic rationality must be considered for steel used for a digester with welded steel sheet structure. Refer to Tables 5.3 and 5.4 for steel and welding materials.

2. *General requirements for digester structure*

Table 5.2 Thickness requirement for steel sheets

Name	Nominal size		Allowable deviation
Steel sheet thickness	≥3 mm	3 mm	−0.1 to +0.3 mm
		3–6 mm	−0.35 to + 0.3 mm
		8–12 mm	−0.5 to +0.3 mm

Table 5.3 Application scope of steel sheets

S/N	Steel No.	Steel standard	Application scope		Remark
			Permissible temperature (°C)	Permissible maximum plate thickness (mm)	
1	Q235-A.F	GB700	>−20	8	It is only used for digester top at −20 to 0 °C
		GB3274	>0	12	
2	Q235-A	GB700	>−20	16	
		GB3274	>0	34	
3	Q245R	GB6654	>−20	34	
4	Q345A	GB1591	>−-20	12	
		GB3274	>−10	20	
		GB6654	>−20	34	

Table 5.4 Welding materials

S/N	Type	Manual arc welding	CO_2	Submerged arc automatic welding	
		Welding rod type	Protective welding wire	Welding wire	Weld flux
1	Q235-A.F	E43	H08Mo2SiA	H08A	HJ300
2	Q235-A	E43	H08Mn2SiA	H08AMnA	HJ300
3	Q245R	E4315	H08Mn2SiA	H08AMnA	HJ300
		E4316			
4	Q345A	E5015	H08Mn2SiA	H10MnSi H10Mn2	HJ500 HJ501
		E5016			
		E5018			

The working pressure of a digester is usually constant or close to being constant. The positive pressure is generally 3–6 kPa, and the negative pressure should be no less than 0.49 kPa. The working temperature is generally 10–60 °C. Negative deviation and corrosion allowance of steel sheets should be taken into account to the additional thickness of the structural steel sheets.

The volume of a digester is determined by the quantity of feedstock or biogas output and volumetric organic load or volumetric biogas production rate of an anaerobic digestion process. The manufacturing difficulty and economical efficiencies of digesters for the same volume may be different if different diameter and height are adopted, and this fact must be considered in biogas plant design. From the aspect of material use, materials can be greatly saved when the diameter-height ratio of a vertical cylindrical digester is 1:1. The wider the steel plates are, the less welding is necessary during digester manufacturing, and the possibility of weld leakage is accordingly reduced. Meanwhile, the manufacturing process speeds up and the labor cost of welding is saved.

Digester bottoms should be made of steel sheets by assembling and welding to form a unity with the digester body. When the inner diameter of digester is smaller than 12.5 m, strip-shaped plates are desirable for the digester bottom. When the inner diameter of digester is no smaller than 12.5 m, bow-shaped edge plates are recommended, as shown in Figs. 5.10 and 5.11.

Fillet welding is adopted at bottom of digester wallboard and bottom plate with continuous welding on both sides. All welds should be firm and reliable and free of cracks, slag inclusion, burning-through and some other defects. Leakage of weld is not allowed for weld assembly where sealing is required. Drill holes of digester wallboards should follow the design requirements. After drilling, surface finish of holes, soldered dots, and burrs should be performed to meet Ra25 texture requirements. During drilling, all layers of steel sheets should be free of deformation and defects. For angle-cutting of digester wallboards, a line with an angle-cutting template shall be marked, and a plasma cutting machine shall be used to cut angles. Sharp burrs on the surface shall be grinded. The steel sheets and angle steel that need bending

Table 5.5 Minimum nominal thickness of digester wall

Tank diameter (m)	Minimum nominal thickness of digester wall (mm)	
	Carbon steel	Stainless steel
D ≤ 16	5	4
16 < D≤35	6	5

shall strictly follow the requirements in their drawings. Radial deviation per meter in the circumferential direction should be no more than 3 mm. Thickness of the steel sheets of digester wall should meet the requirements in Table 5.5.

The digester wall shall be provided with a reinforcing ring as required in the design. The reinforcing ring can be used as a structure and also used for easy installation of heat insulation materials of digester wall. It can also be used to fix the protective layer of the heat insulation materials. In order to prevent deformation at the top of digester wall, edge-wrapped angle steel should be installed.

Self-supporting fixed tops, which include non-reinforced rib tops and reinforced rib tops, are usually used for this kind of digester. A reinforced rib top consists of 4–6 mm thick steel sheets and reinforced ribs (usually made of flat steel), or it consists of an arch frame (made of section steel) and steel plates. The specified thickness of digester tops (excluding corrosion allowance) and specified thickness of the supporting member (excluding corrosion allowance) should be no less than 4.5 mm.

3. *Digester anti-corrosion*

The corrosion resistance consideration of a digester with welded steel sheet structure and its auxiliary facilities shall be based on medium corrosion strength. Rust shall be removed from surface weld joint (excluding galvanized materials) mechanically or manually. The cleaned surface must be free of invisible grease, dirt, oxide scale, rust, and paint coating. The treated surface should be applied with anti-rust primer promptly, generally within 6 h. Anti-rust primers shall be the epoxy or zinc-rich type, such as epoxy zinc-rich paint and epoxy coal asphalt paint. The treated surface should be properly protected to prevent re-rusting and contamination if primer cannot be applied to the surface immediately. Any rust and contamination found on the surface should be re-treated, then paint and anticorrosion are conducted.

Construction method of paint anti-corrosion coatings: After the paint is diluted, it is evenly applied from top to bottom with a roller. The total thickness of the applied membrane is 0.15–0.20 mm, and the coating can be done in two or three steps. Painting on the external wall surface of a digester should match well with the primer. Finishing coat is not required for the external wall when heat insulation layer is available nor for the internal wall of a digester.

4. *Digester Foundation*

The reinforced concrete bottom slab is generally used as digester foundation. This type of foundation is often used in the bearing layer of natural foundation where the bearing capacity is not very large (compared to the upper load), and it is also called a raft foundation. The foundation of a digester should meet the design requirements and specifications. Inspection of installation and acceptance of handover of civil works should be appropriately done, and inspection and acceptance should be well recorded. The allowed deviation of foundation center elevation should be kept within ±20 mm. The allowed deviation of concave/convex degree on foundation surface should be no more than 25 mm. The allowed deviation of outer diameter of a circular foundation should be ±50 mm.

5.3.2 Digester with Spiral Double Fold Hem Structure

A digester with spiral double fold hem structure is commonly known as a "LIPP Tank". This kind of digester is fabricated through the process of spirals, double folds, and hems by special rolling, linking, pressing and molding equipment based on the principle of metal plastic work-hardening and thin shell structure. The spiral double fold hem technology for tank manufacturing has the advantages of short construction period, steel saving, light weight, and a service life of more than 20 years. However, special equipment is required for manufacturing. The steel sheets used is not of general specifications on the market, and the volume of the digester cannot be too large.

In digester manufacturing by the spiral double fold hem technology, the steel sheets form plane connection inside the digester and spiral rising and continuously linking the bars outside the digester by linking the upper and lower layers. Through a molding machine and a linking machine, steel sheets are folded into hooks at the upper and lower steel sheets on the molding machine and are linked together on the linking machine with the folded hooks. The molded circular body rises spirally on the bracket. When the required height is reached, the upper and lower ends shall be cut flat to complete the manufacturing of digesters with spiral double fold hem structure. Although it is a thin-wall structure, such digester has a very large ring-tension strength due to its linking bars with equal spacing (Cai 1997).

1. *Digester rolling equipment*

Equipment for digester manufacturing by the spiral double fold hem technology includes the following parts: an uncoiler that is to unfold coiled steel plates; a molding machine that bends the uncoiled steel sheets and conduct preliminary processing and molding; a bending machine that bends, links, rolls and binds the steel sheets processed by the molding machine to form a spiral cylinder; load-bearing brackets that carry the spiral rising cylinders; a high-frequency stud welding machine that reduces damages to materials during welding. There are mainly two kinds of machines for spiral double fold hem technology, and the model and application scope are shown in Table 5.6. The SM-type model is composed of an uncoiler, a molding machine, and

Table 5.6 Machine for the spiral double fold hem technology

Model	Applicable thickness for steel sheet (mm)	Steel sheet width (mm)	Bar spacing (Mm)	Digester diameter for manufacturing (M)
SM30	2–3	495	380	3.5–18
SM40	3.5–4	495	360	4–18

a flanging machine, and is equipped with conveyor rack during construction (Tang and Wang 2013).

2. *Materials for digester body*

Materials for digester are generally 495 mm wide and 2–4 mm thick galvanized steel sheets or stainless steel-galvanized clad steel plates. For stainless steel-galvanized clad steel plates, the galvanized steel sheets shall meet the strength requirements, and the stainless steel membrane clad layers shall meet the anticorrosion requirements. Because of the existence of screw linking bars, the steel sheets possess a large strength against circumferential tensile stress. In theory, the steel sheet thickness of a digester can be less than 2 mm, but considering structural stability, the sheets to be used shall generally be thicker than 2 mm. Moreover, due to the limitation of linking tightness and compression strength of process machinery, sheets to be used are generally no more than 4 mm. For digesters of considerable sizes, steel sheets with stronger mechanical properties can be selected to make the materials thickness ≤4 mm. The current standard for galvanized steel sheet is GB/T 2518-2008, *Continuously Hot-Dip Zinc-Coated Steel Sheet and Strip*, and the current standard for stainless steel clad steel is GB/T 8165-2008, *Stainless Steel Clad Plates and Strips*.

3. *Digester foundation*

Because of low materials amount requirements for digesters with spiral double fold hem structure, the requirements for foundation bottom slab of such digesters are far lower than those of reinforced concrete digesters. When a foundation bottom slab is poured, a preformed groove of 150 mm wide and 100 mm deep shall be reserved on the bottom slab surface according to the diameter of digester to be made. A certain number of anchor-shaped stainless steel embedded parts shall be evenly placed in the groove according to the digester diameter so that the accomplished digester fits perfectly into the reserved groove. The digester and the embedded parts will be fixed with bolts, and then the groove will be sealed with expanded concrete, asphalt, mat, etc. Finally, the groove will be covered by a protective layer of fine stone concrete, as shown in Fig. 5.12 with details. Other requirements for digester foundation are the same as those for foundation of digester with steel plate welding structure.

4. *Manufacturing process*

The manufacturing sequence for digesters with a spiral double fold hem structure is as follows: starting with the digester top, then rolling the wall of digester from top to bottom, and finally settling and fixing the digester.

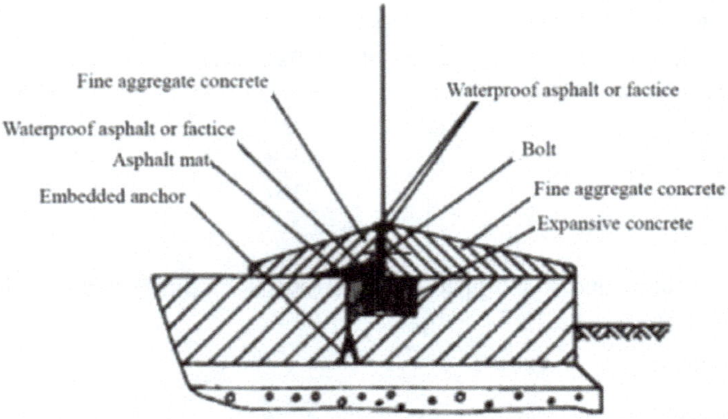

Fig. 5.12 Location and sealing of preformed groove of digester with spiral double fold hem structure (Lin 2000)

(1) Site preparation: Reserve a clearance of about 1 m inside and outside the digester wall and construct a foundation platform to be used as the equipment installation and operation space.
(2) Equipment installation: Install the load-bearing brackets, uncoiler, molding machine and bending machine.
(3) Digester manufacturing, linking and molding. See Fig. 5.13 for the molding process of the linking bar with a cross-sectional view.

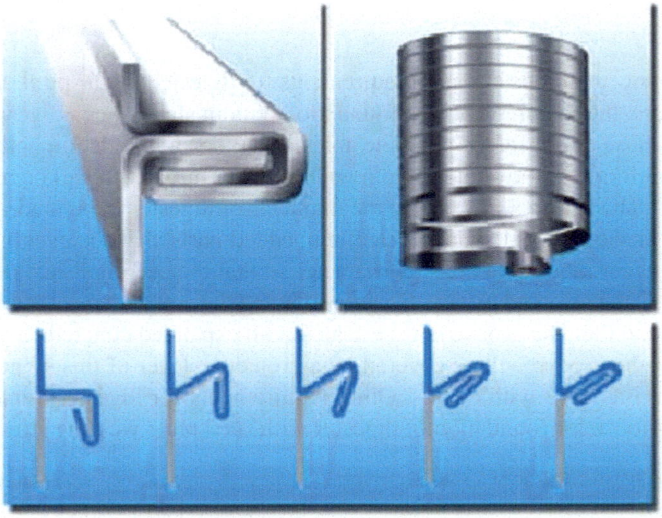

Fig. 5.13 Molding process of the linking bar and its cross section

(4) Installation of digester top. When the digester wall is rolled up to 1 m high, stop linking and rolling. Cut the upper end flat and install the top eaves and the digester top.
(5) Rolling and lifting of the tank. Continue to roll the tank after installation of digester top. lift the digester wall while rolling until reaching the target height;
(6) Settling and fixing the digester wall. Cut the digester wall bottom flat, pull out the equipment inside the digester and settle the digester wall. Fix the digester wall and the embedded parts of the digester bottom by bolts, and then conduct secondary pouring before sealing.
(7) Installation of process pipe holes

    It is quite simple and convenient to prepare process pipe holes and manholes on digesters with a spiral double fold hem structure. It is only needed to open holes in determined positions, and then install the prefabricated process pipe holes and manholes at the openings. The inner laminate is lined with a rubber sealing plate, and the inner and outer laminates are fixed with the digester wall by bolts. In order to prevent water seepage at bolt holes, a special sealant should be applied to the bolt holes during installation.

    Multiple methods are available for preparing process holes on digester top The general method is to clamp a 0.6–0.8 mm thick stainless steel sheet between the upper and lower coping channel steel. A rubber sealing ring is added between the stainless steel sheet and the lower coping channel steel, and the sheets are fixed by bolts at digester top with the special sealant applied on bolt holes and joints. See Fig. 5.14 for details.
(8) Anti-Corrosion

Although digesters with a spiral double fold hem structure made of galvanized steel sheets are anti-corrosive to some extent, the galvanized layer attached to the surface of steel sheets is not enough to resist corrosion from liquid and gas, especially at the opening and the welding joints of fixtures for installation of platform, railings, insulation layers, etc. The galvanized layers on steel sheets get easily damaged, so anti-corrosion treatment is still required after completion and passing the tests. Similarly, thermal insulation structure is also needed for digester with spiral double fold hem technology. Anti-Corrosion treatment for digesters with a spiral double fold

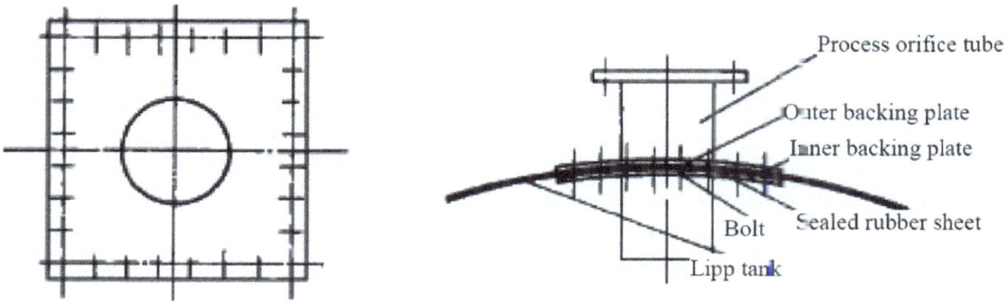

Fig. 5.14 Installation of process pipe hole

hem structure is the same as the anti-corrosion treatment for digesters with welded steel sheet structure.

5.3.3 Digester with an Assembled Enameled Steel Sheet Structure

Such digester has a thin shell structure and is made of pre-fabricated flexible enameled steel sheets assembled by bolt connection and rubber sealing. It is therefore also referred to as "assembled enameled steel sheet digester" or "assembled enameled digester". The baseplate material of an enameled steel sheets is a cold-rolled plate made of low-carbon steel, with a yield strength $\leq$280 MPa and a tensile strength of 270–410 MPa. Enamel glaze is made by high-temperature firing of various inorganic raw chemical materials. Enameled steel sheets are made by calcinations after enamel coating on the surface of the steel sheets. Assembled enameled steel sheet digester have the advantages of good corrosion resistance, short construction period, steel economy, light weight, and easy disassembly. Its disadvantage is that leakage is made possible by bolt connection assembly method, and adjustment of holes opening locations on construction site is not convenient.

1. *Digester body material: enameled steel sheet*

The manufacturing process of an enameled steel sheet digester starts from selection of steel sheets. The steel sheets are then cut, pressed, punched, treated on surface, sprayed with enamel glaze on both sides. Finally, an enamel firing at 900 °C is conducted in an electric furnace. Therefore, the steel sheets to be selected must have the following properties (Tang and Wang 2013):

(1) The surface must be smooth and flat without mechanical damage, delamination, oxide scale or rust spot, and other attachments, and the surface shall not crack during deformation of steel sheet.
(2) Chemical composition of steel sheets influence enamel quality. Cold-rolled low carbon steel sheets and strips must meet the requirements of GB/T13790-2008.
(3) Physical and mechanical properties of metals may change a lot during cold plastic deformation, so the enameled steel sheets for punching or pressing must have good plasticity.
(4) Generally, cold-rolled low-carbon steel sheets and strips used for enamel should have a yield strength less than 280 MPa and a tensile strength 270–410 MPa.

Enameling methods can be classified into wet enameling and dry enameling, with wet enamel glaze being the ingredients for the former and dry enamel powder for the latter. Wet enameling method is more commonly used. Enameling methods can also be classified into manual enameling, mechanical enameling, and electrophoretic enameling according to the operation mechanism. According to the properties and functions of various chemical raw materials, enamel glaze can be classified into matrix agent, fluxing agent, opalizer, adhesion agent, colorant, etc.

(1) Matrix agent is the main component of enamel glaze, accounting for 40–60% of total enamel glaze. Matrix agent generally has high chemical stability, thermal stability, hardness, tensile and compressive strength. The main components of enamel matrix agent are silicon dioxide, boron trioxide, etc.
(2) Fluxing agent promotes melting of enamel glaze and improves its technology and physicochemical properties. The commonly used fluxing agent includes lithium carbonate, soda, sodium nitrate, etc.
(3) Opalizer contributes to the good covering ability of enamel glaze. The commonly used opalizers include titanium oxide, antimony oxide, zirconium oxide, strontium oxide, etc.
(4) Adhesion agent enables enamel glaze to combine firmly with bases, and some common adhesion agent includes cobalt oxide, nickel oxide, copper oxide and antimony oxide.
(5) Colorant. Coloring of enamel is mainly from selective absorption of visible light by the heavy coloring ions or particles of enamel. Colorant gives various colors to enamel glaze. The commonly used colorants include copper oxide, chromium oxide, cobalt oxide, nickel oxide, etc.

Enamel layer thickness is 200–350 μm for main surface of the enameled steel sheet, and 100–150 μm for edges of holes.

2. *Specifications and dimensions of enameled assembly digesters*

For the assembled enameled steel sheet digester, the key to saving material is a reasonable overall design, and unification of the steel sheet specifications is necessary. According to Chinese steel plate specifications, size of enameling equipment, and economy of the overall assembly, the plate size is suggested be length × width = (2–2.8 m) × (1–2 m). Refer to Table 5.7 for digester volume, diameter, and height.

3. *Digester assembly*

Assembly of enameled steel sheet digester are generally proceeded from up to down. That is, digester top is assembled first, followed by the first lap of digester wall, and the next lap downward is then assembled. Special machinery or ordinary scaffolding can be used for installation. For assembling of enameled steel sheets, steel sheets can be overlapped and fastened by self-locking bolts. The overlap surface of steel plates shall be provided with sealant. The surface of self-locking bolts is protected by acid/alkali-resistant polypropylene engineering plastics to avoid any corrosion from liquid mixtures in the digester, and the thread is sealed with thread adhesive. Standard bolts of graded no less than 4.8 with rubber cap can be used. Color galvanizing is required on bolts, nuts and gaskets for anti-corrosion. Neutral weatherproof silicone sealant is used. Connection between the digester top and the wallboard is shown in Fig. 5.15. The bolt sealing of wallboard connection surface is shown in Fig. 5.16, and the bolt sealing at the stirrup is shown in Fig. 5.17.

4. *Digester bottom installation*

Connection between the foundation and the lowest lap of wallboard of an enameled assembly tank is relatively complicated, because the hydrostatic pressure is the

Table 5.7 Specifications/model list of enameled assembly digester

S/N	Diameter (m)	Height (m)	Single-ring plate quantity (Nr.)	Single plate gauge (m)	Enameled plate thickness (mm)
1	3.82	3.6–9.6	5	2.5 × 1.25	3.0
2	4.58	3.6–20.4	6	2.5 × 1.25	3.0–6.0
3	5.53	2.4–20.4	7	2.5 × 1.25	3.0–6.0
4	6.11	2.4–18.0	8	2.5 × 1.25	3.0–6.0
5	6.88	2.4–20.4	9	2.5 × 1.25	3.0–6.0
6	7.64	2.4–18.0	10	2.5 × 1.25	3.0–6.0
7	8.40	2.4–16.8	11	2.5 × 1.25	3.0–6.0
8	9.17	2.4–16.8	12	2.5 × 1.25	3.0–6.0
9	9.33	2.4–15.6	13	2.5 × 1.25	3.0–6.0
10	10.70	2.4–14.4	14	2.5 × 1.25	3.0–6.0
11	11.46	2.4–13.2	15	2.5 × 1.25	3.0–6.0
12	12.22	2.4–12.0	16	2.5 × 1.25	3.0–6.0
13	12.99	2.4–12.0	17	2.5 × 1.25	3.0–6.0
14	13.75	2.4–10.8	18	2.5 × 1.25	3.0–6.0
15	14.51	2.4–10.8	19	2.5 × 1.25	3.0–6.0
16	15.28	2.4–9.6	20	2.5 × 1.25	3.0–6.0
17	16.04	2.4–9.6	21	2.5 × 1.25	3.0–6.0
18	16.81	2.4–9.6	22	2.5 × 1.25	3.0–6.0
19	17.57	2.4–8.4	23	2.5 × 1.25	3.0–6.0
20	18.33	2.4–8.4	24	2.5 × 1.25	3.0–6.0
21	19.10	2.4–8.4	25	2.5 × 1.25	3.0–6.0
22	19.86	2.4–8.4	26	2.5 × 1.25	3.0–6.0
23	20.63	2.4–7.2	27	2.5 × 1.25	3.0–6.0
24	21.39	2.4–7.2	28	2.5 × 1.25	3.0–6.0
25	22.15	2.4–7.2	29	2.5 × 1.25	3.0–6.0
26	22.92	2.4–7.2	30	2.5 × 1.25	3.0–6.0
27	23.68	2.4–7.2	31	2.5 × 1.25	3.0–6.0
28	25.21	2.4–6.0	33	2.5 × 1.25	3.0–6.0
29	28.27	2.4–6.0	37	2.5 × 1.25	3.0–6.0
30	30.56	2.4–4.8	40	2.5 × 1.25	3.0–6.0
31	32.85	2.4–4.8	43	2.5 × 1.25	3.0–6.0
32	35.91	2.4–4.8	47	2.5 × 1.25	3.0–6.0

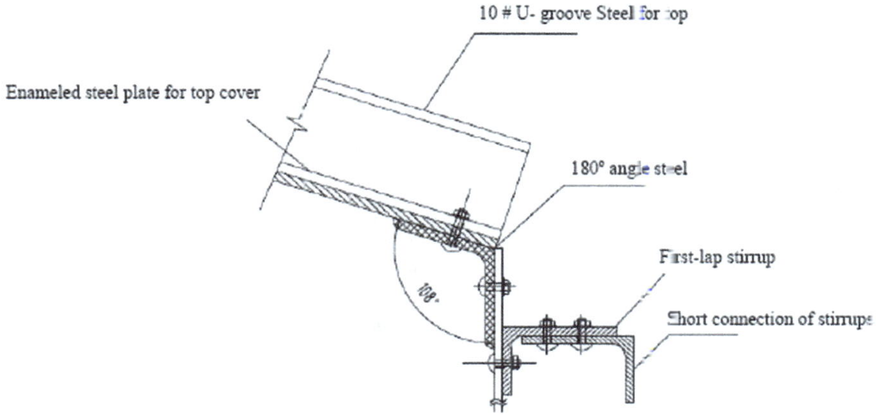

Fig. 5.15 Connection of digester top and wallboard

Fig. 5.16 Bolt sealing of wallboard connection surface

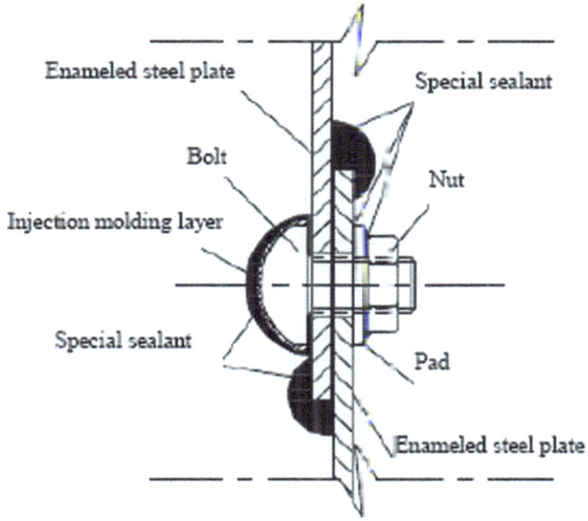

greatest at the connection. Moreover, this connection is also where different materials converge and is therefore susceptible to leakage. The treatment of the connection is as follows: use bolts to fix the lowest lap of wallboard and the foundation stirrups (made of angle steel), then use expansion bolts to fix the foundation stirrups on the digester foundation. Conduct a waterproofing treatment with factice on the contact surface between foundation stirrups and the foundation, then make a reinforced concrete slope protection inside the digester, while making reinforced concrete ring beams outside the digester. Ensure there is no leakage at the bottom and the digester is stable, as shown in Fig. 5.18. Other requirements of digester foundation are the

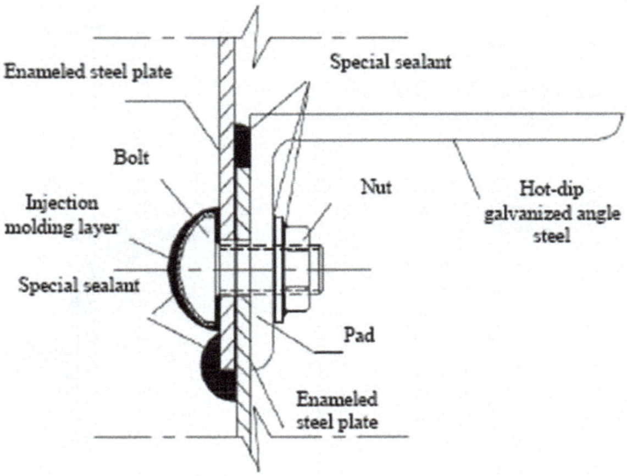

Fig. 5.17 Bolt sealing at a stirrup

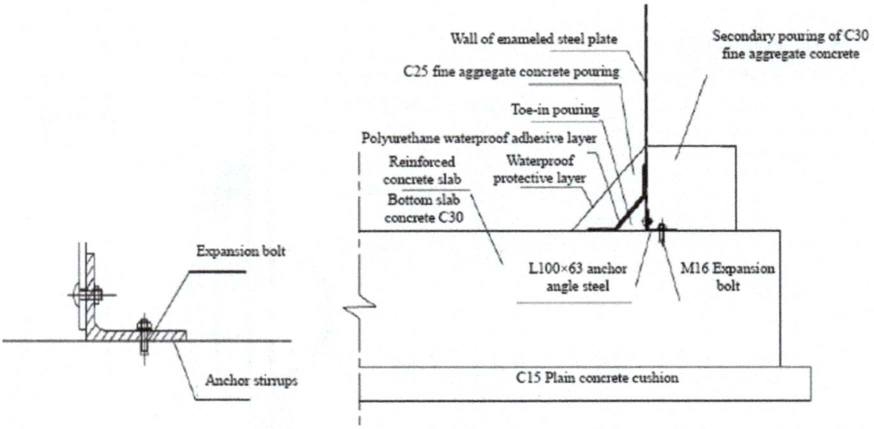

Fig. 5.18 Bottom anchoring of enameled assembly digester

same as requirements for foundation of digester foundation with welded steel sheet structure.

5.3.4 Auxiliary Structures and Facilities of Steel Digesters

Auxiliary structure and facilities of digesters generally include nozzle opening, platform, ladder, and railings directly welded on the wall or top of digesters.

Nozzle openings: openings on digester wall, the distance between the nozzle opening edge and the butt weld lines of the digester wallboard should be more than 8 times the thickness of wallboard and no less than 250 mm. The nominal pressure of steel flanges for nozzle openings should be no less than 0.6 MPa for digester wall and no less than 0.2 MPa for digester top. The equivalent area method is used for reinforcement of openings. Reinforcement is not considered when the opening diameter D $\leq$ 100 mm. The materials and thickness of the reinforcing rings shall be the same as those of the digester body. Manholes on a digester is made for easy access in maintenance, so manhole diameter should be no less than 600 mm. Manhole on a stainless steel digester should be made of stainless steel as well, with a diameter of at least 500 mm.

Platform: Platforms of a digester are mainly used for operation, maintenance and personnel access, and platforms include service platforms, staircase platforms, and assess platforms. Steel grating plates are the most ideal platform plates, but alternatives such as corrugated steel plates and similar anti-skid plates are acceptable. Thickness of steel plates is usually 4.5–6 mm, and appropriate drainage holes should be opened in case of liquid accumulation on platform surface. Service platforms are designed to facilitate maintenance and operation, and the installation location and size should be based on easy access. The design load should be 4000 Pa. The staircase platform is designed for turning and resting, and the design load should be 3500 Pa. The assess platform is designed for personnel to walk, so its width should be no less than 700 mm and railings should be provided, and it should withstand an equivalent uniform load of 2000 Pa.

Staircase: Two kinds of staircases (straight and spiral staircases) are available for digesters. Width of both should be no less than 600 mm. The whole staircase should be able to withstand 5000 N centralized live load. Each step of staircase should be able to withstand 1500 N centralized live load. The step height should be 200–250 mm, and the step spacing must be the even. In case of straight staircases, a safety cage should be set up over 2 m from the ground. In case of spiral staircases, the stepping board shall be made of corrugated steel plates, steel grating plates, or similar anti-skid plates. The ascending angle of a spiral staircase should be no more than 45°, and the minimum width of the stepping board should be 200 mm.

Railings: Railings on a digester is designed for safety protection of personnel. The size of railing columns should be no less than DN40. The spacing of columns should be no more than 1.5 m. The height should be no less than 1.05 m, and the bottom edge of the column should have a baffle to prevent objects and feet from sliding out. Railings should withstand a concentrated load of 1000 N acting at any point and in any direction on top.

5.4 Digester with a Membrane Structure

Such digester is called a covered lagoon in some other countries and is commonly known as a "black-membrane digester" in China. It is an anaerobic digestion reactor consisting of a bottom membrane and a top membrane by sealing the high-quality membranes on an excavated earthwork foundation. Wastewater inlet/outlet, residues extraction pipes, and biogas collection pipes are installed in the digester according to the requirements of anaerobic digestion process. Generally, such digester is built underground or semi-underground and have integrated digestion and biogas storage. The whole digester is completely sealed by the anti-seepage membranes. See Fig. 5.19 for reference.

Digesters with a membrane structure has the advantages of simple construction, a short construction period, low costs, a simple process, easy operation and maintenance, long hydraulic retention time, effective digestion, and a capability of utilizing geothermal heating to heat up and insulate. Anti-seepage membranes can replace conventional digester-construction materials (e.g. concrete, steel plate, brick, etc.) at a low cost to solve the shortage of digester-construction materials. Meanwhile, digesters with a membrane structure can also solve the problems of water seepage/leakage and gas leakage caused by shrinkage and expansion of digesters with a concrete structure due to temperature changes, as well as some easily occurred problems such as steel sheet corrosion, pipeline blockage, equipment damages, and high operating costs.

However, digesters with a membrane structure also have many shortcomings, including a large volume, hydraulic retention period of more than 40 days, a large land occupancy, difficulty to increase temperature, and difficulty to achieve balanced gas production due to digestion at ambient temperature. Therefore, such digesters are not suitable for the northern China region. Due to difficulty in installation of mechanical mixing devices, such digesters are not good at dealing with easily crusting or high fiber-content raw materials. Solid matters are easy to sink to the bottom, the area of which is large, so sludge discharge is also difficult. Protection is required for potential explosion risks from friction and static electricity. Large area increases the possibility of leakage by poor construction quality or man-made damages.

1. *Materials for digesters with a membrane structure*

Such digesters easily leak and has safety risks. Therefore, it is necessary to use membranes with good sealing performances, a high tensile strength, strong anti-aging properties, corrosion resistance, and good anti-seepage effects. Construction

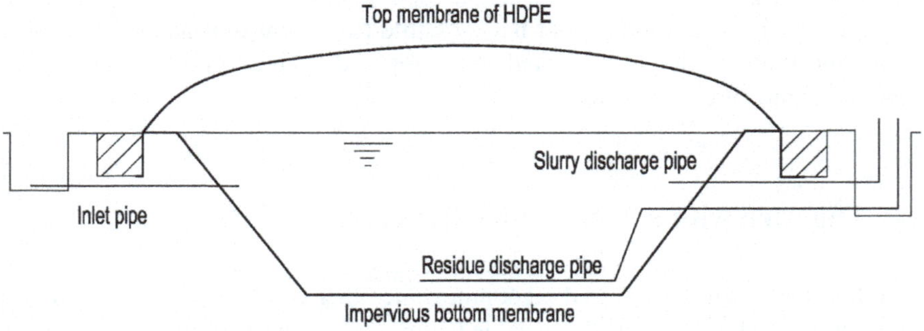

Fig. 5.19 Digester with membrane structure

materials for such digester mainly include HDPE (High Density Polyethylene) membranes, EPDM (Ethylene Propylene Diene Monomer) membranes, PPR (polypropylene random), etc. According to the features of materials and digester construction experiences in china and other countries, environmental-friendly HDPE (High Density Polyethylene) membranes (hereinafter referred to as HDPE membranes) are generally used to construct this type of digester.

When HDPE membranes are used to build the digester, the thickness of bottom and top membrane should be ≥1.5 mm. The bottom should be made of rough-surface membrane, and the top should be made of smooth-surface membrane. In appearance inspection, the cut should be flat and straight without obvious sawtooth trail. Perforation repair points are not allowed. Mechanical (processing) scratches or obvious scratches are not allowed. There should be no more than 10 hard lumps per square meter. When the thickness is ≤2 mm, any hard lump that runs through the membrane is not allowed. Bubbles and impurities are not allowed. Cracks and stratification should not occur. The appearance should be uniform. Knots or missing parts are not allowed.

The tensile properties, 90° tearing strength, puncturing strength, environmental stress cracking resistance, oxidation induction time, and aging and ultraviolet resistance of HDPE membranes should meet the requirements of CJ/T 234-2006, *High Density Polyethylene Geomembrane for Landfills*.

2. *Construction of digesters with a membrane structure*

The basic construction process of such digester is in the following sequence: earthwork excavation → pipeline embedment → anchoring ditch excavation → digester bottom construction (bottom membrane laying) → water leak detection → top membrane laying → gas leak detection → welding edge of bottom membrane and that top of membrane → backfilling of anchorage ditch. When constructing such digester, the civil work location depends on site conditions. Before construction, technical engineers should conduct setting-out on site, and excavators shall work along the setting-out. Digesters are preferably rectangular, and the surrounding slope and digester surface should be flat and uniform. Anchoring ditches shall be excavated to fix the bottom and top membranes to prevent wind blowing.

(1) Pipeline embedment

After the digester body is excavated, blind ditches shall be excavated according to the actual site condition, and intake pipes, outlet pipes, sludge discharge pipes and biogas exhaust pipes shall then be laid. After the bottom and top membranes are laid, the connection between the HDPE membrane and the pipelines should be strengthened and sealed.

(2) Digester bottom construction (bottom membrane laying)

The bottom construction of a digester include the following process: building of a groundwater collection and drainage system, settling of the foundation layer and compact soil protection layer, preparation of the bottom exhaust system, and strengthened sealing of connection of the HDPE membrane and pipelines.

Ground water collection and drainage system: Groundwater collection and drainage system must be set up when groundwater level is high, which is harmful to the stability of the foundation layer at the digester bottom, or when the surface water seepage around the digester is harmful to surrounding slope foundation. Such system should be able to collect and drain groundwater and seeped surface water promptly and effectively, and it should have anti-silting capacity and ensure long-term reliability. Underground blind ditch, gravel diversion layer or geotextiles composite drainage network diversion layer should be used for such system.

Foundation layer and compact soil protection layer: Foundation layer should be flat, compact, and free of cracks, loose soil, surface water, stone, tree roots, and other sharp debris. Compactness should be no less than 93%. Structure of a foundation layer of surrounding slope should be stable, and the compactness should be no less than 90%, with a slope gradient more than 1:2.

Bottom gas exhaust system: Underground soil may produce gas (possibly water vapor) due to temperature changes and some other reasons. Underground gas may cause rising and bulge of the bottom membrane, even breaking the impervious layer at the digester bottom. In order to prevent this situation, underground gas drainage facilities should be provided at the foundation layer or the compact soil protection layer. Gas may be exhausted through gravel-based blind ditches or perforated pipe wrapped with permeable materials such as geotextiles.

HDPE anti-seepage membrane layer: Permeability coefficient of compact soil under an HDPE membrane should be no more than 1×10^{-9} m/s, and the thickness should be no less than 750 mm. Thickness of HDPE membrane should be no less than 1.5 mm. Before installation of HDPE membrane, the protective layer under the membrane should be checked. Flatness error per square meter should not exceed 20 mm. The bottom membrane should be settled after getting welded in factory and fully tested based on the size of the digester to be built. Hot melt welding or extrusion welding should be adopted. The lap width of hot melt welding shall be 100 ± 20 mm, and that of extrusion welding shall be 75 ± 20 mm. The membrane laying work should be done in place at one time. Dragging is not allowed after the laying work. It is necessary to leave spaces for any size change caused by thermal expansion and cold contraction of materials. Appropriate waterproofing and drainage measures should be taken for protective layers under the membrane. Measures shall be taken for aboveground membrane to prevent HDPE membranes from wind damages. Constructors should wear protective gloves. Personnel and tool vehicles should not step directly onto HDPE membranes without protective measures. Protective layers (e.g. non-woven geotextiles) shall be provided on the HDPE membrane. After the digester bottom is constructed, water test is required to ensure there is no leakage at digester bottom.

(3) Laying of top membrane

After the digester bottom has passed the water test, water shall be fully drained from the digester, and the top membrane can be installed. The top membrane should be a 1.5 mm thick HDPE membrane (performance is same as that of the bottom membrane). According to the actual topographic size of the site, the HDPE anti-seepage membrane should be cut and laid sequentially to cover the whole digester. After the top membrane is laid, the bottom and top membranes shall be welded at the edge, embedded and fixed in the anchoring ditch to ensure anchoring is impermeable and will not be pulled out, so as to prevent membrane displacement.

After construction of the whole digester, the gas tightness test should be done to ensure no leakage of digester before material feeding.

(4) Notices in construction

It is strictly forbidden to lay anti-seepage membranes and carry out welding and seaming construction in rainy days. When welding, the base surface should be dry, and the moisture content should be below 15%. The anti-seepage membrane surface should be wiped clean by a dry gauze cloth. Any fire source should not be brought into the construction site. It is strictly forbidden to step onto anti-seepage membranes with nail shoes, high-heels, and hard-sole shoes. Vehicles and other machinery should not roll over the anti-seepage membrane and its protective layer. The welding quality should be guaranteed. The welder should check on the welding quality any time and adjust welding temperature and walking speed according to changes of ambient temperature.

3. *General requirements in operation of digesters with a membrane structure*

After the whole construction of digester is completed, no personnel should not walk on the membrane structure. According to the design requirements, necessary exhaust and emergency combustion treatments should be carried out. Any delay in exhaust may cause the top membrane to be easily pulled out from the anchoring ditch, and a serious delay may cause the top membrane to break and affect the service life of the digester. If the top membrane is damaged and biogas leakage is found, some simple treatments shall be conducted immediately. Such treatments are done using adhesive materials (e.g. adhesive tape) to seal the damaged spots in order to avoid potential safety hazards caused by biogas leakage, and the construction party should be notified promptly for maintenance and treatment.

Because the digesters with a membrane structure have not been used for a long time in China, their service life and safety need to be further explored. Therefore, attention should be paid to monitoring of groundwater and biogas leakage around this type of digester.

4. *Other types of digesters with a membrane structure*

Soft-structure digesters, typically represented by red mud plastic digesters, also occupy a certain part of market in China. Red mud plastic is made as follows: the waste residue ("red mud") after aluminum oxide extracted and mixed it into polyvinyl chloride plastics, which greatly improves the anti-ultraviolet, flame retardant, low temperature resistance, and many other characteristics. One advantage of a soft-structure digester is low cost. The soft-structure digester also has the advantages of anti-corrosion, anti-freezing, anti-seismic, etc. However, its service life is short, which is the main reason that restricts its large-scale application. In addition, the

Fig. 5.20 Digester with red mud plastic

thermal insulation performance of the soft structure membrane is poor, which limits its application in cold regions.

Red mud plastic digesters includes two types: the semi-plastic type and all plastic type. A "Semi-plastic" digester is mainly composed of two parts: a digestion pond and a gas hood. Digestion pond is generally cast on site. It can also be precast or bricked with a flat bottom and straight body. Round cross section types and rectangular (rounded corner) section types are commonly seen. The upper part of the digestion pond is provided with a water sealing groove, which is used to seal the joints between the digestion pond and the gas hood. A fixed hook is provided in water sealing groove to fix the gas hood. The gas hood is made of 0.6–1.0 mm thick red mud plastic membranes, and the gas hood is used to hitch the digestion pond and fix the hood on the hook, and then the water sealing tank is filled with water, which can effectively seal the digester as shown in Fig. 5.20. The construction of an "all plastic" digester is similar to that of digesters with a membrane structure.

5.5 FRP Digester

FRP is fiberglass reinforced plastics made of fiberglass, resin, additives and fillers. FRP is mainly composed of fiberglass and plastics, and its performance is far superior to plastics, featuring in light weight and high strength. Although the specific gravity of FRP is only 1/5–1/4 of carbon steel, its tensile strength is higher than that of carbon steel. FRP has good corrosion resistance and good resistance to common acids, alkalis, salts, various oils and solvents. In addition, FRP is also an excellent electrical and thermal insulation material.

FRP digester is made of fiberglass reinforced plastics (FRP) with glass fiber as the reinforcing material and resin as the matrix. It has the advantages of convenient construction, short construction period, good water and gas tightness, strong corrosion resistance, acid/alkali resistance, freezing resistance, tension/torsion resistance, factory-based and standardized production, no restriction on production environment and construction technology, easy construction treatment in higher-groundwater-level area and quicksand area, little construction difficulty, convenient transportation, long-distance transportation, etc. However, FRP has only about 2% breaking elongation and has brittleness. FRP digesters are slightly higher in construction cost compared to digester with a brick-concrete structure, reinforced concrete structure, or steel-structure. On the other hand, because FRP belongs to petrochemical materials and is affected by oil price, its cost fluctuation is also large.

The main molding methods of FRP include manual plastering, winding, and spraying.

Manual plastering: on the mold where the mold-release agent has been applied, solidified materials (e.g. fiber cloth) are laid manually and applied with resin glue solution until the required thickness is achieved.

Compression molding: Lamellar mixtures made of fibers or mat sheets impregnated in resin paste. Fibers or felt is impregnated with a resin paste to form a lamellar mixture. The lamellar mixture is then placed in a matched mold, and the product is molded at a certain temperature and pressure.

Winding molding: In the condition of controlling tension and presetting line type, wind the continuous filament fiber or fabric (that is impregnated in resin glue) to the core mold or mold to form a finished product. Winding molding is also called continuous fiber winding molding.

Spray molding: spray the prepolymer, catalyst, and staple fiber onto the mold or core mold synchronously to form a finished product.

FRP digesters include FRP household digesters, domes of FRP household digesters, and FRP digesters in biogas plant. FRP household digesters and dome of FRP household digesters are generally made of lamellar molding materials by compression molding, manual plastering, and spray molding, and sometimes they are also made by winding molding process. Small FRP digesters are generally produced by winding molding process.

1. *FRP household digesters and FRP dome*

The volume of an FRP household digester is generally 4–10 m^3. The dome of an FRP digester can match with digesters (6–10 m^3) with digesters of brick-concrete and reinforced-concrete walls. In the past, Manual plastering molding process was generally used, but now most of them are produced by compression molding process. The molded FRP household digesters (Fig. 5.21) and domes of FRP digesters (Fig. 5.22) should be flat and smooth in appearance without obvious scratches, folds, exposed fiber on the outer surface, pinholes, hollow gas bubbles, and heterogeneous and incomplete impregnation defects. The inner surface should be smooth and uniform without obvious gas bubbles. Edges of all components and the connecting parts should be neat, with uniform thickness and no stratification. The whole structure

Fig. 5.21 FRP household digester

Fig. 5.22 Dome of FRP household digester

should comply with regulations of household digester standards, and should meet the requirements of biogas production, biogas storage, convenient feeding, discharging, and maintenance. Connections of the inlet/outlet, movable cover, and hydraulic chamber to the digester body should be strengthened.

Wall thickness of FRP household digesters and their domes should meet the requirements of Table 5.8, and the physicochemical properties of the materials should meet the requirements in Table 5.9.

Table 5.8 Requirements for Minimum thickness of digester wall

Item	FRP digester							FRP cover
Digester volume (m^3)	4	5	6	7	8	9	10	–
Minimum thickness of digester wall (mm)	4.0			5.0		6.0		6.0

Table 5.9 Physicochemical properties of FRP materials

S/N	Item	Performance indicators		
1	Bending strength of structural layer (MPa)	Compression molding technology (M)		≥80
		Winding molding technology (C)		≥150 (circumferential)
		Manual plastering molding technology (S)		≥100
		Spray molding process (P)		≥100
2	Babbitt hardness of surface	≥40		
3	Elastic modulus of structural layer (GPa)	Compression molding technology (M)		≥3
		Winding molding technology (C)		≥10
		Manual plastering molding technology (S)		≥3
		Spray molding process (P)		≥5
4	Resin weight content (%)	Inner lining layer		≥[illegible]
		Structural layer	Compression molding technology (M)	≥25
			Winding molding technology (C)	≥23
			Manual plastering molding technology (S)	≥45
			Spray molding process (P)	≥45
5	Water absorption (%)			≤1.[illegible]

Note SMC is not limited by the inner lining layer

The sealing property of FRP household digesters should satisfy pressurizations up to 8 kPa for 24 h, the corrected pressure drop shall be no more than 3%. The sealing property of FRP dome should be pressurized to 12 kPa, the pressure shall be held for 24 h, the corrected pressure drop shall be no more than 3%.

The cementation materials used for on-site installation of FRP digesters should be consistent with the materials of the digester body, and the installation procedure should be strictly carried out in accordance with the specific and detailed instructions

Fig. 5.23 Aboveground FRP digesters

provided by the product manufacturer. The method of connecting of dome of FRP digester and other structural parts should be strictly carried out in accordance with the specific and detailed construction instructions provided by the dome manufacturer.

2. *Aboveground FRP digester in biogas plants*

FRP digesters in biogas plants are generally made by winding molding process (Fig. 5.23). Its volume can be as much as 2000 m^3. Raw materials for FRP tanks include glass fiber, glass mat, matrix resin, and auxiliary materials (curing agents, accelerating agent, pigments, etc.). FRP digesters generally consist of lining layers and winding layers. When selecting materials for the inner liner, in order to avoid the strain concentration caused by different elastic modulus of glass fibers and resins in the inner liner, the matrix material should be resin with good toughness, elongation, low curing shrinkage, and corrosion resistance. The reinforcing materials should be discontinuous short-cut fiber products with good resin wettability, low coefficient of strain concentration after resin curing, and high resin content. In general, 45 g/m^2 surface mat, 450 g/m^2 chopped mat, and glass cloth of common specifications are used. When materials for winding layer are selected, a reasonable resin curing system should be selected. In such system, there is an appropriate proportional relationship between catalysts and initiators, which varies with ambient temperature and digester diameter. In general, the proportion of catalysts and initiators in resin should be controlled below 4%, and the design of resin curing system should be based on small sample tests. The reinforcing material should be twistless continuous fiber rough yarn of good wettability with resin. In production, 196 unsaturated polyester resin, 400 TEX medium alkali winding yarn, 45 g/m^2 alkali-free FRP surface mat, 0.4 mm alkali-free woven rough yarn, curing agent and accelerating agent are usually used. Epoxy FRP laminate is selected as the thread material for manufacturing joints, and its main properties include a bending strength of 250 MPa, a longitudinal tensile strength of 300 MPa, a transverse tensile strength 200 MPa, an interlaminar tensile strength of 26 MPa and a punching shear strength of 174 MPa. The threaded joints of cylindrical tubes made of the above materials can ensure safe uses (Chen and Li 2010).

The winding molding FRP digesters generally consists of three parts: a sealing head, a sealing bottom and cylinder.

Generally, a sealing head is of conical or vault type, and it can be connected by flange. In order to facilitate transportation, the upper sealing head can be prefabricated in segments in supplier's factor and assembled into a whole on site. The minimum thickness of the sealing head shall be no less than 9.6 mm.

The layered structure of the cylinder body consists of an inner surface layer, an impermeable layer, a structural layer, and an external protective layer. The inner surface layer and the impermeable layer shall be collectively called the inner lining layer, with a total thickness of no less than 4 mm. The inner surface layer is a resin-rich layer with a thickness of 0.25–0.50 mm, and the resin should be corrosion-resistant. The reinforcing material may be corrosion-resistant glass fiber surface mat or organic fiber surface mat, with a resin content of more than 90% in the inner layer. The impermeable layer is reinforced by corrosion-resistant resin and alkali-free glass fiber jet spun yarn or chopped strand mat, with a resin content of more than 70%. The main function of this layer is to protect the inner surface layer and to improve the ability of internal pressure resistance failure of the inner lining, so as to prevents crack diffusion. The structural layer is reinforced by continuous twistless rough yarn. For digesters with different heights, the thickness of the structural layer should meet the minimum strength requirement. Other reinforcing materials such as twistless cloth, non-directional cloth, chopped strand mat or chopped strand can also be diffused during the winding process to provide additional strength. Resin content shall be 25–40% for structural layers reinforced fully by twistless rough yarn. The external protective layer should be able to resist ultraviolet aging and meet some other protection requirements. A resin-rich layer of no less than 0.25 mm thick shall be provided for outer surface of the digester.

The cylinder body of digester should be wound uniformly without obvious color difference. Inner/outer surface of the product shall be free of pinholes, poor impregnation (i.e. fibers are not fully impregnated by resin), scars (including breaks, cracks, scratches), rough appearance (i.e. sharp protrusions or fibers exposure), bubbles (surface bubbles formed by gas accumulation) and other defects. Other appearance and structure requirements are the same as those of FRP household digesters.

5.6 Inspection of Digester

1. *General requirements for digester foundation*

Foundation quality of digester is the most important part of the whole construction of digester. Once foundation problems arise, it is very difficult to rectify foundation deviations. Therefore, quality of the basic construction of digesters shall be ensured. General requirements for foundation design and construction shall be as follows:

(1) Before foundation construction of a digester, the bearing capacity of the groundsill where the foundation is constructed must meet the design requirements, i.e. the bearing capacity per unit area of groundsill > (the overall weight of digester filled with liquid + the foundation weight)/horizontal projection area of the foundation.
(2) Binding of steel bars as well as pouring and curing of concrete during construction of digester foundation must be strictly carried out and recorded in accordance with the regulations. Photographs shall be kept as record of concealed projects.
(3) After pouring of digester foundation is completed, it is necessary to check the foundation size, especially the position of the embedded part (if any). Coordinates deviation from foundation center should be no more than 20 mm, the elevation deviation should be no more than 20 mm, and there should be no protruding angle on the foundation surface along any direction. The concave and convex angles of the foundation surface should not exceed 25 mm when measurement is conducted by taking a line from center to periphery.

2. *Sedimentation observation of aboveground digester*

For settlement observation of digester foundation, observation points should be set up symmetrically and evenly within a 10 m-scope of digester circumference and along the circumferential direction. Refer to Table 5.1 for the quantity of observation points.

Digester foundation settlement is a normal phenomenon during water tests and pressure tests of a digester. Such settlement is acceptable as long as the foundation settlement is within the range specified in Table 5.11.

Settlement observation requirements: (1) The whole process observation shall be carried out before water filling, during water filling, during pressure stabilization after water filling, and after water releasing. (2) Settlement observation shall be carried out at least once a day by a specialist on a regular basis and shall be recorded. The measurement accuracy should be level "II". (3) Water filling should be stopped immediately if abnormal settlement of digester foundation is found in the process of water filling and should not be continued until after treatment. (4) Settlement velocity should be no more than 10–15 mm/d and the lateral displacement should be no more than than 5 mm/d.

Table 5.10 Setting of settlement observation points of digester

Volume (m^3)	Quantity of settlement observation points (Nr.)	Volume (m^3)	Quantity of settlement observation points (Nr.)
1000 and below	4	10,000	12
2000	4	20,000	16
3000	8	30,000	24
5000	8	50,000	24

Table 5.11 Permissible value of foundation deformation

Deformation characteristics of digester foundation	Digester type	Diameter in bottom ring (m)	Permissible value of settlement difference (mm)
Plane inclination (in arbitrary diameter direction)	Vertical fixed-top	$D \leq 22$	0.015D
Non-plane-inclination (non-uniform settlement around tank)	Vertical fixed-top		$\Delta S/I \leq 0.0040$

Note ΔS is the settlement difference (mm) of adjacent measuring points around the digester; I is the spacing (mm) of adjacent measuring points around the tank

Settlement observation method: (1) Make the groundsill foundation solid and fill water to 1/2 of digester height at first if a very small settlement is predicted. Next, carry out settlement observation and compare the observed parameters with those before water filling, followed by a calculation of settlement. Continue to fill water to 3/4 of the digester if the value is not more than the prescribed allowable value, and then calculate the settlement. Next, fill water to the highest water level if the value is still not more than the prescribed allowable value and maintain for 48 h. If there is no obvious settlement, water can be then released. If there is still settlement after 48 h, the water level shall be kept and observed every day until the settlement is stable. (2) For soft groundsill foundation, fill water into the digester at 0.6 m/d and stop filling water when reaching 3 m. Observe and record the value and continue to fill water when settlement decreases. Release filled water of the day when settlement is found to increase and repeat the above steps with a smaller filling rate until the settlement is stable.

3. *Water test and pressure test of aboveground digester*

Before the insulation layer is constructed and backfilled, complete water tests and gas tightness tests should be carried out after the main structure of digester has reached the design strength, and after the waterproof layers and the coating layers are constructed.

A complete water test method of digesters is as follows:

(1) Water filling: water shall be filled for four times. Each filling should be 1/4 of the design water depth. The rising speed of water level in filling process should be no more than 2 m/d. The interval between the two fillings of water should be no less than 24 h. The drop-out value of water level in each filling process shall be measured for 24 h to calculate the water seepage rate. During and after filling, the appearance of digester should be checked. Water filling should be stopped if seepage volume is found too large. Water should not be filled until the reason is identified and treated.

(2) Water level observation: water level should be measured by needle water level gauge for seepage measurement when water is filled to the designed water depth. Readings of the needle water level gauge shall be precise to 1/10 mm. The interval between completion of water filling to the designed depth and the beginning of seepage volume measurement should be no less than 24 h. The interval for measurement of initial reading and final reading of water level should be 24 h. The time for continuous measurement may be determined according to actual situation. If the seepage rate measured on the first day meets the allowable standard, it shall be measured again on the next day. If the seepage rate measured on the first day exceeds the allowable standard and the subsequent seepage rate decreases gradually, then the observation may be extended.

(3) Seepage rate is calculated according to formula 5.1:

$$q = \frac{A_1}{A_2}(E_1 - E_2) \tag{5.1}$$

where

q Seepage rate (L/m^2 d);
A_1 Water surface area (m^2) of digester
A_2 Total wetting area (m^2) of digester;
E_1 The initial reading (mm) of needle water level gauge in digester;
E_2 Final reading (mm) of needle water level gauge 24 h after the measurement of E_1.

In water tests of a digester, only the overflow outlet or the manhole on the digester top is connected to atmosphere, and the expose area is smaller than water surface area of the digester, so the calculation of evaporation volume is not considered in this formula.

(4) According to calculation results from above formula, a complete water test is acceptable if the seepage volume is less than 2 L/m^2 d. On the other hand, a complete water test should be checked, corrected, and measured again if the seepage volume is more than 2 L/m^2 d.

After a test is accepted, the gas tightness test can be carried out. Pressure for gas tightness test should be 1.5 times of the design working pressure of digester, and the pressure drop after 24 h should be less than 3% of the test pressure. The gas tightness test methods are as follows:

First, install and seal the manhole cover at the top of the digester and seal pipes such as gas-guide tubes, overflow pipes, etc. Connect the barometer (with 10 Pa of precision. If a U-tube water pressure meter is used, the scale shall be precise to mm) and use air compressor to inflate air into digester to reach test pressure. After stabilization, the measured barometric value in the digester is the initial reading. Stabilize the pressure and observe for 24 h, and the measured barometric value is the final reading. If the drop-out value of the barometer is less than 3% of the test pressure, the test can be confirmed as "acceptable". If the drop-out value of the barometer is more than 3% of the test pressure, a leakage is indicated and it should be checked and corrected, and the gas tightness test should be carried out again.

4. *Water test and pressure test of underground hydraulic digester*

A complete water test method for underground hydraulic digester is as follows:

Water pressure test method: inject water into the digester and stop adding water when the water level rises to the zero pressure line. Mark the water level line after the digester body is completely wet and observe for 12 h. When there is no obvious change in water level, it means no leakage from the fermentation chamber, and the pressure test can be carried out. In the pressure test, install the movable cover first and then seal it tightly. Connect the U-type water-column barometer. Add water to the digester and stop adding water when the value of U-type water-column barometer rises to the maximum design working pressure. Record the value of U-type water-column barometer. Stabilize the pressure and observe for 24 h. Anti-seepage performance of digester can be confirmed by meeting the requirements when the drop-out value of the barometer is less than 3% of the design pressure.

Gas pressure test method: digester leakage test by adding water is the same as water pressure test method. Once no leakage in the digester wall is confirmed, water should be released from the digester. Tightly seal the feeding port/discharging port and the movable cover. Install U-type water-column barometer. Inflate air into the digester and stop inflation when the value of U-type water-column barometer rises to design working pressure. Turn off the air switch. Stabilize the pressure and observe for 24 h. Whether the anti-seepage performance of the digester has met the requirements can be confirmed when the drop-out value of U-type water-column barometer is less than 3% of the designed pressure.

References

Cai, L., and Q.W. Wang. 1997a. *Lipp technology and its application in wastewater treatment.*

Cai, L., and Q. Wang. 1997b. Lipp technology and its application in sewage treatment plant. *Water Supply and Drainage* 23 (6): 58–61.

Cai, L. 1997. Germany Lipp tank technology in large and medium-sized biogas plant. *China Biogas* 15 (2): 29–32.

CECS 138-2002. *Specification for structural design of reinforced concrete water tank of water supply water supply and sewage engineering*. Standard of China Engineering Construction Standardization Association.

Chen, C. 2012. Application of membrane structure in biogas plant. *China Biogas, China Biogas Society Academic Annual Meeting Proceedings*.

Chen, W.F., and J.F. Li. 2010. Application and development of large fiberglass tank in oil field produced water treatment station. *Enterprise and Technology* 5: 85–86.

CJJ113-2007. *Technical code for liner system of municipal solid waste landfill.* Industry Standard of the People's Republic of China.

CJ/T234-2006. *High-density polyethylene geomembrane for landfills*. Urban Construction Industry Standard of the People's Republic of China.

Deng, L.W. 2015. *Biogas project*. Beijing: Science Press.

Department of Personnel and Labor of the Ministry of Agriculture, the Compiling and Examining Committee of the Training Materials for Agricultural Vocational Skills. 2004. *Biogas production technicians*. Beijing: China Agricultural Publishing House.
Fang, G.Y., E.P. Chang, and C.D. Cai. 1986. Red mud plastic biogas digester. *Solar Energy* 1: 6–10.
Hao, S.Y. 2006. *Manual of design and calculation of reinforced concrete water tank*. Beijing: China Construction Industry Press.
HG/T20679-1990. *Design code for external corrosion protection of chemical equipment, piping*. Industry Standard of the People's Republic of China.
HG/T3983-2007. *Filament wound glass-fiber reinforced thermosetting resin chemical resistant large tanks made on site*. Industry Standard of the People's Republic of China.
GB/T12754-2006. *Prepainted steel sheet and strip for building*. People's Republic of China.
GB/T14978-2008. *Continuously hot-dip aluminum-zinc alloy coated steel sheet and strip*. People's Republic of China Standard.
GB/T17643-2011. *Geosynthetics–polyethylene geomembrane*. People's Republic of China Standard.
GB/T18772-2008. *Technical standard for the environmental monitor on the disposal site of landfilled domestic waste*. Standards of the People's Republic of China.
GB/T2518-2008. *Continuously hot-dip zinc-coated steel sheet and strip for building*.
GB/T25832-2010. *Hot rolled steel plates and strips for porcelain enameling*. People's Republic of China.
GB/T28904-2012. *Steel strips for steel-aluminum composite*. People's Republic of China for Steel-Aluminum Composite.
GB/T4750-2002. *The collection of standard design drawings for household anaerobic digesters*. Standard of the People's Republic of China.
GB4752-2002. *Operation rules for construction of household anaerobic digesters*. Standard of the People's Republic of China.
GB50003-2011. *Code for design of concrete structures*. Standards of the People's Republic of China.
GB50010-2010. *Code for structural design of concrete*. Standards of the People's Republic of China.
GB50069-2002. *Structural design code for special structures of water supply and waste water engineering*. Standards of the People's Republic of China.
GB50128-2005. *Code for construction and acceptance of vertical cylindrical steel welded storage tank*. Standards of the People's Republic of China.
GB50341-2003. *Code for design of vertical cylindrical welded steel oil tanks*. Standards of the People's Republic of China.
GB50924-2014. *Code for construction of masonry structures engineering*. Standards of the People's Republic of China.
GB/T7690.3-2013. *Reinforcements—Test method for yarns—Part 3: Determination of breaking force and breaking elongation for glass fibre*. Industry Standards of the People's Republic of China.
GB/T8165-2008. *Stainless steel composite steel plate and steel strip*. Standard of People's Republic of China.
GB/T8237-2005. *Liquid unsaturated polyester resin for fiber reinforced plastics*. Industry Standard of the People's Republic of China.
Lin, W.H. 2000. Architectural design and construction of Lipp tank technology in large and medium-sized biogas plant. *China Biogas* 18 (2): 24–27.
NY/T1702-2009. *Technology specifications of biogas digester for domestic sewage treatment*. Agricultural Standard of the People's Republic of China.
NY/T2597-2014. *Collection of standard design drawings of biogas digester for domestic sewage treatment*. Agricultural Industry Standard of the People's Republic of China.
NY-T1220[1].1-2006. *Technical code for biogas engineering-Part 1: Process design*. Agricultural Standard of the People's Republic of China.

NY-T1220[1].3-2006. *Technical code for biogas engineering-Part 3-Construction and acceptance*. Agricultural Standard of the People's Republic of China.

NY/T1699-2016. *Technical specifications for household anaerobic digesters of fiberglass reinforced plastics*. Agricultural Standard of the People's Republic of China.

SH3046-1992. *Petrol-chemical design specification for vertical cylindrical steel welded storage tanks*. Industry Standard of the People's Republic of China.

Tang, Y.F., and Y.X. Wang. 2013. *Design and application of large and medium-sized biogas*. Beijing: Chemical Industry Press.

Xia, B.S., Q.C. Hu, and L. Song. 2008. Technical analysis of the design drawing of the biogas digester for domestic sewage treatment in villages and towns. *Agricultural Engineering* 24 (11): 197–201.

Xiong, Z.Q. 2012. Construction of the biogas digester for domestic sewage treatment. *Agricultural Disaster Research* 2 (9–10): 46–48.

Xu, Y., Y.F. Yang, and P. Zhu. 2005. *Spherical tank and large storage tank*. Beijing: Chemical Industry Press.

Zhang, H.W., W.N. Huang, and Q.H. Zeng. 2010. The application of red mud plastics in the biogas plant. *China's Biogas* 28 (5): 24–26.

Zhou, M.J., R.L. Zhang, and J.Y. Xu. 2009. *Practical biogas technology*. Beijing: Chemical Industry Press.

Chapter 6
Biogas Cleaning and Upgrading

As a mixture of gases, biogas contains not only methane and carbon dioxide, but may also contain other impurities in small amounts. Some impurities can affect the equipment for biogas utilization by causing problems such as corrosion and mechanical wear. Therefore, biogas has to be purified to meet relevant standards and requirements before it is used. Biogas cleaning includes dewatering, removal of hydrogen sulfide, and removal of other impurities. The process of further removing carbon dioxide is called biogas upgrading.

6.1 Biogas Characterization and Quality Standards

The degree and process of biogas cleaning depend on both the original biogas composition and quality standards of the target products.

6.1.1 Biogas Composition

Biogas has complicated compositions that vary with the type of biogas feedstocks and the relative amount of the each feedstock. The compositions also depend on anaerobic digestion conditions, digestion stages and operation methods (Al Seadi et al. 2008).

For instance, biogas produced from straw, animal manure, living organic waste and industrial wastewater each has different components. When a digester is in the stage of digestion at normal temperature, the ratio of components in biogas has a typical range of 50–70% methane, 30–50% carbon dioxide. Other than that, biogas also contains impurities such as a small amount of carbon monoxide, hydrogen, hydrogen sulfide, oxygen, nitrogen, and particulate matter. Biogas is usually composed of the following constituents, as listed in Table 6.1.

L. Deng et al., *Biogas Technology*,
https://doi.org/10.1007/978-981-15-4940-3_6

Table 6.1 Typical composition of biogas

Compound	Volume ratio (%)
Methane	50–75
Carbon dioxide	25–50
Nitrogen	0–4
Oxygen	0–3
Hydrogen sulfide	0.03–0.5
Hydrogen	2–7

6.1.2 Influence and Harm of Impurities in Biogas

Carbon dioxide takes up a large proportion in biogas impurities. While other impurities take a smaller fraction, they still can exert negative impacts on biogas utilization. The effects of various impurities are briefly described below (Arthur et al. 2013; Deng et al. 2015).

Carbon dioxide

Apart from methane, carbon dioxide is the most concentrated component in biogas. It is formed during decomposition of different types of substrates used for biogas production. The transformation of substrates into biogas is a complex process that involves several steps and different types of microorganisms. Carbon dioxide is formed in different steps and acts as an electron acceptor for the methane-producing bacteria. The carbon dioxide will decrease the volumetric energy content of biogas. If high volumetric energy content is important (e.g. when the gas is used as a vehicle fuel or subjected to grid injection), carbon dioxide is considered an impurity that should be removed. For other applications such as power and heat generation, carbon dioxide does not usually cause problems. However, carbon dioxide will form carbonic acid together with water that has condensed in these processes.

Water

Since water is always present during anaerobic digestion, some of it will evaporate in the digester and thus be present in biogas. Biogas leaving the digester is therefore always saturated with water. The water content depends on pressure and temperature inside the digester. Water in raw biogas can cause many problems. It can, for example, increase fluid resistance of biogas in pipelines. Another negative effect is that water lowers the energy content of biogas and can thus affect heat value of biogas. The acid solution produced from mixing of water, carbon dioxide and hydrogen sulfide can cause corrosion and blockage of metal pipelines, valves, flow meter, compressor, gas holder, and engines. Depending on the temperature and pressure downstream from the combustion stage, water may condense and cause problems in downstream heat exchangers and exhaust components.

Hydrogen sulfide

Another common impurity in biogas is hydrogen sulfide. Other sulfur-containing impurities may also be present in raw biogas, but hydrogen sulfide is most common. Concentration of hydrogen sulfide has a lot to do with digestion substrate and process. Biogas produced from high-protein or sulfate substrate has a higher content of hydrogen sulfide.

Hydrogen sulfide is a colorless, poisonous and inflammable gas with a strong smell of rotten eggs. It can be easily oxidized with strong oxidation-reduction reaction potential. It is also highly soluble in organic solvent. Moreover, hydrogen sulfide readily reacts with alkali and produces metal sulfide or hydrosulfide.

Under ambient temperature, hydrogen sulfide can react with some metallic oxides such as ferric oxide and zinc oxide to produce metal sulfides. It can also react with metal ions (except ammonium and alkali metal) in the liquid form and produce sulfides with low solubility.

The presence of hydrogen sulfide during biogas utilization can lead to corrosion in compressors, gas holders, and engines since it forms sulfuric acid when oxidized in the presence of water. As a result, combustion of biogas containing hydrogen sulfide will lead to emissions of sulfur dioxide, and when combined with vapor in the combustion products, sulfur dioxide will form sulfurous acid, which causes corrosion to metal surfaces of the low temperature parts of combustion equipment.

Hydrogen sulfide is also highly toxic and can result in serious health risks.

Hydrogen sulfide is a strong nerve poison. It strongly stimulates human's mucosa. It can cause conjunctivitis of eyes and is easily absorbed by lungs and stomach.

Hydrogen sulfide can combine with alkaline substances in tissues to form sodium sulfide, causing damages to eyes and respiratory tract. Systemic poisoning from exposure to hydrogen sulfide is caused by cytochrome oxidase inhibition, bio oxidation blockage, and asphyxia in tissue hypoxia. Once hydrogen sulfide enters the bloodstream, it combines with hemoglobin to form non-reducible hemoglobin, which results in poisoning symptoms. Health effects of human exposure to different concentrations of hydrogen sulfide are shown in Table 6.2 (Zhang et al. 2005).

Oxygen and nitrogen

Since biogas is formed under anaerobic conditions, neither oxygen nor nitrogen is usually present in biogas. However, both can be found if air enters the digester system by chance. If oxygen is present, it will be slowly consumed by facultative anaerobe. Nitrogen in raw biogas can thus be a sign of de-nitrification or air leakage inside digesters. It is more common for nitrogen to be found in landfill gas, and there may also be small amounts of oxygen, due to the fact that the extraction of landfill gas can cause a low pressure in the landfill, which results in air being sucked in. Oxygen can lead to formation of flammable mixtures with methane and oxygen content in biogas, and thus oxygen has to be carefully controlled.

Table 6.2 Harm of different concentrations of hydrogen sulfide on human body

Concentrations of hydrogen sulfide (mg/m^3)	Contact time	Poisoning reaction
0.035		Start to smell
0.4		Odor becomes obvious
4–7		Smell a moderate odor of rotten eggs
30–40		Strong but bearable smell
70–150	1–2 h	Respiratory tract and eye irritation; olfactory fatigue may occur and results in inability to smell
300	1 h	8 min of eye irritation; prolonged exposure causes emphysema
760	15–60 min	Emphysema, bronchitis, pneumonia; headache, dizziness, gait instability, nausea, vomiting, and difficulty in urinating caused by long contact time
1000	Several seconds	Acute poisoning appears soon; respiratory paralysis
1400	Instantly	Instantaneous coma and respiratory paralysis

Ammonia

Ammonia is an impurity often found in raw biogas; it is formed in the digester during hydrolysis of substrates containing proteins, such as livestock farm waste and slaughterhouse waste. High levels of ammonia in the digester can cause inhibition of methane production in the digester.

Volatile organic compounds

Volatile organic compounds in biogas are mainly alkanes, siloxanes and halogenated hydrocarbons. The type of compounds and their concentrations depend on the substrate used for biogas production.

Siloxanes are compounds used in products such as fire retardants, shampoos and deodorants. If siloxanes are present in the substrate entering the digester, they will also be found in small amounts in the biogas due to the fact that some of these compounds evaporate. Temperature in the digester will determine how much siloxane evaporation into the biogas will occur. Low-molecular-weight siloxanes will evaporate to a greater extent than others. Siloxanes can also be found in landfill gas. During combustion of siloxanes, the resulting silica and microcrystalline quartz deposit on spark plugs, valves and cylinders, causing surface corrosion and damage to engines. Siloxane oxide formed in combustion is insoluble and will form unwanted depositions on combustion equipment.

Halogenated hydrocarbons are hydrocarbon molecules containing chlorine, bromine or fluorine. They can be present in raw biogas due to volatilization of halogen-containing material in landfills. Halogenated hydrocarbons can cause corrosion and acidification when combusted, due to formation of acids.

Particles

Particles are often present in raw biogas. In many cases, particulates form nuclei onto which water drops condense. Particles will deposit in the compressor and gas holders and cause blockage. Particles can cause equipment wear due to their abrasive properties.

6.1.3 Quality Demands on Biogas for Utilization

How clean biogas needs to be depends on how biogas is to be utilized. Biogas can be used for domestic fuels, boiler fuels, power generation or fuel for vehicles. There are corresponding quality requirements on biogas for different purposes. Equipment has certain threshold requirements for gas components. In general, it can be said that the cleaner biogas is, the lower the maintenance cost is, although a higher cleaning cost is resulted. Gas cleaning can thus sometimes be a compromise between cleaning and maintenance costs. When evaluating biogas composition and its impurities, one cannot look at each impurity individually since different impurities also affect each other.

(1) **Quality demands on biogas for domestic fuel**

Biogas is commonly used as domestic fuel. Major appliances include biogas stoves lamps, rice cookers and water heater. Biogas should be dewatered, desulfurized before it is stored and distributed. Purified biogas should meet the following quality demands: greater than 18 MJ/m^3 of lower calorific value, less than 20 mg/m^3 of hydrogen sulfide content in biogas, and lower than 35 °C of biogas temperature (Zhou et al. 2009).

(2) **Quality demands on biogas for boiler fuel**

Boilers are used for heat production. Compounds in biogas that can cause problems in a boiler are hydrogen sulfide, particles and siloxanes. In the cooling system of a boiler, the flue gas is cooled down and water in the gas condensates. Hydrogen sulfide will form sulfuric acid with water, which may then cause corrosion. Particles and siloxanes can also introduce problems since they can introduce partial blockage of tubes in a boiler.

Industrial boilers are sometimes certified for using raw biogas. For small boilers, hydrogen sulfide together with condensed water may cause corrosion. Due to their larger size of parts, large boilers have fewer issues with particles and siloxanes than small boilers. Once a boiler accommodates to a specific biogas composition, this gas composition should be retained unless the boiler is equipped with an oxygen or carbon monoxide sensor for flue gas. With such sensors, the boiler can tolerate more variations in biogas compositions.

(3) **Quality demands on biogas for power production**

Biogas generator sets using internal combustion engines can adapt to different biogas compositions. For biogas composition requirements for biogas generators in China, refer to the standard NY/T 1223-2006 "Biogas Generator Sets". The lower calorific value of biogas is not less than 14 MJ/Nm^3 (equivalent to methane volume fraction not less than 30%); hydrogen sulfide content is less than or equal to 200 mg/Nm^3 (when methane volume fraction is from 30 to 50%), 250 mg/Nm^3 (when methane volume fraction is from 50 to 60%), or 300 mg/Nm^3 (when methane volume fraction is at 60% or more); the relative pressure of biogas inlet is 2000–2400 Pag.

Fuel cells can also be used for power production from biogas. Different types of fuel cells use different fuels and have different sensitivities to impurities in the gas. High-temperature fuel cells (e.g. molten carbonate fuel cells (MCFCs)) can use methane from biogas as a fuel. However, for fuel cells that operate at lower temperatures (e.g. proton exchange membrane (PEM) fuel cells), biogas has to be catalytically reformed to hydrogen before used as a fuel. Compounds such as hydrogen sulfide, halogenated hydrocarbons, ammonia, and siloxanes that are toxic to fuel cells should be removed from biogas (Zhou et al. 2014).

(4) **Quality demands on biogas for natural gas grid injection**

At present, since there is no quality standard for biogas-sourced natural gas in China, here we mainly refer to the standard for natural gas of gas field and oil field sources, GB 17820 "natural gas" (see Table 6.3), which is applicable to natural gas transported through pipeline after pre-treatment. ClassIand Class II natural gases are mainly used as domestic fuels. Class III natural gas is mainly used as industrial raw materials or fuels. In addition, under the premise of meeting national safety and health standards, the supply and demand sides can create contracts and agreements to determine specific technical requirements for natural gas that does not belong to the above three categories.

(5) **Quality demands on biogas for vehicle fuel**

Table 6.3 Natural gas characteristics

Technical parameters	ClassI	Class II	Class III
Higher calorific value/(MJ·m^{-3})	≧36.0	≧31.4	≧31.4
Total Sulfur/(mg·m^{-3})	≦60	≦200	≦350
Hydrogen sulfide/(mg·m^{-3})	≦6	≦20	≦350
Carbon dioxide/(%)	≦2.0	≦3.0	–

Note The above parameters refer to GB 17820-2012 "Natural Gas". (1) The standard reference condition of gas volume in this standard is 101.325 kPa, 20 °C; (2) Under the conveying condition, when the temperature of the buried pipe top is 0 °C, the ice dew point should not be higher than −5 °C; (3) water dew point pressure of the natural gas to be injected into gas grids should be the highest delivery pressure.

Table 6.4 Quality standard of compressed natural gas for vehicles

Characteristic parameter	Technical index
High calorific value/($MJ \cdot m^{-3}$)	≧31.4
Total sulfur mass concentration/(by sulfur $mg \cdot m^{-3}$)	≦200
Hydrogen sulfide mass concentration/($mg \cdot m^{-3}$)	≦15
Carbon dioxide volume ratio/(%)	≦3.0
Oxygen volume ratio/(%)	≦0.5
Water dew point/(°C)	In a particular geographic area where cars are driven, under the highest operating pressure, water dew point should not be higher than −13 °C; when the minimum temperature is lower than −8 °C, water dew point should be 5 °C lower than the lowest temperature

Note The standard reference condition for gas volume in the standard is 101.325 kPa, 20 °C

When biogas is used as a vehicle fuel, water removal, hydrogen sulfide removal, dehalogenation, and decarbonation are required. Biogas is used as vehicle fuels with different standards in different countries. Switzerland has a specific standard for vehicle biogas fuels (SJIS 1999), which strictly regulates the content of methane, hydrogen sulfide and water. The United States (International SAE 1994), Switzerland (ISO 2006), and Germany (DIN 2008) each has issued corresponding standards. The United Nations Economic Commission for Europe (UNECE 1958) also covers relevant standards. Relatively mature standards are developed by the European Standards Committee (Deng and Chen 2007; Yin et al. 2009).

As of now, there is no standard for biomethane upgrading from biogas purification for vehicle gas. The only reference is the standard "Car Compressed Natural Gas" (GB 18047-2000) (see Table 6.4). The compressed natural gas for vehicles must meet the main requirements in Table 6.4.

6.2 Biogas Cleaning Techniques and Equipment

6.2.1 Water Removal

In the anaerobic digestion process, due to evaporation, there is a certain amount of water in biogas. In mesophilic or thermophilic digestion, the moisture content in biogas is especially higher. Vapor in digesters is usually saturated, which means that the relative humidity is 100%, but the absolute moisture content is also related to temperature. In general, the saturated moisture content of 1 m^3 of dry biogas is 35 g at 30 °C and 111 g at 50 °C. Therefore, in order to protect biogas utilization equipment from severe corrosion and damage, and to meet the requirements of downstream cleaning equipment, vapor in biogas must be removed (Zhou et al. 2009).

Water removal technology of biogas mainly includes condensation method, adsorption method, absorption method, etc.

6.2.1.1 Techniques for Water Removal

(1) Condensation method

The condensation method is the easiest way to remove water from biogas. This method can be used for any flow rate of biogas. Saturated vapor pressure changes with biogas temperature, and the condensation method uses this property to separate water from biogas by means of cooling or compression. Biogas produced in mesophilic or thermophilic digestion can be cooled down, and the condensed water can be removed by cooling in a condenser in a heat exchange system.

During biogas transportation by pipeline, vapor is condensed into water due to temperature drop. Therefore, a condenser is usually installed at the lowest point of the biogas pipeline to remove the condensed water in the pipeline.

Besides water, other impurities such as water-soluble gases and aerosols are also removed during condensation. Study (Urban et al. 2007–2008) has found that this method can reach a dew point of 3–5 °C. Under the initial moisture content of 3.1% (volume ratio), 30 °C, ambient pressure, moisture content can be reduced to 0.15% (volume ratio).

Compressing biogas before cooling can further increase efficiency. This method has better water removal effect, but it cannot fully meet the requirements of biogas to be fed into natural gas pipeline network. This can be compensated by downstream adsorption cleaning technology (swing pressure adsorption and hydrogen sulfide removal adsorption).

(2) Adsorption

The adsorption method refers to absorption of moisture in biogas by solid surface force (van der Waals force and dispersive force) when the biogas passes through a solid adsorbent, thereby achieving the purpose of drying. According to the nature of the surface tension, adsorption can be categorized into chemical adsorption (cannot be regenerated after water removal) and physical adsorption (renewable after water removal). Commonly used adsorbent materials are silica gel, activated alumina, molecular sieves, and composite desiccants.

The physical adsorption method can achieve a dew point of −90 °C (Ramesohl et al. 2006). The adsorption device is installed on a fixed bed and can be operated under normal pressure or at the pressure of 600–1000 kPa, which is suitable for water removal of small-flow biogas. Usually two units run in parallel one for absorption and the other for regeneration. In biogas water removal engineering, the condensation method and the adsorption drying method are usually combined. Water is partially removed by the condensation method, and then the water removal method is used for a more thorough water removal.

The feature of physical adsorption is that the heat released during the adsorption process generally includes heat of vapor condensation and heat released by the adsorbent due to wetting from water. The amount of heat released during the entire adsorption process is small, and the adsorbent material can be regenerated by increasing the temperature or reducing the pressure. The adsorption process is reversible, and the dewatering performance by physical adsorption far exceeds that by the solvent absorption method. The physical adsorption method can obtain biogas with extremely low dew point, and the method is not sensitive to changes in temperature, pressure and flow rate; the apparatus is simple and easy to operate. Corrosion and blistering are not likely to occur. Due to the good drying effects, this method is suitable for all biogas utilization purposes.

(3) Absorption

The absorption method is to remove water in biogas by using a water removal absorbent in countercurrent contact with biogas. Water removal absorbents are generally hydrophilic. Commonly used dehydrating absorbents are calcium chloride, lithium chloride, and glycolic compounds (ethylene glycol diethylene glycol triethylene glycol, etc.) (Ryckebosch et al. 2011).

At present, more absorbents in application are in the form of glycol compounds. When ethylene glycol is used as an absorbent, it can be heated to 200 °C to volatilize impurities to realize regeneration of alcohol (Weiland 2003). Literature shows that water removal by ethylene glycol can reach a dew point of −100 °C (Schonbucher 2002). From the economic point of view, this method is suitable for water removal of large-flow (500 m^3/h) biogas, so the absorption method can be used as a pretreatment method for biogas upgrading.

6.2.1.2 Devices for Water Removal

(1) Biogas water collector

Water removal of household biogas digesters usually uses a biogas water collector that is similar to biogas condenser on pipelines in structure, only that the former is simpler. It is categorized into manual drainage water collector and automatic drainage water collector. Manual drainage water collectors can be made of ground glass bottles and rubber stoppers. The stopper should be bored with two holes, into which two glass elbow adaptors with an inner diameter of 6–8 mm are inserted. The rubber stopper is then plugged into the mouth of glassware. Tighten the stopper and ensure that it doesn't leak.

The horizontal ends of the two elbow adaptors are respectively connected to biogas pipes. When the condensate level approaches the lower end of the elbow, unscrew the cork and pour out the water. The main components of the automatic drainage water collector are the same as the manual drainage water collector, except that the upper

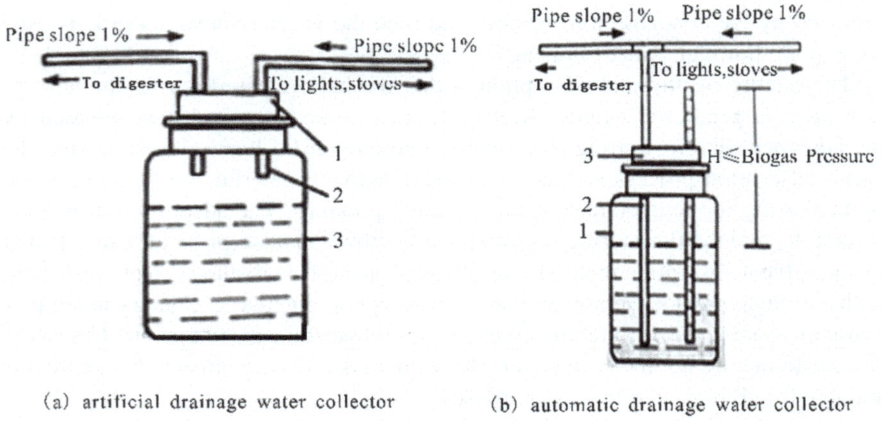

Fig. 6.1 Biogas water collector. 1. Rubber plug 2. Glass pipe 3. Glass bottle 4. Water overflow hole

end of one straight pipe is connected with a tee. Two horizontal ends of the tee are connected to biogas pipelines. The upper end of the other straight pipe discharges to the atmosphere as the water overflow drainage hole. The water overflow hole should be lower than the tee pipe, otherwise the condensate will block the pipe when the gas production is low (Fig. 6.1).

(2) Gas-water separator

In gas-water separators, horizontal and vertical filter screens are installed in the device. It is best to refill the packing. The filter screen or packing can be made of stainless-steel wire mesh, copper wire mesh, polyethylene wire mesh, polytetrafluoroethylene mesh or ceramic lacy, ring, etc.

When biogas enters from the lower part of the device in a tangential manner with a certain pressure, it rotates under the centrifugal force, and then passes through a vertical filter followed by a horizontal filter. Moisture in biogas is separated from dry gas, and vapor is condensed. Water droplets formed in the gas-water separator flow down along the inner wall, accumulating at the bottom of the device to be periodically removed (Fig. 6.2).

When designing gas-water separator, the following design principles should be followed. The amount of biogas entering the separator should be calculated according to the average daily biogas output. The biogas pressure in the gas-water separator should be greater than 2000 Pa, and the pressure loss of the separator should be less than 100 Pa. The superficial velocity in the separator is preferably between 0.21 and 0.23 m/s. The biogas inlet pipe shall be arranged tangential to the gas-water separator cylinder, and the gas-water separator shall be provided with a liquid accumulation bag and a wastewater blow-off pipe. The flow rate in the inlet pipe of the gas-water separator should be 15 m/s, and the flow rate in the outlet pipe should be 10 m/s (Zhou et al. 2009).

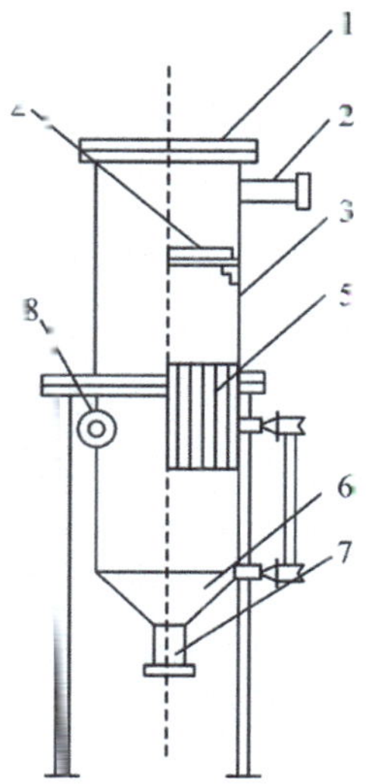

Fig. 6.2 Gas-water separator. 1. Blocking plate 2. Outlet pipe 3. Cylinder 4. Flat filter 5. Vertical filter 6. Shell cover 7. Wastewater blow-off pipe 8. Biogas inlet pipe

(3) Condensate traps

Condensate traps are similar to the condenser of urban pipeline gas. At the lowest point of the biogas pipeline, a condensate trap must be installed to discharge the condensed water in the pipeline periodically or automatically. Otherwise, the resistance in the biogas pipeline may be increased, which may affect the stability of the biogas transmission and distribution system. It must be easy to operate, and the condensate traps must be installed in freeze-proofing areas. The diameter of condensate traps should be 3–5 times the diameter of inlet pipes, and the height of a trap should be 1.5–2.0 times its diameter. The specifications of condensate traps at different biogas flow rates are shown in Table 6.5. The condensate traps need to be drained regularly. The condensate traps can be classified into manual and automatic drainage based on different drainage methods (Fig. 6.3) (Zhou et al. 2009; Deng et al. 2015).

(4) Freeze dryer

Freeze dryer is a kind of equipment for drying biogas by using the principle of cooling condensation. Biogas is cooled and compressed by a freeze dryer, so moisture in biogas is removed to achieve the purpose of biogas drying. A freeze dryer mainly

Table 6.5 Specifications of condensate traps

No.	Outer diameter of condensate traps	Diameter of inlet and outlet	Suitable biogas flow
1	Φ600	DN 150–200	Biogas flow ≧1000 m^3/d
2	Φ500	DN 100–150	Biogas flow = 500–1000 m^3/d
3	Φ400	DN 50–100	Biogas flow ≦500 m^3/d

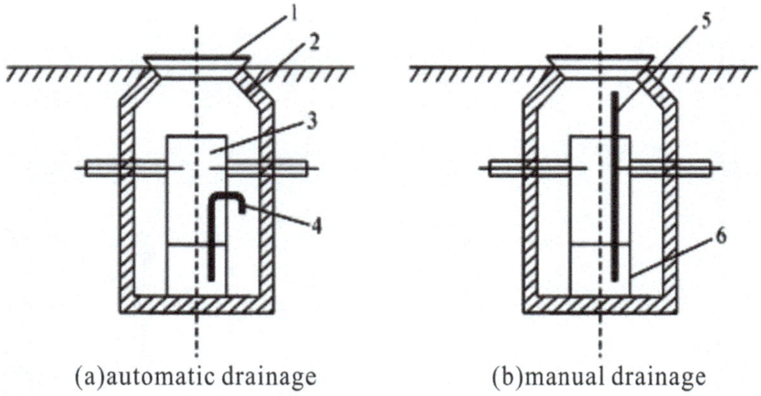

Fig. 6.3 Condensate traps. 1. Well cover 2. Collecting well 3. Condensate traps 4. Automatic drainage pipe. 5. Drainage pipe 6. Drainage valve

consists of three parts: a heat exchange system, a refrigeration system, and an electrical control system. The hot and humid compressed biogas from the compressor is first pre-cooled by a heat exchanger, and then the pre-cooled biogas is cooled again in the refrigerant circulation of the freeze dryer, following which this cooled gas exchanges heat with cold biogas discharged from the evaporator. The temperature of the compressed biogas is further reduced. After the previous stages, the compressed biogas enters the evaporator and exchanges heat with the refrigerant, and the temperature of the compressed biogas is lowered to 0–8 °C, at which moisture in the biogas is removed. Condensed water in the compressed biogas is separated by a condenser and discharged through an automatic drainage. The dry and low-temperature biogas enters the heat exchanger for heat exchange, and leaves after the temperature rises. Freeze dryer is mainly used in the water removal unit of ultra-mega biogas plants. Commonly used freeze dryers are classified into air-cooled and water-cooled according to the cooling methods of condensers. Depending on the intake gas temperature, these condensers are also classified into high-temperature gas intake type (below 80 °C) and ambient-temperature gas intake type (about 40 °C). Based on the working pressure, the condensers are classified into ordinary type (0.3–1.0 MPa) and medium or high-pressure type (1.2 MPa or more). Since water removal by condensation is relatively economical, it is generally necessary to remove water partially by passing gas through a gas-water separator or a condensate trap prior to the freeze dryer treatment.

6.2.2 Hydrogen Sulfide Removal

The methods of hydrogen sulfide removal can be generally classified into two categories: the direct hydrogen sulfide removal and the indirect hydrogen sulfide removal.

The direct hydrogen sulfide removal is a direct separation and removal of H_2S from biogas. The indirect hydrogen sulfide removal refers to reduction or suppression of H_2S production in produced biogas by using specific methods. According to different principles of hydrogen sulfide removal, the direct hydrogen sulfide removal from biogas can be further classified into physical, chemical and biological methods. Based on the physical state of desulfurizer, hydrogen sulfide removal is divided into wet methods and dry methods.

6.2.2.1 Techniques for Hydrogen Sulfide Removal

(1) **Wet hydrogen sulfide removal methods**
(1) Water washing method

Water washing method uses water to spray to and wash biogas to remove hydrogen sulfide. At a temperature of 20 °C and a pressure of 1.013×10^5 Pa (1 barometric pressure), 1 m^3 water can dissolve 2.3 m^3 of hydrogen sulfide. When the hydrogen sulfide content in biogas is high and the gas volume is large, it is suitable to use water washing method for hydrogen sulfide removal, and in the meanwhile, a part of the carbon dioxide can be removed to increase methane content in the biogas.

(2) Sodium carbonate absorption method

Sodium carbonate absorption is a commonly used wet hydrogen sulfide removal method. When sodium carbonate solution absorbs acid gas, the pH of sodium carbonate does not change very quickly, which ensures the operational stability of the system. In addition, sodium carbonate solution absorbs H_2S faster than CO_2, and can partially and selectively absorb H_2S. This method is usually used to remove a large amount of CO_2 from gas and can also be used to remove CO_2 and H_2S from natural gas and biogas.

The gas containing H_2S is in countercurrent contact with the $NaCO_3$ solution in the absorption tower, and the solution is generally sprayed from the top of the tower with a 2–6% $NaCO_3$ concentration. When it reacts with H_2S rising from the bottom of the column, $NaHCO_3$ and NaHS are formed. After absorbing H_2S, the solution is sent back to the regeneration tower, and $NaCO_3$ is regenerated by steam heating under reduced pressure. In other words, H_2S is released, and $NaCO_3$ is regenerated. Hydrogen sulfide removal reaction and regeneration reaction are reversible reactions.

$$Na_2CO_3 + H_2S \rightarrow NaHS + NaHCO_3 \tag{6.1}$$

The solution flowing out of the regeneration tower is recycled to the absorption tower. The concentration of H_2S in the gas discharged from the top of the regeneration tower can reach 80% or more, suitable for producing sulfur or sulfuric acid. The sodium carbonate absorption method has a simple process and is inexpensive, so it is suitable for treating gases with high H_2S content. One disadvantage is that the hydrogen sulfide removal efficiency is not high enough, generally 80–90%, whereas the steam and power consumption is large due to difficulty in regeneration.

(3) Sodium hydroxide absorption method

Sodium hydroxide absorption method is another commonly used wet hydrogen sulfide removal method. The main reaction which uses aqueous sodium hydroxide solution as the absorbent has two steps. The first step is to use sodium hydroxide to react with hydrogen sulfide to produce sodium sulfide and water, by which removal of hydrogen sulfide is achieved. The second step is to oxidize sodium sulfide to sodium hydroxide and elemental sulfur to complete regeneration of sodium hydroxide. During the hydrogen sulfide removal process, biogas enters the bottom of the alkali absorption tower and contacts with the absorption liquid in a countercurrent fashion. The cleaned biogas is discharged from the top of the tower. Due to mass transfer conditions such as flow rate and flow volume, hydrogen sulfide cannot be completely dissolved in the alkali solution, and sodium hydrosulfide is easily formed during the absorption process. Sodium hydrosulfide reacts with oxygen and forms harmful substances such as thiosulfate and sulfate. These harmful components will accumulate in the absorbent, and it is therefore necessary to replenish fresh lye in time before each use. This method was used in the second-phase project of Gaobeidian Wastewater Treatment Plant. Biogas of this plant was initially absorbed by 35% NaOH solution. During operation it was found that the alkaline solution quickly crystallized during the cycle and blocked the pump. The operator thus further reduced the concentration of the alkaline solution and found that alkaline solution of 16–20% can avoid precipitation of NaOH crystals, while the H_2S removal efficiency cannot satisfy the needs. Furthermore, generation of harmful substances requires continuous make-up and replacement of the absorbent solution, which increases the treatment cost as well as the possibility of secondary pollution. At the same time, requirement on equipment anti-corrosion is relatively high, so this method is rarely used in hydrogen sulfide removal from biogas.

(4) Ammonium hydroxide method

Hydrogen sulfide is an acid gas, and when hydrogen sulfide is absorbed by alkaline ammonia water, a neutralization reaction occurs (Formula 6.2).

$$H_2S + NH_4OH \rightarrow NH_4HS + H_2O \tag{6.2}$$

The first step is the dissolution of hydrogen sulfide in the gas into ammonia water, which is a physical dissolution process. The second step is neutralization of dissolved hydrogen sulfide and ammonium hydroxide to form ammonium hydrogen sulphate, which is a chemical absorption process. The regeneration method is to blow air into a solution containing ammonium hydrogen sulfide to generate a reverse process of the absorption reaction and desorb the hydrogen sulfide gas.

After replenishing fresh ammonia water, the desorbed ammonia hydroxide solution is then used for absorption. Hydrogen sulfide generated during regeneration must be re-treated to avoid environmental pollution.

(5) Alcohol amine absorption method

Hydrogen sulfide removal by alcohol amine absorption has been industrialized since 1930 and is the most widely used method in the gas cleaning industry. The method has a simple process, reliable performances, low costs with availability, and high cleaning efficiency. There are six main amines used: monoethanolamine, diethanolamine, triethanolamine (TEA), methyldiethanolamine (MDEA), diglycolamine, and diisopropanolamine. The main reaction (taking monoethanolamine as an example) is shown in Eqs. 6.3 and 6.4.

$$2RNH_2 + H_2S \rightarrow (RNH_3)_2S \tag{6.3}$$

$$(RNH_3)_2S + H_2S \rightarrow 2RNH_3HS \tag{6.4}$$

The above reaction is reversible, in which the forward reaction (absorption) is dominant at a lower temperature (20–40 °C) and the backward reaction (desorption) is dominant at a higher temperature (above 105 °C). The classical alcohol amine solution is a good solvent for absorbing hydrogen sulfide with the advantages of low price, strong reaction, good stability, and easy recovery. The disadvantages are easy bubbling, corrosion, no selectivity to hydrogen sulfide and carbon dioxide, degradation in the presence of organic sulfur, high vapor pressure, and large solution loss.

The basic process of hydrogen sulfide removal by alcohol amine is mainly composed of four parts (Liu 2015).

① Acid gas absorption

Through the separator of the gas inlet, liquid and solid impurities in the mixed gas are first removed. The mixed gas containing hydrogen sulfide and carbon dioxide enters the bottom of the absorption tower, and the absorbent solution is sprayed from the top downward. The mixed gas contacts with the alcohol amine solution from bottom to top, wherein H_2S and CO_2 are absorbed into the liquid phase while other gas components are discharged from the top of the absorption tower, and the liquid part is discharged from the bottom of the absorption tower.

② Flashing

An alcohol amine solution which has absorbed hydrogen sulfide and carbon dioxide is often referred to as rich solution. Rich solution flows out from the bottom of the absorption tower and then depressurizes into a flash tank. Combustible hydrocarbons dissolved and entrained in the rich solution are flashed out. The flash gas generated in the flash tank can be used as the fuel gas of the device.

③ Heat exchange

After flashing, the rich solution enters the lean/rich solution heat exchanger through a filter and exchanges heat with the regenerated heated alcohol amine (referred to as lean solution). The heated rich solution then enters the low-pressure regeneration tower from the top.

④ Absorbent regeneration

When rich solution enters the regeneration tower, part of the acidic component is first separated by flash evaporation at the top of the tower, and then flows from top to bottom. During its flow, the rich solution is in contact with the heated vaporized vapor in the re-boiler. It uses hot steam to strip off other acidic components in the solution. The solution left in the regeneration tower is a residual acid gas containing only a small amount unstripped gas which is called a lean solution. The hot lean solution transfers heat to the rich solution that has not entered the regeneration tower through the lean/rich solution heat exchanger, recovers part of the heat, and then the lean solution which has been further cooled to the appropriate temperature is sent to the top of the absorption tower by a solution circulation pump to complete regeneration and circulation of the absorbent liquid.

Hydrogen sulfide removal by alcohol amine absorption is one of the mainstream processes for large-scale plants of hydrogen sulfide removal from natural gas. The amine method can simultaneously remove hydrogen sulfide and carbon dioxide, and it is suitable for hydrogen sulfide removal and upgrading in ultra-mega biogas plants.

⑤ Liquid phase catalytic oxidation

Research on this kind of method began in the 1920s, and the number of methods has grown to more than 100 kinds, of which more than 20 kinds have industrial application value. The liquid phase catalytic oxidation method has the following characteristics: ① High hydrogen sulfide removal efficiency, which can reduce the sulfur content of cleaned gas to less than 10 ppm (13.3 mg/m^3), or to even less than 1–2 ppm (1.33–2.66 mg/m^3). ② The removed hydrogen sulfide can be further converted into elemental sulfur without secondary pollution. ③ Removal can be operated at room temperature, and can be operated under pressurization. ④ Most desulfurizer can be regenerated, so this method has low operation costs. However, when sulfur content in the raw gas is too high, the H_2S/HS-reaction in the liquid phase immediately slows down due to a decrease of pH of the solution, thereby affecting the mass transfer rate of absorption and increasing the cost of the device. (Liu 2015).

(2) **Dry hydrogen sulfide removal methods**

When a solid is used as an absorbent to adsorb and remove hydrogen sulfide, it is called dry hydrogen sulfide removal methods. Factors affecting hydrogen sulfide removal include the compositional properties of the gas-solid phase, pressure, and contact time. Dry hydrogen sulfide removal is a gas-solid mass transfer process. In order to achieve a good hydrogen sulfide removal efficiency, a reasonable design of biogas flow rate in the absorption tower is needed.

In the hydrogen sulfide removal process, the biogas containing hydrogen sulfide enters the tower from one end and exits the tower from the other end. If the load of the packing layer is high at the inlet end and low at the outlet end, packings in the tower cannot be uniformly and fully utilized and the utilization rate of the packings is low. At the same time, uneven load will also result in uneven distribution of elemental sulfur amongst the packings or even concentrated accumulation, and the pressure drop along the packings will increase, which further affect the gas flow rate and hydrogen sulfide removal efficiency.

In addition, the continuous replacement of packings increase production cost. Therefore, in plants with large biogas production rates, dry hydrogen sulfide removal has significant limitations. At the same time, due to the characteristics of gas-solid mass transfer, dry hydrogen sulfide removal is generally suitable for applications where the amount of biogas is small, and the concentration of hydrogen sulfide is low.

Since the reaction process is simple, dry hydrogen sulfide removal is a simple and low-cost hydrogen sulfide removal method compared with wet hydrogen sulfide removal. Common dry hydrogen sulfide removal methods include molecular sieve, activated carbon, iron oxide, zinc oxide, etc. (Zhou et al. 2009; Li 2007; Deng et al. 2015).

(1) Molecular sieve method

Molecular sieve is a kind of aluminosilicate micro porous crystal composed of silicon oxide and aluminum oxide tetrahedron that form the main structure, as well as metal cations such as Na^+, K^+, Ca^{2+} and Li^+ in the crystal lattice. Molecular sieves can take advantage of the micro porous structure and its large specific surface area to adsorb H_2S. Due to the strong polarity inside the pores of the molecular sieve, it has preferential adsorption capacity for polar molecules and unsaturated molecules. Molecular sieves can be used to separate molecules with different polarities, different degrees of saturation, different molecular sizes, and different boiling points. Due to their special structure, molecular sieves have higher adsorption capacities and thermal stability, as well as a wider application range than other adsorbents.

Molecular sieves are available in the form of both natural and artificial synthetic zeolites. Natural zeolites are mostly converted from volcanic tuffs and tuffaceous sedimentary rocks in marine or lacustrine environments. Common natural zeolites include clinoptilolite, mordenite, erionite and chabazite. According to their different crystal structures, artificial synthetic zeolites are classified into different molecular sieve types, such as 3A molecular sieves, 4A molecular sieves, 5A molecular sieves, 10X molecular sieves, 13X molecular sieves, 13XAPG molecular sieves, etc. Different artificial synthetic molecular sieves are suitable for different applications. Molecular sieves are mainly used for the cleaning of low H_2S concentration gas, and modification of molecular sieves can improve the efficiency of hydrogen sulfide removal. By molecular sieve adsorption, the sulfur content in the gas can be reduced to less than 0.4 ppm. Under a temperature of 200–300 °C with steam, the saturated molecular sieves after adsorption can be regenerated. Using high-temperature steam to regenerate molecular sieves has the problem of large capital investment, which limits its application.

(2) Activated carbon adsorption

Activated carbon is a hydrophobic adsorbent and a porous carbon-containing medium. It can be prepared by high-temperature carbonization and activation of many kinds of carbon-containing materials such as coal and coconut shell. Carbon is not the only component of activated carbon. In terms of elemental composition, 80–90% is composed of carbon, which explains its hydrophobic properties. As a commonly used solid desulfurizer, activated carbon is characterized by a large adsorption capacity, acid and alkali resistance, and good chemical stability. Activated carbon can easily desorb, and there is no change in its crystal structure during desorption and regeneration at a high temperature. Moreover, it has high thermal stability and retains its original adsorption performance after repeated adsorption and desorption operations. The activated carbon used for separating inorganic sulfide (H_2S) has generally the same number of micropores and macropores, the average pore diameter of which ranges from 8 to 20 nm. When the activated carbon is used to remove sulfide, the activated carbon contains certain amount of water in order to optimize the activity level of adsorption. Therefore, activated carbon can be activated by steam. In order to improve the hydrogen sulfide removal capacity of activated carbon, some modifications on ordinary activated carbon are necessary. The commonly used modifiers are metal oxides and salts, such as ZnO, CuO, $CuSO_4$, Na_2CO_3, Fe_2O_3 and etc. According to the hydrogen sulfide removal mechanism, removal by activated carbon can be classified into three types: the adsorption method, the oxidation method and the catalytic method. Due to hydrogen sulfide removal reaction, elemental sulfur gradually deposits on the surface of activated carbon, which needs to be regenerated after a certain sulfur quantity is accumulated.

(3) Iron (III) oxide method

The removal of hydrogen sulfide from fuel gas by iron oxide is an old method that developed in the 1840s with its birth in the city gas industry. Hydrogen sulfide removal by iron (III) oxide at ambient temperature used back then is still widely used at the moment. Hydrogen sulfide removal by iron (III) oxide at medium-temperature developed in modern times has been used in some industrial plants, and hydrogen sulfide removal by iron (III) oxide at high-temperature has also been reported.

The reaction formula of the hydrogen sulfide removal process by iron (III) oxide is as follows:

$$\text{hydrogen sulfide removal}\quad Fe_2O_3 + H_2O + 3H_2S \rightarrow Fe_2S_3 \cdot H_2O + 3H_2O \tag{6.5}$$

$$Fe_2O_3 + H_2O + 3H_2S \rightarrow 2FeS + S + 4H_2O \tag{6.6}$$

$$\text{regeneration}\quad 2Fe_2S_3 \cdot H_2O + 3O_2 \rightarrow 2Fe_2O_3 \cdot H_2O + 6S \tag{6.7}$$

$$4FeS + 3O_2 \rightarrow 2Fe_2O_3 + 4S \tag{6.8}$$

Hydrogen sulfide removal by iron (III) oxide is a commonly used dry hydrogen sulfide removal method. Hydrogen sulfide in biogas is removed by chemical reaction on the surface of solid iron (III) oxide. The smaller the flow rate of biogas in the reactor, the longer the contact time, and the more fully the reaction proceeds, and the better the hydrogen sulfide removal effect. In general, the optimum reaction temperature is 25–50 °C.

When the mass fraction of iron (III) sulfide in the desulfurizer reaches 30% or more, the hydrogen sulfide removal effect becomes poor because the surface of the iron (III) oxide is covered with a layer of elemental sulfur. When the desulfurizer fails and no longer function continuously, it is necessary to contact the deactivated desulfurizer with air to oxidize Fe_2S_3 to regenerate the desulfurizer.

After many repeated uses, it is necessary to replace iron (III) oxide or iron (III) hydroxide. If iron (III) oxide covers a layer of wood chips, the same quantity of iron (III) oxide has a larger specific surface area and a lower density, which can increase the H_2S absorption rate per unit mass of the desulfurizer, and about 100 g of iron oxide wood pieces can absorb 20 g of H_2S (Song et al. 2007).

Iron (III) oxide is abundant, cheap and easy to obtain. The method of iron (III) oxide has the advantages of high removal efficiency (greater than 99%), low investment, and simple operation. The disadvantages include being sensitive to water, high hydrogen sulfide removal cost, heat release from regeneration, risk of burning in tower, decrease of reaction surface area as the number of regeneration increases, and toxic dust release.

(4) Zinc oxide method

The zinc oxide method for hydrogen sulfide is one that the above-mentioned iron (III) oxide is changed to zinc oxide. The reaction mechanism and behaviors of hydrogen sulfide removal by zinc oxide have long been recognized. Zinc oxide also has a combined partial conversion and absorption function, which can convert the organic sulfur fraction such as CO_S and CS_2 into hydrogen sulfide, and hydrogen sulfide can be further absorbed and removed.

Due to the difficulty to dissociate generated ZnS, the high efficiency of hydrogen sulfide removal, and low biogas sulfur content of less than 0.1×10^{-6} mg/m^3 after hydrogen sulfide removal, removal by zinc oxide has been applied to fine hydrogen sulfide removal processes. Compared with the iron (III) oxide method, the zinc oxide method has higher hydrogen sulfide removal efficiency and faster H_2S adsorption.

The hydrogen sulfide removal capacity of zinc oxide increases as temperature increases, but removal of H_2S can be carried out at a relatively lower temperature (200 °C). The method is suitable for treating gas with low concentration of hydrogen sulfide. According to the experimental research in industrial scale hydrogen sulfide removal rate can reach 99%. However, after zinc oxide participates in hydrogen sulfide removal, it is generally impossible to recover its hydrogen sulfide removal ability by a simple method, and the current price of zinc oxide is not as inexpensive as that of iron oxide. The reaction of hydrogen sulfide with zinc oxide is shown in 6.9.

$$H_2S + ZnO \rightarrow ZnS + H_2O \tag{6.9}$$

3. **Hydrogen sulfide removal by biological process**

Hydrogen sulfide removal by biological process includes biological hydrogen sulfide removal with and without oxygen.

The method of biological hydrogen sulfide removal with oxygen is to inject air into a digester or a separate hydrogen sulfide removal tower; hydrogen sulfide and oxygen in the air react to generate elemental sulfur or sulfate with microorganisms as catalysts. The reaction is shown in Eqs. 6.10 and 6.11.

$$2H_2S + O_2 \rightarrow 2H_2O + 2S \tag{6.10}$$

$$2H_2S + 3O_2 \rightarrow 2H_2SO_3 \tag{6.11}$$

Depending on the type of energy needed, microorganisms capable of converting hydrogen sulfide can be divided into photosynthetic bacteria and colorless sulfur bacteria. Photosynthetic bacteria require a large amount of radiant energy in the conversion process, which is difficult to achieve economically and technically. Because when micro particles of sulfur form in wastewater, the wastewater will become turbid, and light transmittance will be greatly decreased, thereby affecting the hydrogen sulfide removal efficiency.

Amongst the microbiota of colorless sulfur bacteria, not all can be used to oxidize hydrogen sulfide. Some sulfur bacteria accumulate sulfur in the interior of the cell and the growth of the bacteria will cause sludge bulking in the reactor. This can cause trouble for the separation of elemental sulfur. If sulfur cannot be separated in time, there will be further oxidation problems affecting hydrogen sulfide removal efficiency. Therefore, during the operation of the hydrogen sulfide removal unit, reaction conditions must be strictly controlled to control the dominant growth of such microorganisms.

If air is directly introduced into the upper part of the digester, and hydrogen sulfide removal takes place in the upper part of the digester and the foam layer of the wall, this type of hydrogen sulfide removal is called in-digester biological hydrogen sulfide removal. Hydrogen sulfide removal reaction is catalytically oxidized by thiobacillus, so some mechanical structures are usually installed at the top of digesters to facilitate propagation of bacteria. Since the product is acidic and corrosive, and the reaction is dependent on a stable foam layer, the hydrogen sulfide removal reaction is preferably carried out in a separate reactor.

The biological hydrogen sulfide removal outside a digester is to inject air to a separate hydrogen sulfide removal tower, passing biogas, air, and nutrient solution through the packed beds with a large specific surface area. Hydrogen sulfide is converted by the microorganism attached to the packing, by which biogas is cleaned. The generated elemental sulfur or sulfate still exists in the liquid phase.

As shown in Fig. 6.4, the outside-digester biological hydrogen sulfide removal reactor is somewhat similar to a scrubber in which porous packing is installed for microorganisms to grow onto. Auxiliary facilities such as sewage tanks, pumps and blowers are equipped as well. The sewage tank holds the spray liquid containing alkali and nutrients, and the liquid digestate can be used as a spray liquid.

The packing is sprayed periodically, and the spray has the function of washing away acidic products and providing nutrients to microorganisms. Under the condition

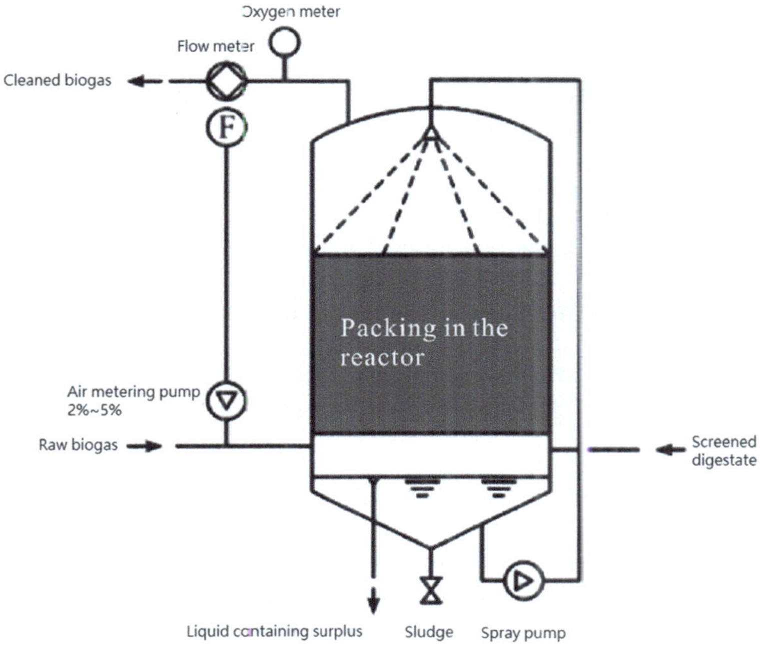

Fig. 6.4 Biological hydrogen sulfide removal process

of sufficient air, hydrogen sulfide removal efficiency can reach as high as 99% (Yin and Zhang 2017). As to rinsing, air/water mixed pulse and intermittent flushing is used. The deposition of elemental sulfur should be avoided when spraying stops (Yang et al. 2010, 2013).

The advantages of biological hydrogen sulfide removal with oxygen are that this method require neither catalysts nor chemical sludge disposal. It produces little biological sludge, consumes low energy, and recovers elemental sulfur with high efficiency. The disadvantage is that if too much oxygen is injected into the system, there is a risk of explosion.

Biological hydrogen sulfide removal without oxygen is a new biological hydrogen sulfide removal process developed by BIOMA (Deng et al. 2009). The process uses nitrate and nitrite in the aerobic post-treatment of digested effluent as electron acceptor and H_2S in biogas as electron donor to react with microorganisms, achieving simultaneous removal of nitrogen and hydrogen sulfide. The main reactions are as follows:

$$5S^{2-} + 2NO_3^- + 12H^+ \rightarrow 5S^0 + N_2 + 6H_2O \quad (6.12)$$

$$5S^0 + 6NO_3^- + 2H_2O \rightarrow 5SO_4^{2-} + 3N_2 + 4H^+ \quad (6.13)$$

$$5S^{2-} + 8NO_3^- + 8H^+ \rightarrow 5SO_4^{2-} + 4N_2 + 4H_2O \tag{6.14}$$

Given the conditions that NO_x–N (the sum of NO_3^-–N and NO_2^-–N) concentration in the influent is 270–350 mg/L, hydrogen sulfide content in biogas is 1273–1697 mg/m^3, hydraulic retention time is 0.985–3.72 d, and the unloaded tower residence time is 3.94–15.76 min, the NO_x–N removal efficiency can reach 96.4–99.9% with an effluent NO_x–N concentration of 0.114–110.6 mg/L. The hydrogen sulfide removal rate is 96.4–99.0% with the hydrogen sulfide concentration of about 100 mg/m^3.

This process has the following advantages: nitrogen in wastewater and hydrogen sulfide in biogas are simultaneously removed; nitrogen removal from liquid digestate does not require additional carbon source; hydrogen sulfide removal from biogas does not require the addition of oxygen nor desulfurizer; cost is low (Deng et al. 2009).

6.2.2.2 Devices for Hydrogen Sulfide Removal

Hydrogen sulfide removal from biogas mainly uses iron (III) oxide methods and the biological process. The following section mainly introduces three devices for hydrogen sulfide removal based on iron oxide methods and the biological process.

(1) Small device for hydrogen sulfide removal based on iron (III) oxide methods

Biogas generated by household digesters or small biogas plants has a small flow and a relatively low concentration of hydrogen sulfide. Usually, a small device containing iron oxide is used to directly remove hydrogen sulfide. This small device consists of desulfurizer and a vessel (Fig. 6.5). The volume of a small device is generally not less than 2.5 L, and the ratio of height to diameter is 2:1–3:1.

The vessel shall be made of material with pressure resistance (not less than 10 kPa), corrosion resistance, and high temperature resistance (greater than 110 °C). ABS plastic is generally used for high temperature resistance. Wall thickness of any part of the vessel shall be greater than 2 mm, and there shall be no defects such as pores and cracks. The cover shall have a gasket. The gasket shall be wear-resistant and elastic.

The pressure drop generated by the first use of the vessel should be less than 200 Pa. The inlet and outlet of the vessel adopts PVC hoses or PE pipes which match the biogas transmission pipe. The length of the hoses or pipes should be no less than 20 mm, and there are 3 sealing joints or wire ports. When a hose is connected, the buckle should be fastened and no gas leakage is allowed. The weight of the small device for hydrogen sulfide removal is generally greater than 2000 g.

Iron (III) oxide desulfurizer particles with cylindrical shapes are usually used, the diameter (φ) of which is usually from 4 mm to 6 mm with a length (L) of 5 mm to 15 mm. The sulfur capacity of the desulfurizer in the first run should not be less than 12%, and the cumulative sulfur capacity should not be less than 30%, and the bulk

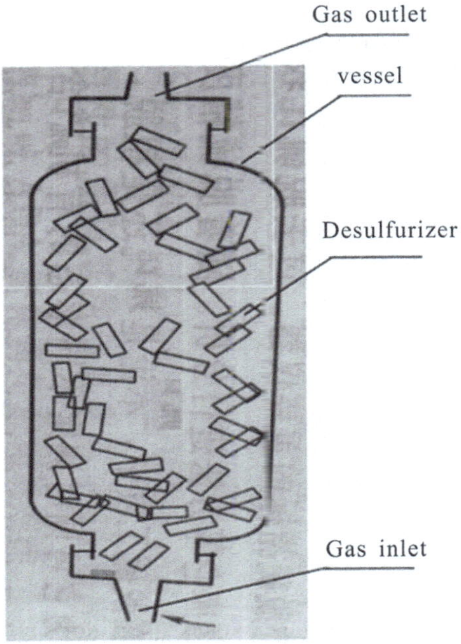

Fig. 6.5 A small device for hydrogen sulfide removal

density of desulfurizer should be 0.60–0.80 kg/L. The desulfurizer should be easy to replace, and there should be no short circuit in the flow of biogas in the device.

When the small device for hydrogen sulfide removal is used independently, it should have a clear indication of the direction of entry and exit. When the hydrogen sulfide removal capacity reaches saturation, the desulfurizer needs to be discarded and oxidized in the air. Preferably it is exposed to sunlight until the color changes from black to orange, yellow, or brown. Once oxidation is done, the desulfurizer is put back into the vessel. Desulfurizer regeneration cannot be carried out in the vessel. The desulfurizer should not be used for more than 3 times. Desulfurizer must be replaced after uses for some time. The timing of replacement is determined by hydrogen sulfide content of the biogas to be processed.

(2) Tower for hydrogen sulfide removal based on iron (III) oxide methods

Tower for hydrogen sulfide removal based on iron (III) oxide methods is the main device for sulfide removal used in large and medium biogas plants in China. If its desulfurizer is made from iron (III) oxide, the hydrogen sulfide removal device usually takes the form of a tower (Fig. 6.6). Hydrogen sulfide in biogas can be removed by single-stage or multi-stage devices. The appropriate type should be adopted according to the specific situation, the concentration range of H_2S, and the degree of hydrogen sulfide removal.

One-stage hydrogen sulfide removal is suitable for biogas with an H_2S content of 2 g/m^3 or less, whereas biogas with an H_2S content of 2–5 g/m^3 should better use two-stage hydrogen sulfide removal. Biogas with an H_2S content of 5 g/m^3 or more

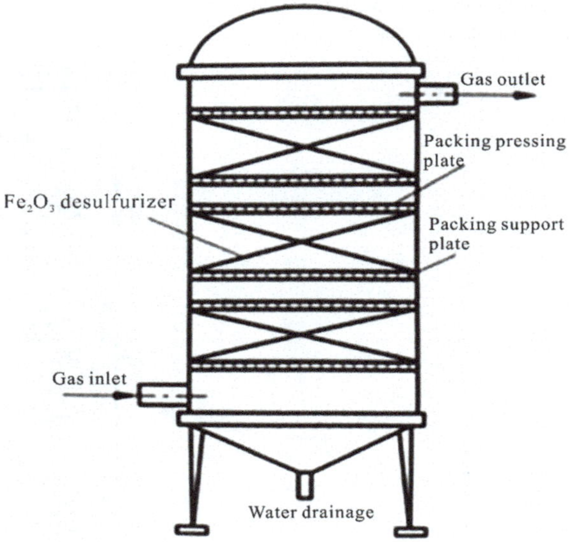

Fig. 6.6 Schematic diagram of the tower for hydrogen sulfide removal by iron (III) oxide

needs a three-stage hydrogen sulfide removal. If the H_2S content is above 10 g/m^3, it is best to use wet methods for crude removal first, followed by fine removal with iron (III) oxide. The tower for hydrogen sulfide removal is a packed bed composed of a tower body, two shell covers, an inlet and outlet gas pipe, an inspection hole, a sewage hole, a bracket and an internal wooden grille.

In order to prevent the condensed water from deposition at the top of the tower that soaks the desulfurizer, a certain thickness of crushed aluminum silicate fiber cotton or other porous filler can be laid on the top desulfurizer to block the condensed water. Depending on the amount of biogas to be treated, the tower can be classified into single-layer beds or bunk beds. Generally, when the bed height is about 1 m, a single bed is taken. If the height is greater than 1.5 m, a bunk bed is adopted (Zhou et al. 2009).

① Space velocity

Space velocity refers to the volume of biogas that can be treated by per unit volume of desulfurizer per hour, in units of h^{-1}. Its expression is:

$$V_{SP} = V_m/V_t \tag{6.15}$$

where

V_{SP} space velocity of biogas, h^{-1}
V_m biogas flow, m^3/h
V_t desulfurizer volume, m^3

It is not difficult to see from the equation above that space velocity is one of the important parameters which indicates the performance of the desulfurizer. Different desulfurizers have different activities. In selecting space velocity, a comprehensive consideration should be given according to factors such as the concentration of H_2S in biogas, the operating temperature, and the elevation of the workspace.

The higher the space velocity is selected, the shorter the contact time of biogas and desulfurizer. That is, the contact time t_j is the reciprocal of the space velocity.

$$t_j = 1/V_{SP} \tag{6.16}$$

where

t_j contact time, s

Under normal temperature and pressure, when the concentration of H_2S in biogas is less than 3 g/m^3, t_j is 100 s, and the space velocity is 36 h^{-1}. If the H_2S content is 4–8 g/m^3, t_j is 450 s (the equivalent space velocity is 8 h^{-1}). If biogas contains a small amount of oxygen, the space velocity can be increased to a suitable extent.

② Linear velocity

Linear velocity refers to the speed at which biogas passes through the bed of the desulfurizer, and the value is the ratio of the bed height to the contact time.

$$U_S = H_{ch}/t_j \tag{6.17}$$

where

U_S linear velocity, mm/s
H_{ch} bed height, mm
t_j contact time, s.

Linear velocity of biogas passing through the tower is a key parameter in designing the size of the device. If the linear velocity is too low, biogas is in a stagnant state. As the linear velocity increases, the gas flow enters the turbulent zone. The gas film thickness is reduced to a greater extent, thereby increasing the activity of the desulfurizer. Linear velocity is preferably 0.020–0.025 m/s when choosing cylindrical desulfurizers.

③ Bed height

The filling height of each layer of particle desulfurizer is 1.0–1.4 m. When the height of the bed exceeds 1.5 m, a double layer can be used. Layered filling is beneficial to avoid bias flow or partial short circuit and to improve hydrogen sulfide removal effect.

④ Ratio of height to diameter

Hydrogen sulfide removal by iron (III) oxide is a chemical adsorption process. During each stage of adsorption, the bed of hydrogen sulfide removal can be divided into spare zone, working zone and saturation zone. The working zone is the part of the bed that performs hydrogen sulfide removal. The elevation of the working zone is related to the activity of the desulfurizer and space velocity. According to the test results of TTL type desulfurizer, when the space velocity is 50 h^{-1} or less and the concentration of H_2S is 1–3 g/m^3, the height of the bed is 3–4 times the tower diameter.

⑤ Desulfurizer replacement time

The replacement time of desulfurizer is proportional to the technical performance of the desulfurizer (the working sulfur capacity and the amount of desulfurizer), and inversely proportional to H_2S concentration and biogas flow rate. For small, medium and large biogas plants, it is necessary to consider the size of the device for hydrogen sulfide removal, the size of the land area, and maintenance inconvenience caused by frequent replacement of desulfurizer. Generally, the renewal frequency of desulfurizer is every 6 months. The amount of desulfurizer can be calculated by the following formula.

$$G = \frac{t \cdot c \cdot V}{s \cdot 1000} \quad (6.11)$$

where

G the amount of the desulfurizer filled in device, kg
t time of desulfurizer use, d.
s sulfur capacity of desulfurizer, %
C H_2S content in biogas, g/m^3
V biogas flow, m^3/d

⑥ Structure and material of the tower

The towers are usually manufactured by welding Q235-A or Q235-A.F steel plates. The inner surface of the tower should be coated with two layers of anti-rust paints or epoxy resins. The outer surface should be coated with 1 to 2 layers of anti-rust paints. Seal of the shell cover is made of asbestos rubber or neoprene. The temperature monitoring uses a WNG-12 type glass thermometer with right-angle metal protector, and it is located at a height of 100 mm from the bottom of the bed.

The liquid level is displayed at the lower part of the tower below the inlet by plexiglass tube to prevent the condensed water from accumulating at the bottom and from affecting normal operation of the desulfurizer. The observation window can be placed 50 mm above the bed, using plexiglass to observe the changes in bed and well control regeneration time of the desulfurizer.

(3) Tower for hydrogen sulfide removal outside the digester based on biological process

There are two types of towers for hydrogen sulfide removal outside the biogas digester based on biological process. One is to directly blow air into the tower, and the other is to aerate in the regeneration tank outside the tower. When oxygen is not allowed to mix into biogas or biological natural gas, aeration in regeneration tank outside the tower is generally used. Biological hydrogen sulfide removal without oxygen also takes this form.

Main design parameters of biological hydrogen sulfide removal outside the digester include:

Biological hydrogen sulfide removal load: 8–10 m^3 biogas/(h m^3 packing).

Gas-water ratio (biogas: spray solution): (5–10): 1.

Air addition amount: 2–5% of the volume of biogas treated or 1–1.5 mg/L of dissolved oxygen in the spray solution.

Process temperature: 25–35 °C.

pH of the spray solution: greater than 6.0.

The following section mainly introduces the biological hydrogen sulfide removal system aerated in a regeneration tank outside the tower. The system does not only have a tower, but also must be equipped with a rich solution tank, a regeneration tank, a sedimentation tank, a lean solution tank, a circulating pump, a blower, metering pumps, a heating unit, an automatic dosing unit, and a water replenishment unit. Details of these components are described as below (Fig. 6.7) (Yin and Zhang 2017).

① The tower body

The height-to-diameter ratio of the tower is (8–10):1, and a tower can be divided into 3–5 sections. The lower section collects the absorption liquid and overflows to the rich solution tank. The middle section is the packing layer. Because H_2S is corrosive, the selected packing needs to be corrosion-resistant and evenly layered to

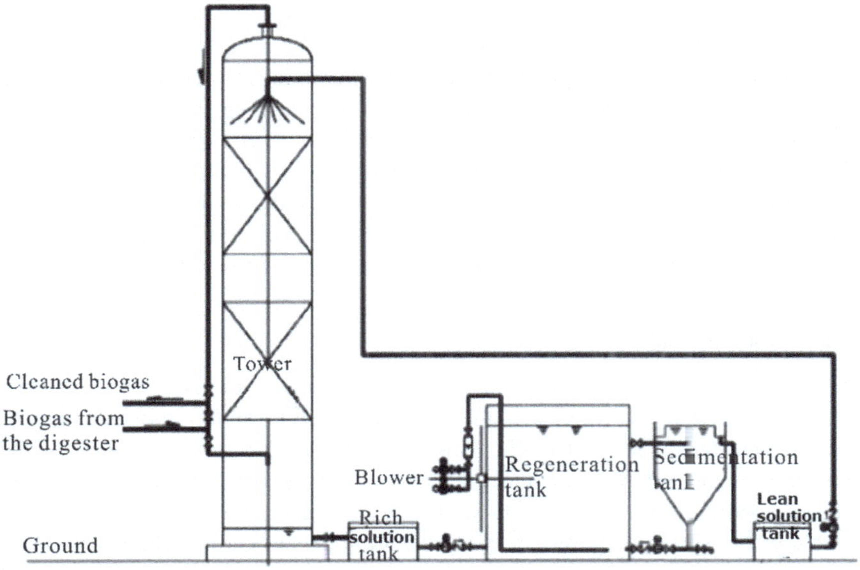

Fig. 6.7 Process flow chart of hydrogen sulfide removal outside the digester based on biological process (Yin and Zhang 2017)

ensure uniform contact between circulating solution and biogas for improved gas-liquid contact efficiency, and the packing should also be able to withstand sufficient strength.

The absorbent liquid is evenly sprayed onto the packing via the upper water distributor. The upper part is equipped with a demister to prevent generated elemental sulfur from entering and blocking the pipe. A water distributor for backwashing the demister is arranged in the absorption tower, and the elemental sulfur accumulated in the demister can be regularly washed out.

② Rich solution tank

After absorbing H_2S, the HS^--rich absorbent liquid flows into the rich solution tank, which buffers and ventilates the system so as to ensure that the circulating pump will drive the solution in the rich solution tank into the regeneration tank as a part of safe operation. The nutrient salts required for the colorless sulfur bacteria such as magnesium sulfate, dipotassium hydrogen phosphate, urea, and the supplementary alkali are all added through the rich solution tank. The hydraulic retention time of the rich solution in the tank is 0.5–1.0 h.

③ Regeneration tank

The regeneration tank has an aeration unit, a heating unit, an automatic dosing unit, and a water supply unit to provide the circulating solution with dissolved oxygen, growth temperature, suitable pH and nutrition required by microorganisms. The HS^--rich absorbent liquid enters the regeneration tank and is then converted to elemental sulfur (S) and OH^- by microorganisms. The regeneration tank is aerated with a blower to provide required oxygen for microbial growth and metabolism. Since the regeneration reaction will generate sulfur foam, a sulfur foam trough that encircles the bioreactor is required at the upper reactor. A liquid discharge pipe with a diameter of 100–200 mm is set at the upper part of the regeneration tank through which regenerated solution directly enters the sedimentation tank. Since the regeneration tank will produce sludge, a sludge pipe with a diameter of 200 mm is required at the bottom. The hydraulic retention time of the regeneration tank is 0.5–2.0 h.

④ Sedimentation tank

The role of a sedimentation tank is mainly to precipitate the liquid rich in sulfur from the regeneration tank and to separate the sludge and water, and the sludge containing sulfur is then discharged out of the system. The hydraulic retention time of the sedimentation tank is 0.5–2 h. An inclined tube can be arranged inside to strengthen the sedimentation effect and prevent sludge from entering the absorption tower to block the filler. The lower part of the sedimentation tank is a cone, which serves as a sludge storage tank and is equipped with a sludge discharge pipe with a diameter of 200 mm. It is necessary to regularly discharge sludge from the sedimentation tank and tap water should be regularly replenished.

⑤ Lean solution tank

The supernatant liquid layer from the sedimentation tank is returned to the lean solution tank. The lean solution tank acts as a buffer and a gas release unit. The hydraulic retention time of the lean solution tank is 0.5–1.0 h.

⑥ Circulating solution unit

The lean solution obtained after the rich solution regeneration is pumped to the top of the tower by a circulation pump to form circulating solution. The circulating solution connects the tower, the regeneration tank, the sedimentation tank, the circulating pump dosing device and the heating device. The sprinkler unit allows full contact between the biogas and the circulating solution to promote conversion of H_2S from gas phase to liquid phase. Sludge discharged from the bottom of the sedimentation tank is transferred into the digestate storage tank.

⑦ Automatic dosing unit

Microbial growth requires certain nutrients, which are automatically added in the form of nutrient solution. The nutrient solution is added to the circulating solution, and the dosage is precisely controlled by the metering pump. Since some of the HS^- is converted to sulfate, pH of the circulating solution gradually decreases. Therefore, it is necessary to continuously add alkali to neutralize sulfate ion and alkali is added in the same form as the nutrient solution.

⑧ Structure and material of tower

The tower can be welded by Q235-A or Q235-A.F steel plates. The inner surface of the tower should be coated with epoxy resin. The outer surface should be coated with 1–2 mm of anti-rust paint. It is best to use acid-resistant glass fiber-reinforced plastic. All fillers and attachments are anti-corrosion materials such as plastic and stainless steel S316. The material of biogas pipeline is PE, and the material of air pipe can be galvanized steel.

6.2.3 Oxygen and Nitrogen Removal

Oxygen and nitrogen can be removed by activated carbon, molecular sieves, or membranes. Oxygen and nitrogen are also partially removed during hydrogen sulfide removal processes or biogas upgrading processes. However, it is difficult to separate oxygen and nitrogen from biogas. If there are demanding requirements for oxygen and nitrogen (for instance, when biogas injected into natural gas network or used for vehicle gas) in biogas, the biogas can hardly be acceptable unless oxygen and nitrogen content is low enough. One of the most advantageous methods available as of now are catalytic removal using palladium/platinum catalysts and chemisorption with copper.

6.2.4 Removal of Other Trace Gases

Trace gases in biogas include ammonia, siloxanes and benzene gases, the occurrence of which in agricultural biogas plants is rare. These trace gases are often removed during the cleaning process of hydrogen sulfide removal and water removal described above. For example, ammonia is easily soluble in water, so it is usually removed during the water removal stage of biogas. Siloxane and benzene gases can be absorbed by organic solvents, strong acids, strong bases, silica gel, or activated carbon, or removed at low temperatures.

6.3 Biogas Upgrading

The main purpose of biogas upgrading is to remove CO_2 to obtain high-purity methane. Biogas upgrading technology is mostly derived from decarbonization technology of natural gas and syngas from ammonia synthesis. Since the treatment capacity of biogas is much smaller than that of natural gas or synthetic ammonia conversion gas, more attention should be paid to miniaturization and energy saving in the selection of decarbonization technology. At present, widely used biogas upgrading technology mainly includes six kinds: pressure swing adsorption, water washing, organic solvent physical absorption, organic solvent chemical absorption, high pressure membrane separation, and low temperature upgrading. The details are as follows.

6.3.1 Pressure Swing Adsorption (PSA)

The principle of pressure swing adsorption (PSA) is to use the difference in size and physical properties of gas molecules, as well as the difference in adsorption capacity, adsorption rate, and affinity to the adsorbent by different gas components. The adsorption capacity varies with pressure. Separation of the mixed gas by adsorption is achieved during pressurization, and regeneration of the adsorbent is completed at the time of reduced pressure, and gas separation is achieved this way.

The amount of absorbed component is affected by pressure and temperature. When pressure increases, the absorbed amount also increases. On the other hand, when pressure decreases, adsorption is done to a lesser extent and this amount decreases. When temperature rises, the amount of adsorption decreases. When temperature decreases, the amount of adsorption increases. Commonly used adsorbent materials are activated carbon, zeolite, and molecular sieves (mostly carbon molecular sieves). Commercial application of pressure swing adsorption (PSA) began in the 1860s.

Apart from CO_2, other gas molecules such as H_2S, NH_3 and H_2O can also be adsorbed. In real world engineering, H_2S and H_2O should be removed before biogas enters the adsorption tower. Part of N_2 and O_2 can also be adsorbed together with CO_2. From data provided by large upgrading stations, about 50% of N_2 will be discharged with the off-gas. The purity of biomethane obtained by pressure swing adsorption is greater than 96% (Beil et al. 2012).

Biogas upgrading stations usually arrange 4–6 adsorption towers in parallel, and complete adsorption (adsorption of vapor and carbon dioxide), decompression, desorption (that is, desorption by a large amount of raw gas or product gas), and pressurization. The operating pressure range of pressure swing adsorption is 1×10^5 to 1×10^6 Pa. Most pressure swing adsorption systems pressurize biogas to 4×10^5 to 7×10^5 Pa, so the pressure loss is about 1×10^5 Pa. After hydrogen sulfide and water are removed, the biogas passes to an adsorption tower packed with molecular sieves. Operating temperature range is 5–35 °C (Berndt 2012).

Most of the CO_2 will be adsorbed onto the surface of molecular sieve, while most of the CH_4 will pass through molecular sieve, and only a small amount of CH_4 can be adsorbed. After the product gas leaves the adsorption tower, packing is desorbed by releasing pressure. By increasing the number of flushing and desorption cycles of the feed gas or the product gas, as well as the number of recycling exhaust generated by upstream compressor, methane concentration can further increase, which in turn increases the upgrading cost. According to the data provided by manufacturers and plant operators, the actual operating load is 40–100% of the rated load (Radlinger 2010). Figure 6.8 shows a four-tower PSA process.

In the mid 1980s, power consumption of pressure swing adsorption was 0.35 kWh/m^3 of raw biogas. By 2012, the power consumption was reduced to 0.16–0.18 kWh/m^3 of raw biogas. A biogas upgrading station reported that at an absolute pressure of 3×10^5 Pa, biogas with a methane content of 65% requires electricity consumption of 0.17 kWh/m^3, while biogas with a methane content of 55% requires electricity consumption of 0.18 kWh/m^3.

Another biogas upgrading station reported a power consumption of 0.19–0.23 kWh/m^3 under the same conditions. When the absolute pressure is 5.4×10^5 Pa, the upgraded methane has a methane content of 96–97% and a treatment capacity of 1000 m^3/h (Radlinger 2010).

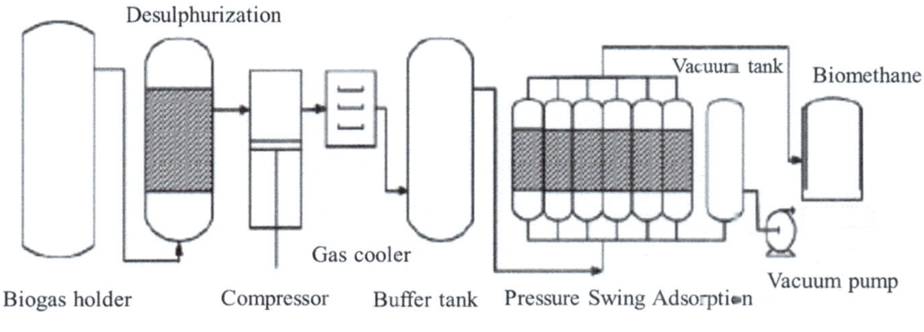

Fig. 6.8 Flow chart of pressure swing adsorption process (Sun et al. 2014)

In previous systems, methane recovery rate was generally 94% (methane loss rate was 6%). In the new system, methane recovery is generally 97.5–98.5% (methane loss rate is 1.5–2.5%). On the one hand, the off-gas is utilized more efficiently. On the other hand, a higher methane concentration can be obtained. By obtaining 17% to 18% of methane in the off-gas, methane concentration of the upgraded biogas is as high as 99% or more (Qi and Guo 2010).

Since the off-gas contains a significant amount of CH_4, it must be oxidized. Sulfur content in the off-gas is not large, so "catalytic oxidation" and "non-flame oxidation" are often used as off-gas treatment technologies in large plants. If the concentration of methane in the off-gas is low enough, regenerative thermal oxidation (RTO) can also be used.

6.3.2 *Water Scrubber*

Water scrubbing is a method of separating biogas by using different solubility of methane and carbon dioxide in water (see Fig. 6.9). Solubility of carbon dioxide in water is larger than that of methane. Therefore, biogas can be removed by a water scrubber. There are usually two types of water scrubbing processes: single-pass absorption and regenerative absorption. The latter uses recycled water, while the former does not. Water scrubbing is a reversible absorption process based on van der Waals force, which belongs to physical absorption. Low temperature and high pressure can increase the absorption rate.

In the water scrubbing process, biogas is first pressurized to 4×10^5–8×10^5 Pa and then sent to the scrubber tower (MT-Biomethan GmbH 2011). Biogas in the scrubber tower is in contact with the countercurrent water flow from bottom to top, and CO_2 and H_2S are dissolved in water in order to be separated from CH_4. CH_4 is discharged from the upper end of the scrubber tower and further dried to obtain biomethane.

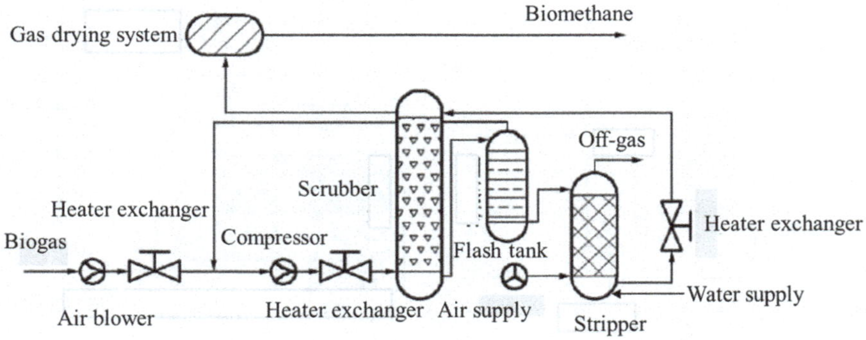

Fig. 6.9 Flow chart of the water scrubbing process (Sun et al. 2014)

Under pressure, a part of CH_4 is also dissolved in water, so water discharged from the bottom of the scrubber needs to enter a flash tower, and the CH_4 and some CO_2 dissolved in water are released from water by depressurization. The released gas is mixed with the feed gas again to participate in the scrubber separation.

Water discharged from the flash tower enters the desorption tower and is regenerated by using air, steam or inert gases (Krich 2005). When H_2S content in the biogas is high, it is not advisable to use the air blow-off method to regenerate water, because the air blow-off method will produce elemental sulfur pollution and block the pipelines. In this case, regeneration can be carried out by steam or inert gases, or hydrogen sulfide removal pretreatment of biogas can be carried out prior to regeneration.

In addition, another problem with air stripping is the increase in concentration of oxygen and nitrogen in the biomethane (Anneli 2009). In places where water resources are relatively cheap, fresh water can always be used without regeneration, which simplifies the system and improves the upgrading efficiency.

Water scrubbing method is highly efficient, and it requires no complicated operation management. A single scrubber tower can upgrade CH_4 concentration to 95% (Schomaker 2000), and the loss ratio of CH_4 can also be controlled at a relatively low level (<2%). When H_2S is lower than 300 ppm, it can be removed at the same time. And because of the use of water as an absorbent, it is also a relatively inexpensive upgrading method. The water scrubbing method is more economical when it is not necessary to regenerate water, and the processing capacity can be adjusted by changing pressure and temperature.

The main problem of the pressurized water scrubbing method is that the investment is large, and the operating cost is high. The microorganisms will grow on the surface of packings in the scrubber tower to form a biofilm, thereby causing clogging of packings. Therefore, it is necessary to install an automatic flushing device or solve the problem by chlorination sterilization (Tynell 2005). Although the water scrubbing process can simultaneously remove H_2S, in order to avoid corrosion of the compression equipment used in the upgrading stage, H_2S should be removed before CO_2 removal. In addition, since the product gas is in a state of water saturation, dehydration of the product gas is required (Zheng et al. 2013).

Power demand for the water scrubbing method is 0.20–0.30 kwh/m^3 of raw biogas. At 5×10^5 Pa (absolute pressure), the current demand values provided by technology providers are 0.22 kwh/m^3 (large upgrading station) and 0.25 kwh/m^3 (small upgrading station). The average power demand recorded by plant operators is 0.26 kwh/m^3 (Wellinger et al. 2013).

According to different sizes of upgrading stations, the demand for water is about 1–3 m^3/day. That is, the upgrading of 1 cubic meter of raw biogas per day corresponds to 2.1–3.3 L of water. In addition, the water consumption reported by the technology providers is less than 1–2 m^3/day (Ohly 2010).

The methane recovery rate is 98.0–99.5% (methane loss is 0.5–2%. The methane loss provided by a technology provider is 1%, or even reduced to 0.5%. A factory operator has actually recorded a methane recovery ranging from 98.8 to 99.4% (Wellinger et al. 2013).

Due to dilution of the air in the desorption tower, methane concentration in the off-gas is much lower than 1%, but the off-gas still contains CH_4. Therefore, the off-gas must be discharged after treatment. Since off-gas usually includes sulfur and CH_4 of a low concentration, regenerative thermal oxidation (RTO) is commonly used for off-gas treatment in large-scale applications for water scrubber (Beil et al. 2012).

6.3.3 Physical Absorption by Organic Solvents

Physical absorption by organic solvents is a physical absorption process (Fig. 6.10). The physical absorption method with organic solvents is similar to the water scrubbing method, with a major difference that the absorbent of the former method is organic solvent. The solubility of CO_2 and H_2S in organic solvents is greater than that in water. Therefore, the amount of liquid phase circulation used for upgrading the same amount of biogas is smaller, together with a smaller power consumption and a lower upgrading cost.

Typical physical absorbents are propylene carbonate (PC method), polyethylene glycol dimethyl ether (Genosorb method), low temperature methanol, and N-methylpyrrolidone. Another common physical absorbent is Selexol®, whose main component is dimethyl polyethylene glycol (DMPEG). Water and halogenated hydrocarbons (mainly from landfill biogas) can also be removed by absorption with Selexol, the recycling of which is usually carried out with steam or inert gases (Song et al. 2007).

One of the features of organic solvent physical absorption method is that CO_2, H_2S, and H_2O can be simultaneously absorbed in the scrubber, and NH_3 can also be absorbed, but formation of unfavorable intermediate products should be avoided. Table 6.6 lists solubility of each gas during biogas upgrading (25 °C, tetraethylene glycol dimethyl ether as absorbent). There is no SO_2 in raw biogas, the source of which is H_2S combustion. Concentration of methane in the product gas ranges from 93 to 98%.

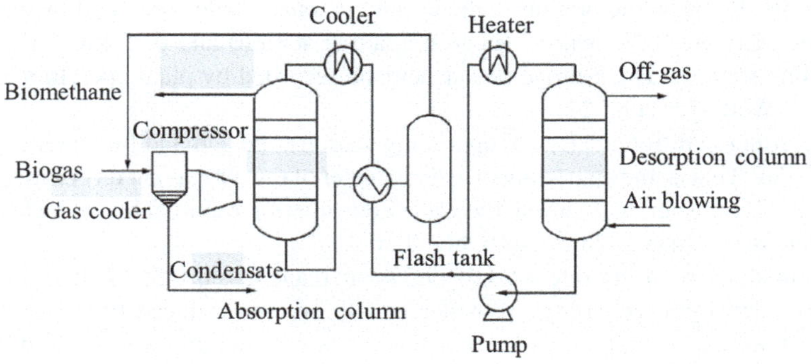

Fig. 6.10 Flow chart of the process of physical absorption by organic solvent (Sun et al. 2014)

Table 6.6 Solubility of selected gas (25 °C, tetraethylene glycol dimethyl ether as adsorbent, unit cm^3/g)

CH_4	CO_2	H_2S	NH_3	O_2	H_2	SO_2
0.2	3.1	21	14.6	0.2	0.03	280

Raw biogas is pressurized to 4×10^5–8×10^5 Pa before entering the absorption column. In current applications, the operating pressure is typically 6×10^5–7×10^5 Pa. The condensed water generated during pressurization needs to be discharged from the system. The operating temperature in the absorption column is 10–20 °C. Organic solvent scrubbers differ from water scrubbers in design and operation. For water scrubbing, thorough hydrogen sulfide removal is generally not required.

Organic solvent absorption, however, requires hydrogen sulfide removal and drying to a much greater extent. After water and hydrogen sulfide removal, the product gas is then processed by entering from the top of the scrubber. To reduce methane loss, two flash towers are typically employed. Desorption can be achieved by stripping with heating (about 40–80 °C) in an absorption column (Girod 2009) or by thermal decoupling (heat exchanger/cooling compression/off-gas treatment) without any external heat source. The actual operating load is 50–100% of the rated load (Wellinger et al. 2013).

The power consumption is around 0.23–0.33 kWh/m^3 of raw biogas. In latest upgrading stations, the expected electricity consumption is about 0.23–0.27 kWh/m^3 of raw biogas. The heat demand of 0.10–0.15 kWh/m^3 of raw biogas can be reduced by increasing the heat recovery rate of the plant (Wellinger et al. 2013).

Since the off-gas contains 1–4% of CH_4 (CH_4 recovery rate is 96–99%), off-gas cleaning is a must-do. Since the off-gas also usually contains H_2S, a typical off-gas treatment technology is regenerative thermal oxidation (RTO), which is a general method for off-gas treatment in large upgrading stations (Beil et al. 2012).

6.3.4 Chemical Absorption by Organic Solvents

Chemical solvent absorption is commonly used to remove CO_2 from biogas. Nowadays, a wide range of chemical solvents are used, such as various alcohol amines and alkali solutions. The absorbent liquid in the absorption tower reacts with CO_2 to separate out CO_2. The rich solution that absorbs CO_2 enters the desorption column, and CO_2 is then released by heating so that regeneration of the absorbent and separation of CO_2 are eventually achieved.

Chemical absorption of biogas by organic solvent, commonly referred to as "amine scrubbing", is a method of separating CO_2 and CH_4 using an amine solution (Beil et al. 2012). Different alkanolamine solutions can be used for separating CO_2 by chemical absorption processes, and different equipment manufacturers require different ethanolamine-water mixtures as absorbents. Commonly used amine solutions

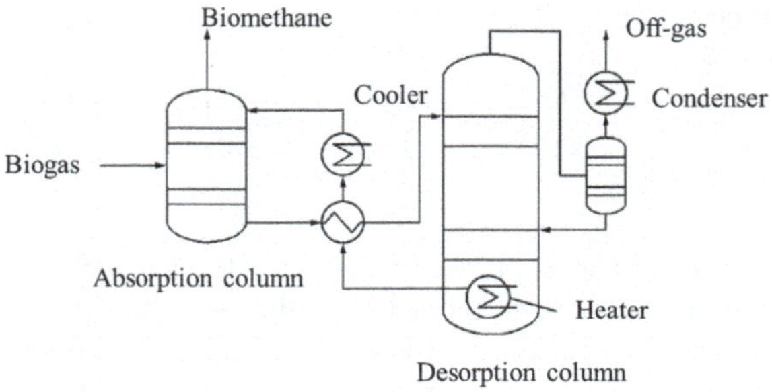

Fig. 6.11 Flow chart of chemical absorption by organic solvents (Sun et al. 2014)

are ethanolamine (MEA), diethanolamine (DEA), methyldiethanolamine (MDEA) (Foo 2009; Zhang et al. 2011), in which N-methyldiethanolamine (N-MDEA) has chemical stability, slight corrosion, and good selectivity. The reaction equation of alkanolamine and CO_2 are as follows.

$$2RNH_2 + CO_2 \rightarrow RNHCOONH_3R \tag{6.19}$$

$$2RNH_2 + CO_2 + H_2O \rightarrow (RNH_3)_2RCO_3 \tag{6.20}$$

$$(RNH_3)_2RCO_3 + CO_2 + H_2O \rightarrow 2RNH_3HRCO_3 \tag{6.21}$$

Since the 1970s, the chemical absorption of amine solutions has been used for the separation of CO_2 and H_2S from acid gases (Diez et al. 2012). Figure 6.11 is a flow chart of the process of chemical absorption by organic solvent.

Apart from CO_2, H_2S can also be absorbed during the amine scrubbing process. However, in most practical applications, there will be a hydrogen sulfide complete removal step prior to sending the gas to the absorber in order to reduce energy requirements of the regeneration process. Methane concentration of the obtained product gas is over 99%. Since N_2 cannot be absorbed, N_2 in the original biogas will reduce the quality of the product gas, but other upgrading methods also have such problems. In addition, entry of oxygen should be avoided as oxygen can introduce side reactions that degrade the amine (Girod 2009).

Since CO_2 is chemically reactive with amine solution after getting absorbed, the absorption process should be carried out under a relatively low pressure. Generally, only a slight pressure increase from existing biogas pressure is required. Regeneration process of the amine solution is difficult and requires a temperature of 160 °C. Therefore, the operation process requires a large amount of heat, and this high energy consumption is considered as a disadvantage of this process. In addition, due to its loss from evaporation, the amine solution has to be often replenished in the process of absorption (Petersson 2009). If the biogas contains a high concentration of H_2S, it needs to be desulfurized in advance, otherwise it will cause chemical absorption poisoning (Hara 1997).

The power consumption of amine scrubbing is about 0.06–0.17 kWh/m^3 of raw biogas. One manufacturer has provided a power consumption warranty value of 0.09 kWh/m^3 (methane concentration in the original biogas is 65%) and 0.11 kWh/m^3 (methane concentration in the original biogas is 55%), both values corresponding to product gas pressures of 5 × 10^3–1.5 × 10^4 Pa and description temperatures of 135–145 °C (Wellinger et al. 2013). Another manufacturer has reported a power consumption of 0.17 kWh/m^3, with corresponding product gas pressure of 2.5 × 10^5 Pa and a desorption temperature of 120–130 °C. Desorption process heat demand is 0.4–0.8 kWh$_{th}$/m^3 raw biogas (Boback 2012; Burmeister et al. 2009).

Compared with other purification methods, the amine scrubbing method has a large absorption capacity, with a methane recovery rate as high as 99.9% (Petersson 2009) and a small methane loss (<0.1%). The off-gas usually does not require further treatment. The solvent is recyclable, and the operation cost is low. However, the disadvantages include a large investment, high regeneration temperatures, and large energy consumptions. Moreover, decomposition and poisoning happen when O_2 or some other substances are present, and precipitation with salt, foaming, and corrosion are also likely to occur by this method.

6.3.5 *Membrane Separation*

Membrane separation, also known as gas permeation, is based on different permeabilities of gas compounds through polymer membranes (Beil et al. 2012). The main features of the membrane method are that there is no phase change, the devices are simple, and the sizes of the devices can be large or small. The three different fluids in the membrane system are the feed (raw biogas), the permeate (CO_2-rich) gas and the retentate (CH_4-rich gas). The pressure gradient formed by different partial pressures of each compound between the retentate and permeate side is defined as the driving force of the system. Increasing feed (retentate) side pressure and lowering permeate side pressure can achieve high flux rates. Figure 6.12 is a flow chart of the membrane separation process. Most of the separation membranes currently used are polymer membranes, mainly including cellulose derivatives, polysulfones, polyamides, polyesters, polyolefins, and silicon-containing molecular membranes.

A membrane module is typically composed of a membrane component and a shell. Some membrane modules have only one component, but most of them have

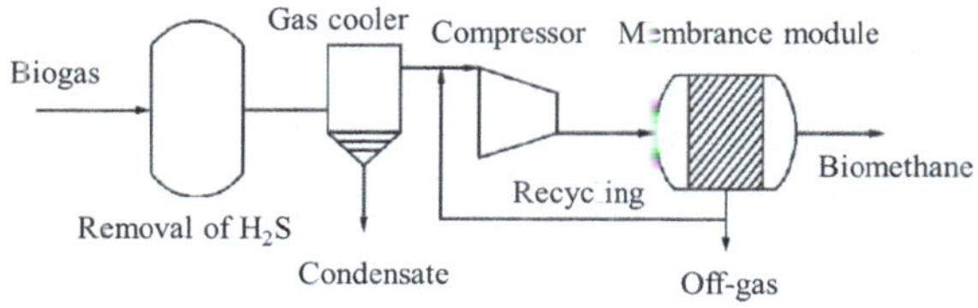

Fig. 6.12 Flow chart of membrane separation process (Sun et al. 2014)

multiple components. Gas membrane modules mainly include plate and frame type, spiral type, and hollow fiber type (Liang 2015).

(1) Plate and frame type

The plate and frame membrane module is also called the flat membrane module, mainly because it is a stacking of many plates and frames. The structure is simple, and the diaphragm can be replaced separately, which greatly reduces the equipment investment and operation cost.

(2) Spiral coil type

The spiral coil membrane module is made of flat membranes. Multiple layers membrane sheets are wrapped around a porous tube to form a membrane roll and the roll is put into a shell, producing a spiral wound separator. This model has the advantages of compact structure and low price.

(3) Hollow fiber type

The hollow fiber membrane is a very fine hollow membrane tube. A hollow fiber membrane module is produced by having a large number of hollow fiber membrane tubes into a U-shaped cylindrical pressure vessel. It has advantages of high packing density and no need of external support.

In order to extend the life of the membranes and achieve optimum separation efficiency, the raw biogas should be dried and completely desulfurized. Dust and aerosols must be separated before the feed enters the membrane (Beil et al. 2012). The feed was pressurized to 7×10^5–2×10^6 Pa in the 1980s and 1990s (system pressure back then was greater than 2×10^6 Pa). A complete hydrogen sulfide removal was conducted before or after pressurization, and then the feed was allowed to enter the membrane module (Baumgarten 2012). Pressure loss in the system is approximately 1×10^5 Pa. In the internal membrane modules, CO_2 can pass through the membrane while most CH_4 is not permeated. In most practical applications, there are at least two stages of membrane separation systems (see Fig. 6.12) (Harasek et al. 2008). The permeated gas still contains some methane, so the off-gas should be recycled (such as a second-stage membrane separation) or undergo one more stage of membrane separation.

Due to different operating pressures, circulating flow rates, and membrane materials, the power consumption by membrane separation was about 0.18–0.35 kWh/m^3 of raw biogas (Burmeister et al. 2009) in the past decade. In the most up to date separation system, power consumption is significantly lower than 0.35 kWh/m^3 of raw biogas. A membrane supplier used to report an electricity consumption of smaller than 0.2 kWh/m^3 of raw biogas (operating pressure 1×10^6 Pa–2×10^6 Pa) (Baumgarten 2012). Some manufacturers identified the power consumption to be around 0.29–0.35 kWh/M^3 of raw biogas, depending on the composition of the feed gas and the degree of methane recovery (97% of biomethane after upgrading) (Harasek 2012). In a large upgrading station located in Germany, power consumption is 0.20 kWh/m^3 of raw biogas.

It was reported in literature that methane recovery rate from membrane separation was 85–99% (1–15% loss of methane) (Burmeister et al. 2009). An economic methane recovery range was defined as 95–96% in previous years. Further increasing methane yields is possible, but it will lead to increased recirculation rates and increased electricity consumption.

Because the off-gas flow contains a significant amount of CH_4, it is necessary to oxidize the off-gas by a treatment step. Suitable treatment methods are comparable to those of the PSA systems.

In general, membrane separation technology is relatively reliable and simple to operate. Such technology is also able to remove H_2S and water at the same time, and to obtain pure CO_2 as a by-product. However, it is necessary to balance the purity and recovery of CH_4, and the membrane separation method requires multiple steps to avoid a large loss of CH_4.

6.3.6 *Cryogenic Upgrading*

Cryogenic upgrading uses the difference in boiling point of methane (−161 °C) and sublimation point of carbon dioxide (−78.48 °C). Since the sublimation point of carbon dioxide is higher than that of methane, carbon dioxide is first condensed and separated. General steps for low temperature separation are as follows:

(1) First, temperature is lowered to 6 °C, at which part of H_2S and siloxane can be partly removed by catalytic adsorption.
(2) After this pretreatment, the raw gas is pressurized to 18×10^5 to 25×10^5 Pa.
(3) Temperature is then lowered to −25 °C, at which the gas is dried and the remaining siloxane can also be condensed.
(4) Hydrogen sulfide removal is performed.
(5) Temperature is lowered to −50 to −59 °C, carbon dioxide is liquefied and then removed.

The expected loss of methane is 0.1–1%, and the actual loss of methane is limited to 2%. The power consumption is 0.18–0.25 kWh/m^3 of raw biogas (Wellinger et al. 2013).

The presence of CH_4 affects the sublimation temperature of CO_2, which requires higher pressures or lower temperatures to condense CO_2 into liquid or dry ice. The GPP biogas upgrading system developed by the Netherlands Gas Treatment Service B.V. uses this system to separate CO_2 from biogas by condensing CO_2 in biogas into liquid. After phase change into liquid, CO_2 can be used as a greenhouse gas fertilizer or further converted into dry ice (Petersson 2009).

Cryogenic upgrading can obtain CO_2 and CH_4 with extremely high purity, and it can also easily obtain liquefied biomethane. However, cooling biogas to below −80 °C requires a large amount of energy, and the separation equipment is also complicated, resulting in high costs (Marco et al. 2013).

In addition to the six mainstream methods above, there is also a latest hydrate separation process called hydrate-based gas separation. Hydrate refers to non-stoichiometric cage crystal materials formed by small gas molecules (such as N_2, O_2, CO_2, H_2S and CH_4) and water at a certain temperature and pressure, and such materials are also known as cage hydrate. Different gases form hydrates at different temperatures and pressures. The non-hydratable gas component and the easily hydratable component are respectively enriched in the gas phase and the hydrate phase, and therefore separation is realized.

6.3.7 Comparison of Key Parameters and Economic Input of Biogas Upgrading Technologies

In Sweden, pressurized water scrubbing method is the most commonly used method. In Germany, pressure swing adsorption is more widely used. In the Netherlands, pressurized water scrubbing, pressure swing adsorption, and membrane separation technology are all common. The economic comparison of biogas upgrading technologies is shown in Table 6.7.

Biogas upgrading cost is related to methane content of the feed gas and the product gas. Upgrading cost of raw biogas with high methane content is cheaper. This is mainly due to the increase of energy output, and total cost is related to higher energy levels. One can reduce costs by increasing efficiency (reducing energy consumption per unit of output) and using raw biogas with high calorific value to reduce actual energy consumption.

A short downtime, smooth equipment operation, and normal equipment performance under conditions of rated load and fully loaded raw biogas are the standards for highly efficient biogas upgrading plants. Fast responses are necessary, so good network services provided by technology providers is critical. In addition, remote monitoring can also be used. When the operation is interrupted, technicians can directly identify the issues and immediately take necessary remedial measures based on the type of problem occurred, thus avoiding time delay and saving long-distance travel of technicians from office to site.

The several CO_2/CH_4 separation methods above have been used in other industries for decades and are the most advanced biogas upgrading methods. The trend in biogas upgrading is to reduce energy consumption, increase recovery, and reduce methane emissions. Main measures taken include reducing pressure of the product gas to reduce power consumption, lowering temperature of the desorption process (amine scrubbing), developing a CO_2/CH_4 membrane with higher selectivity, and combining several technologies (such as membrane and cryogenic upgrading) into one.

Table 6.7 Economic comparison of biogas upgrading technologies

Items	PAS	Water scrubber	Physical absorption by organic solvent	Chemical absorption by organic solvent	Membrane separation	Cryogenic upgrading
Treatment range (m^3/h)	350–2800	300–1400	250–2800	250–2000	250–750	
Power consumption (kWh/m^3 BG)	0.16–0.35	0.20–0.30	0.23–0.33	0.06–0.17	0.18–0.35	0.18–0.25
Heat consumption (kWh/m^3 BG)	0	0	0.10–0.15	0.4–0.8	0	0
Reactor temperature (°C)	–	–	40–80	106–160	–	–
Operation pressure (10^5 Pa)	1–10	4–10	4–8	0.05–4	7–20	10–25
Product pressure (10^5 Pa)	2	5	6.5	1.15	7	–
Methane loss (%)	1.5–10	0.5–2	1–4	~0.1	1–15	0.1–2.0
Methane recovery rate (%)	90–98.5	98–99 5	96–99	~99.9	85–99	98–99.9
Off-gas treatment requirement	Yes	Yes	Yes	No need	Yes	Yes
Complete hydrogen sulfide removal water need	Yes	No need	No need	Yes (depending on manufacturers)	Recommended	Yes
	No need	1–3 m^3/d	No need	Need	No need	No need
Chemical reagent	No need	No need	Need	Need	No need	No need

Note BG refers to biogas. Electricity cost (12–18 Euro cents/kWh_{el}), heat cost (3–5 Euro cents/kWh_{th}), water cost 5 Euro/m^3 (including sewage treatment fee). All values are based on full load operation

References

Al Seadi, T., D. Rutz, H. Prassl, M. Köttner, T. Finsterwalder, S. Volk, and R. Janssen. 2008. *Biogas handbook*. Esbjerg: University of Southern Denmark Esbjerg.

Anneli, P., and W. Arthur. 2009. Biogas upgrading technologies-developments and innovations. IEA Bioenery.

Arthur, W.J., M. Patrick, and B. David. 2013. *The Biogas Handbook: Science, Production and Applications*. Philadelphia: Woodhead Publishing limited.

Baumgarten, G. 2012. Gewinnung von Biomethan mit Hilfe hochselektiver Membranen. Proceedings of 2 VDI-Konferenz Biogas–Aufbereitung und Einspeisung, Frankfurt, 27–28 June 2012.

Beil, M., W. Beyrich, U. Holzhammer, and T. Krause. 2012. Biomethan. http://www.nachwachsende-rohstoffe.de [accessed 15 October 2012].

Berndt, S.S.A., and V. Eichenlaub. 2012. Effizienzsteigerung der Druckwechseladsorption–Entwicklung, Status 2012 und weitere Potentiale. Presentation at 2 VDI-Konferenz Biogas–Aufbereitung und Einspeisung, Frankfurt, 27–28 June 2012.

Boback, R. 2012. Biogasaufbereitung mit dem aminselect-verfahren. Proceedings of 2 VDI-Konferenz Biogas–Aufbereitung und Einspeisung, Frankfurt, 27–28 June 2012.

Burmeister, F., R. Erler, F. Graf, W. Köppel, S. Petzold, and J. Senner. 2009. Stand des DVGW-Forschungsprogramms Biogas. Energie l wasser-praxis, 06/2009, 66–71.

Deng, L.W., and Z.A. Chen. 2007. The current state of biogas engineering in Europe. *China Biogas* 25 (5): 23–31. (In Chinese).

Deng, L.W., H.J. Chen, Z.A. Chen, Y. Liu, X.D. Pu, and L. Song. 2009. Process of simultaneous hydrogen sulfide removal from biogas and nitrogen removal from swine wastewater. *Bioresource Technology* 100 (23): 5600–5608.

Deng, L.W., Y. Liu, X.D. Pu, and L. Song. 2015. Biogas Plants. Beijing: Science Press. (In Chinese).

Diez, R., F. Lampe, R. Rieger, and C. Riemann. 2012. Application of Pura Treat®R+ for the upgrading of biogas. Gas for Energy, 2011, 22–28.

Girod, K., H. Lohmann, and W. Urban. 2009. Technologien und Kosten der Biogasaufbereitung und Einspeisung in das Erdgasnetz. Ergebnisse der Markterhebung 2007–2008. Fraunhofer UMSICHT, Oberhausen.

Foo, K.Y., and B.H. Hameed. 2009. An overview of landfill leachate treatment via activated carbon adsorption process. *Journal of Hazardous Materials* 171 (1): 54–60.

GB 17820-2012. Natural Gas/ National Standards of the People's Republic of China. (In Chinese).

GB 18047-2000. Compressed Natural Gas as Vehicles Fuel/National Standards of the People's Republic of China. (In Chinese).

Hara, K.T.A., A. Kudo, and T. Sakata. 1997. Change in the product selectivity for the electrochemical CO_2 reduction by adsorption of sulfide ion on metal electrodes. *Journal of Electroanalytical Chemistry* 434(1–2): 239–243.

Harasek, M., A. Makaruk, and M. Miltmer. 2008. Application of Gas Permeation for Biogas Upgrade-Operationa Experiences Of Feeding Biomethane into the Austrian Gas Grid.

Harasek, M., and J. Szivacz. 2012. Biogas–Aufbereitung mit Membranen–5 Jahre Betriebserfahrungen mit industriellen Anlagen. Proceedings of 2 VDIKonferenz Biogas–Aufbereitung und Einspeisung, Frankfurt, 27–28 June 2012.

Krich, K., D. Augenstein, and J. P. Batmale, et al. 2005. Biomethane from Dairy Waste—A Sourcebook for the Production and Use of Renewable Natural Gas in California. USDA Rural Development.

Li, L. X. 2007. Biogas purifying technology in large and medium scale biogas-project. Guangxi University. (In Chinese).

Liang, Z. X. 2015. Process design of biogas purification to produce vehicles gas fuel by membrane separation. Guangxi University. (In Chinese).

Liu, W. G. 2015. Key techniques and application for biological removal of hydrogen sulfide from biogas. Wuhan University of Technology. (In Chinese).

Marco, S., M. Thomas, and W. Matthias. 2013. Transforming biogas into biomethane using membrane technology. *Renewable and Sustainable Energy Reviews* 17 (1): 199–212.

NY/T 1223-2006. Biogas-powered Generating Sets/ Agricultural industry standards of the People's Republic of China. (In Chinese).

NY/T 1220.2-2006. Technical code for biogas engineering Part 2: Design of biogas supply/Agricultural industry standards of the People's Republic of China. (In Chinese).

Qi, Y., and X.Z. Guo. 2010. *System design and construction operation of biogas plants*. Beijing: Posts & Telecom Press. (In Chinese).

Radlinger, G. 2010. Zwei Verfahren, drei Hersteller–Aufbereitungsprojekte aus der Sicht eines Betreibers. Proceedings of 1 VDI-Kongress–Biogas–Aufbereitung und Einspeisung. VDI Wissensforum GmbH, Düsseldorf.

Ramesohl, S., F. Hofmann, and W. Urban. 2006. Analyse und bewertung der nutzungsmöglichkeiten von biomasse. Study on behalf of BGW and DVGW.

Ryckebosch, E., M. Drouillon, and H. Vervaeren. 2011. Techniques for transformation of biogas to biomethane.*Biomass and Bioenergy* 35: 1633–1645.

Schönbucher, A. 2002. *Thermische Verfahren-stechnik: Grundlagen und Berechnungs-methoden für Ausrüstungen und Prozesse*. Heidleberg: Springer.

Schomaker, A.H.H.M., A.A.M. Boerboom, and A. Visser. 2000. *Anaerobic Digestion Of Agro-industrial Wastes: Information networks-technical summary on gas treatment*. Nederland: Nijmegen.

Song, C.H., B. Xiao, X.Y. Shi, and X.Y. Yang. 2007. Review on biogas upgrading technology. *China Biogas* 25 (4): 23–27. (In Chinese).

Sun, J., G. Li, and Z.B. Chen. 2014. Biogas purification technology for producing vehicle gas. *Modern Chemical Industry* 4: 141–146. (In Chinese).

Tynell. 2005. Microbial Growth on pall-rings: a problem when upgrading biogas with the technique absorption with water wash. Master's Thesis, Link pings Universitet.

Ohly, N. 2010. Biogasaufbereitung mit der Druckwasserwäche. Proceedings of 1 VDI-Kongress–Biogas–Aufbereitung und Einspeisung. VDI Wissensforum GmbH, Düsseldorf.

Urban, W., K. Girod, and H. Lohmann. Technol-ogien und kosten der biogasaufbereitung und einspeisung in das erdgasnetz. tesults of market surbey, 2007–2008.

Weiland P. 2003. Notwendigkeit der Biogasaufbereitung, Ansprüche einzelner Nutzung srouten und Stand der Technik. Presentation at FNR Workshop 'Aufbereitung von BIogas'.

Wellinger, A., J. Murphy, and D. Baxter. 2013. *The biogas handbook: science, production and applications*. UK: Woodhead Publishing Limited.

Yang, D., C.Z. Liang, W.G. Liu, H.L. Liu, and Y.P. Hao. 2010. Design and operation of bio-desulfurization project of liuminying biogas station. *China Water and waste-water* 26 (14): 56–62. (in Chinese)

Yang, S.G., J.H. Li, and G. Li. 2013. *Gas Biofuel Technology and Engineering*. Shanghai: Shanghai Science and Technology Press. (In Chinese).

Yin, B., L.M. Chen, and Q.P. Kong. 2009. Research on purification technology for vehicle biogas. *Modern Chemical Industry* 29 (11): 28–33. (In Chinese).

Yin, Y.F., and W. Zhang. 2017. Design and operation condition analysis of biogas bio-desulfurization process. *Environmental Science and Technology* 30 (2): 12–16. (In Chinese).

Zhang, J.J., S.L. Zhao, R.X. Zhao, P. Li, Z.B. Cao, and W.W. Shi. 2011. Research on Modern CO_2 Absorption Process. *Modern Chemical Industry* 40 (1): 88–91. (In Chinese).

Zhang, Q.G. 2005. *Biogas Technology and Its Application*. Beijing: Chemical Industry Press. (In Chinese).

Zheng, G., and Q.G. Zhang. 2013. Review on biogas upgrading technologies for producing biomethane. *Transactions of the Chinese Society of Agricultural Engineering* 29 (17): 1–8. (In Chinese).

Zhou, M.J., R.L. Zhang, and J.Y. Lin. 2009. *Practical Technologies on Biogas*, 2nd ed. Beijing: Chemical Industry Press. (In Chinese).

Zhou, W., and H.Y. Zhu. 2014. Application status and development trend of biogas fuel cells. *Electronic World* 17: 130–132. (In Chinese).

Chapter 7
Biogas Storage

Under normal conditions with a balanced feed and a stable digestion temperature, digesters can produce biogas continuously. However, the use of biogas cannot be balanced, meaning that biogas consumption can hardly match immediate biogas on-site production. Therefore, gasholders are needed for storage of the biogas that cannot be used up. Gasholders are in essence buffer devices that adjust peaks and fluctuations in biogas production and solve the contradiction of uniform biogas production and variable biogas consumptions. In other words, the buffer devices store the surplus biogas produced at the moment to cover large consumptions of biogas at some other time in future.

Common gasholders include hydraulic gas storage chambers, low-pressure water-sealed gasholders, and low-pressure waterless gasholders. Although not as common, in some cases, medium pressure or sub-high pressure gasholders are used as buffer devices. Selection of gasholder type should take into consideration biogas plant size, location, biogas utilization, cost and other relevant factors. Household digesters, as well as small and medium underground biogas plants can use hydraulic gas storage chamber. In regions where water does not freeze in winter, biogas plants can adopt water-sealed gasholders, which can ensure a stable pressure and can easily adjust pressure. In regions where water freezes in winter, biogas plants can adopt waterless gasholders to avoid the problem of freezing inside gasholders. Long-distance biogas supply and storage of bio-methane after biogas upgrading can choose buffers of medium pressure or sub-high pressure.

Gasholders are the sources of risks in biogas plants. Relevant regulations should be followed to keep the gasholders at safe distances from buildings. Gasholders should stay away from high density residential areas, large-scale public buildings, important supply warehouses, telecommunication and transportation hubs, and other important facilities. Meanwhile, gasholders must be equipped with safety protection devices which can prevent overfills and over-discharging of gas.

L. Deng et al., *Biogas Technology*,
https://doi.org/10.1007/978-981-15-4940-3_7

7.1 Biogas Storage Pressure and Gasholder Volume

Volumes and storage pressures differ with utilization purposes of the biogas. Biogas used for power generation, water boilers, or household fuel usually adopts low-pressure holders with the pressure under 10 kPa (gauge pressure). Low-pressure gasholders can meet the input pressure requirements for biogas end users. Biogas used for upgrading to bio-methane usually applies medium pressure or sub-high pressure gasholders for more convenient pipeline transportation and increased storage capacity due to the high pressure of upgraded bio-methane. The pressure of medium pressure holders should be under 0.4 MPa (gauge pressure), and the pressure of sub-high pressure gasholders should be under 1.6 MPa (gauge pressure).

Biogas used for households can have a gasholder volume of 50–60% of daily average supply volume. Biogas used for power generation with power gensets that work continuously can have a gasholder volume of 10–30% of daily consumption by power gensets. Biogas used for power generation with power gensets that work intermittently can have a gasholder volume larger than the total consumption by power gensets in a single working period. Biogas used for upgrading and compression should have a gasholder volume of 10–30% of daily consumption.

7.2 Hydraulic Gas Storage Chambers

Household digesters along with small and medium-scale underground biogas plants usually do not require construction of separate gasholders for biogas storage. The chamber at the top of a digester functions as a biogas storage. The maximum biogas pressure in the chamber is determined by the difference between the liquid level of the effluent outflow in the hydraulic chamber and the liquid level of the digestion chamber. The greater the difference is, the higher the pressure is, but usually the maximum pressure does not exceed 12 kPa (gauge pressure); the amount of gas stored in the gas tank is controlled by the volume of the hydraulic chamber, so that the larger the volume of the hydraulic chamber, the larger the volume of biogas storage (Fig. 7.1).

(1) When the biogas outlet pipe is closed, with the generation of biogas in the digestion chamber, biogas pressure in the chamber increases and the liquid level of the digestion chamber decreases. The liquid is then forced into the hydraulic chamber through the outflow pass. When the hydraulic chamber is filled up, digested effluent overflows. As biogas production continues, the liquid level in the digestion chamber continues to drop, and when the liquid level drops to the upper end of the outflow pass, the additional biogas will enter the hydraulic chamber from the outflow pass and escape from the digested effluent overflow chamber.

(2) When biogas is used, the amount of biogas in the digestion chamber reduces, and the pressure decreases. Liquid in the digestion chamber is replenished by

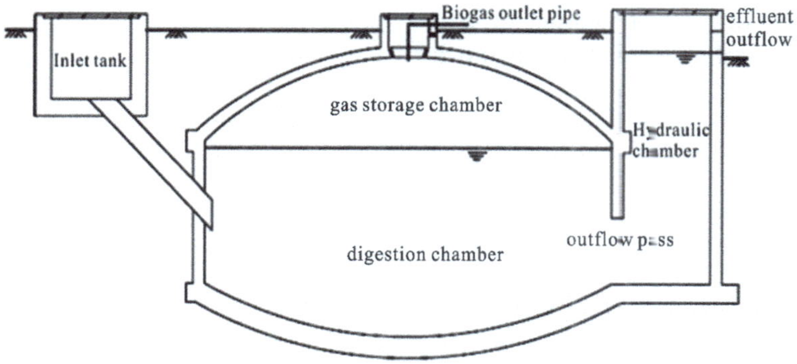

Fig. 7.1 Structure of underground digesters

reflux from the hydraulic chamber through the outflow pass, and the amount of reflux is equal to the volume of the biogas used. Liquid level in the digestion chamber rises back. When the liquid levels of the hydraulic chamber and the digestion chamber are the same, the pressure of the gas storage chamber is 0 Pa (gauge pressure).

7.3 Water-Sealed Gasholder

A water-sealed gasholder (wet gasholder) generally consists of a water-sealed tank and a floating drum, and the floating drum is placed inside the water-sealed tank. The water-sealed tank is filled with clean water that functions as a seal, and the floating drum serves as a biogas storage space. Rises or falls of floating drum in the water-sealed tank through the guiding frame achieve the purpose of gas storage or gas supply. When biogas flows into the gasholder, the floating drum placed in the water sealing tank rises; when biogas is discharged from the gasholder, the floating drum falls. The water-sealed gasholder is a low-pressure gas storage tank. The gas storage pressure is generally 2000–5000 Pa. When there are special requirements, the storage pressure can also be set to 6000–8000 Pa. The pressure can be varied by adjusting the counterweights added to the floating drum.

The volume of the water-sealed gasholder can be calculated according to Eq. 7.1, and the pressure can be calculated according to Eq. 7.2.

$$V = \frac{\pi d^2}{4}(H - h) \tag{7.1}$$

where

V effective volume of the gasholder, m^3;
d diameter of the floating drum, m;

H height of the floating drum, m;
H water sealing height to ensure no gas leakage, m.

$$P = \frac{4}{9.81\pi d^2}\left[(G + G_1) - \frac{\gamma G_2}{7.85}\right] \quad (7.2)$$

where

P gasholder pressure, Pa;
d diameter of the floating drum, m;
G added counterweight, kg;
G_1 weight of the floating drum (including its accessories), kg;
7.85 specific gravity of steel;
G_2 weight of the part of floating drum in the water, kg;
γ water density, kg/m^3.

Water-sealed gasholders have simple structures and are easy to construct. They require no power, and they are reliable in operation. However, in northern China, water-sealed gasholders need heat preservation. The water-sealed tank, the floating drum, and the guide rail are in contact with each other all year round, so they are susceptible to abrasion and corrosion; a gasholder has a large weight and takes a large construction land, and the overall construction cost is relatively high.

1. *Types of water-sealed gasholders*

Water-sealed gasholders can be classified into direct ascending water-sealed gasholders without external guides, direct ascending water-sealed gasholders with external guides, and spirally ascending water-sealed gasholders, as shown in Figs. 7.2, 7.3, and 7.4.

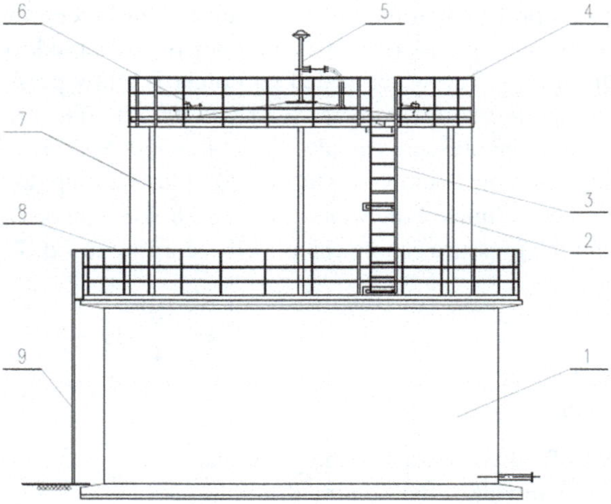

Fig. 7.2 Direct ascending water-sealed gasholders without external guides. 1—Water sealing tank; 2—floating drum; 3—floating drum ladder; 4—floating drum railing; 5—biogas empty pipe; 6—maintenance manhole; 7—guides rail; 8—handrail around water sealing tank; 9—water sealing tank ladder

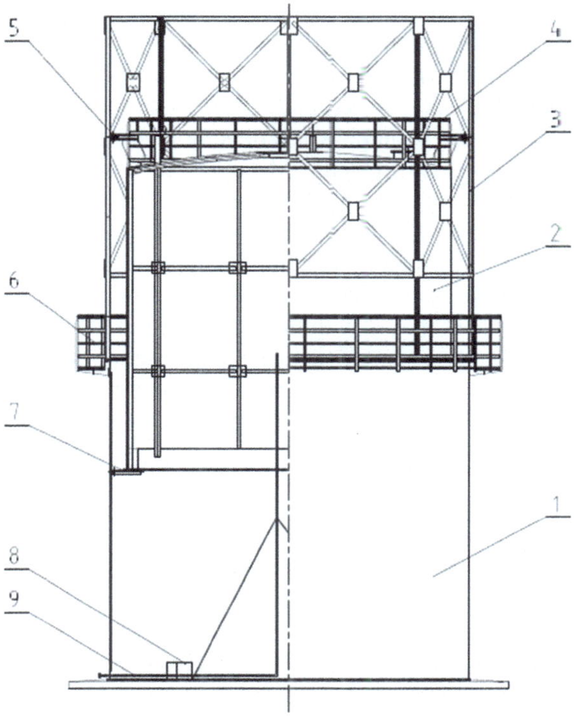

Fig. 7.3 Direct ascending water-sealed gasholder with external guides. 1—Water sealing tank; 2—floating drum; 3—external guide frame; 4—handrail of floating drum; 5—upper guide wheel; 6—handrail of water sealing tank; 7—lower guide wheel; 8—pier for drum; 9—gas inlet pipe

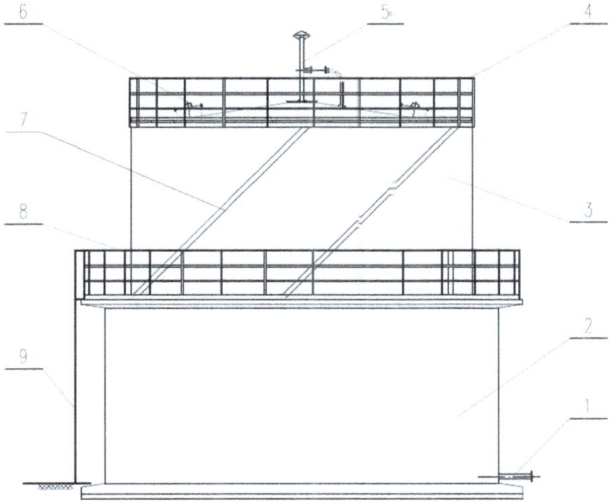

Fig. 7.4 Spirally ascending water-sealed gasholder. 1—Gas inlet pipe; 2—water sealing tank; 3—floating drum; 4—handrail of floating drum; 5—biogas empty pipe; 6—maintenance manhole; 7—guides wheel; 8—handrail of water sealing tank; 9—ladder of water sealing tank

The principles and auxiliary facilities of the three types of water-sealed gasholders are basically the same. Their main difference lies in the installation of guide rails, the upper guide wheels, and the movement of the floating drum. Guide rails of the direct ascending water-sealed gasholders without external guides and spirally ascending water-sealed gasholders are directly welded on the floating drum, and the upper guide wheel is installed on the upper edge of the water sealing tank. The guide rail moves with the floating drum while the upper guide wheel is fixed.

The guide rail of direct ascending water-sealed gasholders with external guides has a mesh structure fixed on the upper edge of the water sealing tank. The upper guide wheel is welded on the floating drum, and the guide rail is fixed. The upper guide wheel moves with the floating drum. The guide rail of the direct ascending water-sealed gasholder is linear, resulting in linear rising and falling of the floating drum. The guide rail of the spirally ascending gasholder is a 45° inclined spiral, and the movement of the floating drum spirals up and down.

2. *Structures of water-sealed gasholder*

1) Size

When the volume of the water-sealed gasholder is determined, the external dimensions depend on the diameter height ratio (D: H). The diameter height ratio refers to the ratio of diameter to height. Diameter D refers to the inner diameter of the water sealing tank of a gasholder. Height H refers to the total height of the gasholder when the floating drum rises to the highest point. For the direct ascending gasholder and the spirally ascending gasholder without external guide frame, D:H = 1.0–1.65; for the direct gasholder with external guide frame, D:H = 0.8–1.2.

The height of the water sealing tank of a water-sealed gasholder is generally not more than 10 m. Some common geometries of water-sealed gasholders can be seen in Table 7.1 (Fig. 7.5).

Table 7.1 Geometry of the commonly used water-sealed gasholders

No.	Nominal volume	Water sealing tank			Floating drum			
	V_0 (m^3)	D (mm)	h (mm)	V (m^3)	D_1 (mm)	h_1 (mm)	f (mm)	V_1 (m^3)
1	25	4400	2800	42.5	3600	2500	260	27
2	50	5400	3300	75.5	4600	3000	300	52.4
3	100	6700	4050	142.7	5900	3750	400	107.7
4	150	8000	4050	203.5	7200	3750	480	162.4
5	200	8800	4300	261.4	8000	4000	540	214.7
6	250	9500	4550	322.4	8700	4250	580	268.1
7	300	10,000	4800	376.8	9200	4500	620	317.8
8	400	11,000	5300	503.4	10,200	5000	660	429.7
9	500	11,800	5800	640	11,000	5500	720	548.8

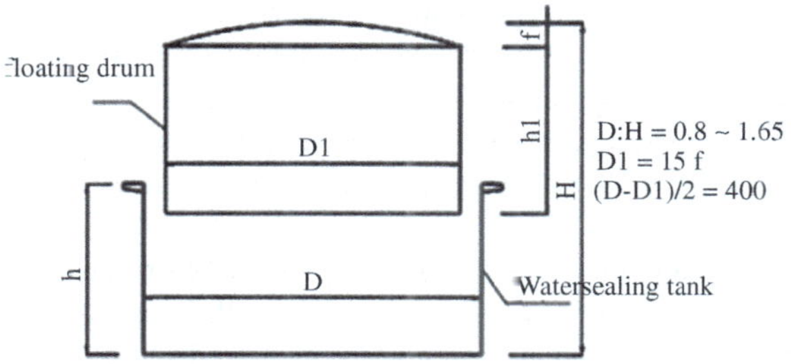

Fig. 7.5 Schematic diagram of water-sealed gasholder

Table 7.2 Minimum thickness the wall of steel-structure water-sealing tanks

Diameter of the tank D (m)	Minimum thickness of the wall of tank (mm)
D < 12	4
12 ≤ D < 16	4.5
D > 16	6

2) Water sealing tank

The water sealing tank mainly serves to support the floating drum and to seal biogas. The structure of the water sealing tank is mostly made of reinforced concrete or welded-steel plates. Design parameters of water-sealed tanks in the form of reinforced concrete can be calculated according to the relevant formula of reinforced concrete tanks; design parameters of the steel-structure water-sealing tank can be calculated according to the strength formula of the steel-structure digester walls, the minimum thickness of which must meet the requirements in Table 7.2.

The water sealing tank of water-sealed gasholders is preferably built above ground to facilitate the access of gas pipes and the drainage of pipelines. If a water sealing tank is to be set underground, full considerations of safety as well as the ease of operation of pipeline drainage and the water sealing tank are necessary.

Support piers of the water-sealing tank need to be provided to ensure that the floating drum is not pressed into the biogas inlet pipe when it is lowered onto the piers. Generally, 4–8 piers are installed, and they need to be evenly arranged with their upper surface all kept at the same level.

3) Floating drum

A floating drum is the main structure of a water-sealed gasholder. The drum is equipped with a guiding system consisting of a guide rail and a guide wheel that maintain smooth movements of the drum. The inner wall of the water sealing tank is equipped with a lower guide rail, and the lower rail is usually made of steel channels. The upper platform of the water sealing tank is equipped with guiding wheels. The bottom of the floating drum has lower guiding wheels, and its outer wall has an upper rail. Guiding rails are usually light-weight rails. The maximum lifting speed V_{max} of the drum is ≤0.9–1.2 m/min.

The maximum pressure of water-sealed gasholders is determined by the weight of its floating drum. When the weight of a floating drum is less than the upward lifting force generated by biogas in the floating drum, the pressure of the gasholder is achieved by adding a counterweight to the floating drum. The weights are usually made of C10 concrete, and they are forged by even pouring of concrete into the counterweight grooves at the bottom of the drum. At the same time, a certain number of concrete blocks of about 20 kg should be produced and placed into a counterweight plate at the top of the drum to balance the floating drum. Modifying Eq. 7.2, we can obtain a formula for calculating the weight of counterweights (7.3).

$$G = \frac{9.81\pi d^2}{4} P + \frac{\gamma G_2}{7.85} - G_1 \tag{7.3}$$

where

- G weight of counterweights, kg,
- P gasholder pressure, Pa;
- d diameter of floating drum, m;
- G_1 weight of the floating drum on the gasholder (including all the supporting part on the floating drum), kg;
- 7.85 specific gravity of steel;
- G_2 weight of the part of the floating drum soaked in water, kg;
- γ (density of water), kg/m^3.

4) Foundation

Water-sealed gasholders are sensitive to uneven settlement. Excessive settlement will cause the guide wheel to get stuck, which will affect the flexibility of the drum and the sealing effect of the water sealing tank. Therefore, water-sealed gasholders require a rigid foundation, and pile foundations are generally used when constructing water-sealed gasholders on soft foundations.

3. *Material for making water-sealed gasholders*

Water sealing tanks can be made of steel plates or reinforced concrete, and the floating drum is usually made of steel. For low pressure water-sealed gasholders with a volume of less than 300 m^3, reinforced concrete can also be used for the drum. Steel produced by Bessemer process should not be used as building materials for water-sealed gasholders.

4. *Subsidiary structures and facilities*

The accessorial structure and facilities of water-sealed gasholders generally include biogas inlet pipelines, relief pipelines, ladders, piers for drum, water pipes, overflow pipes, rails, drum manholes and inspection wells.

A valve must be installed at the biogas inlet line outside the water sealing tank of the water-sealed gasholder, and the drain port should be installed at the lowest point of the pipeline. The height of the biogas inlet pipe in the water-sealing tank must be at least 150 mm higher than the overflow nozzle. The biogas inlet pipe must be installed in a way that neither the pipe will affect the movement of the floating drum nor the movement of the drum will damage the biogas inlet pipe.

The water-sealed gasholder shall be equipped with a manual relief line and an automatic relief line at the same time. Diameter of the relief pipe is usually not less than 50 mm. The manually controlled relief line is equipped with a ball valve, which is usually closed and opened during maintenance. The relief line is used to discharge the biogas in the floating drum. Valves should not be installed on the automatic relief pipes. The lower side of the relief pipe is usually 350 mm higher than the bottom of the floating drum. When the drum rises to the height where the automatic relief pipe nozzle is separated from the water surface of the water sealing tank, biogas is discharged from the automatic relief line to prevent the floating drum from being ejected from the water sealing tank due to pressure accumulation.

Ladders and handrails are required for water sealing tanks and floating drums of water-sealed gasholders. A passage platform must be set at the top of the water sealing tank.

7.4 Waterless Gasholder

Waterless gasholders here refer to low-pressure waterless gasholder (dry gasholder), which can be classified into the rigid structure waterless type and the flexible structure waterless type. Rigid structure gasholders have a rigid outer casing and a piston or a flexible foldable gas sac that can move up and down for storing biogas, whereas flexible structure dry gasholders mainly consists of double-layer membranes. The outer membrane plays a role of protecting and stabilizing gasholders from external wind pressure, snow pressure, and inner membrane pressure. The inner membrane is used for storing biogas, which is called double membrane gasholder. The main advantage of waterless gasholders is that they do not have water sealing structures, and their operation is not influenced by weather. The main disadvantage, however, is that the sealing oil and the flexible membrane are prone to aging.

7.4.1 Rigid Structure Waterless Gasholder

Rigid-structure waterless gasholders are generally classified into the piston type and the gas sac type according to their internal composition.

1. *The piston type waterless gasholder*

In the piston type waterless gasholder, the rigid casing and piston are sealed by oil or rubber membrane. The main components are a rigid casing, a piston system, a sealing system, pressure balance devices, etc. Auxiliary facilities mainly include

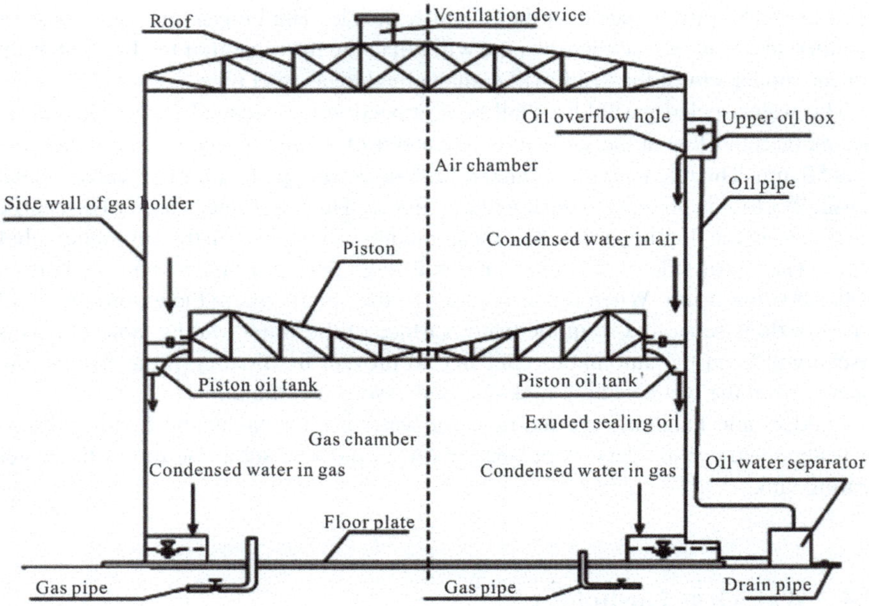

Fig. 7.6 Schematic diagram of a piston-type waterless gasholder with oil seals (Tang and Wang 2013)

platforms for gasholder operation and maintenance, stairs, top ventilators, piston limiters, handrails, manholes, etc. (Fig. 7.6).

The piston-type waterless gasholder with oil seals has a cylindrical or a regular polygon piston, the size of which is slightly smaller than the inner size of the rigid outer casing. The piston moves up and down with increase and decrease of gas storage, and the lifting speed is about 1.5 m/min. The gas pressure is 2.5–8.0 kPa, and the pressure is adjusted by the counterweights set on the piston. The sealing oil fills the oil grooves on the piston and the gaps in between the rigid outer casings, and the oil seals by balancing the oil pressure and the biogas pressure at the lower piston. During the movement of the piston, the sealing oil gradually flows along the gap between the piston and the inner wall of the rigid outer casing. The sealing oil is stored in the oil groove at the bottom of the gasholder together with the condensed water from biogas, and the oil is recycled after a oil-water separator removes the accumulated water (Fig. 7.7).

The piston-type waterless gasholder sealed by a rubber membrane is also called the curtain-type waterless gasholder. The piston moves up and down with increase and decrease of gas storage. The piston lifting speed is about 5 m/min, and the gas storage pressure can reach 10.0 kPa. The rigid casing and the piston are sealed by a rubber film that owns the advantages of corrosion resistance, strong flexibility, folding repeatability, and good sealing performances. The rigid outer casing acts as a protective layer for the rubber film.

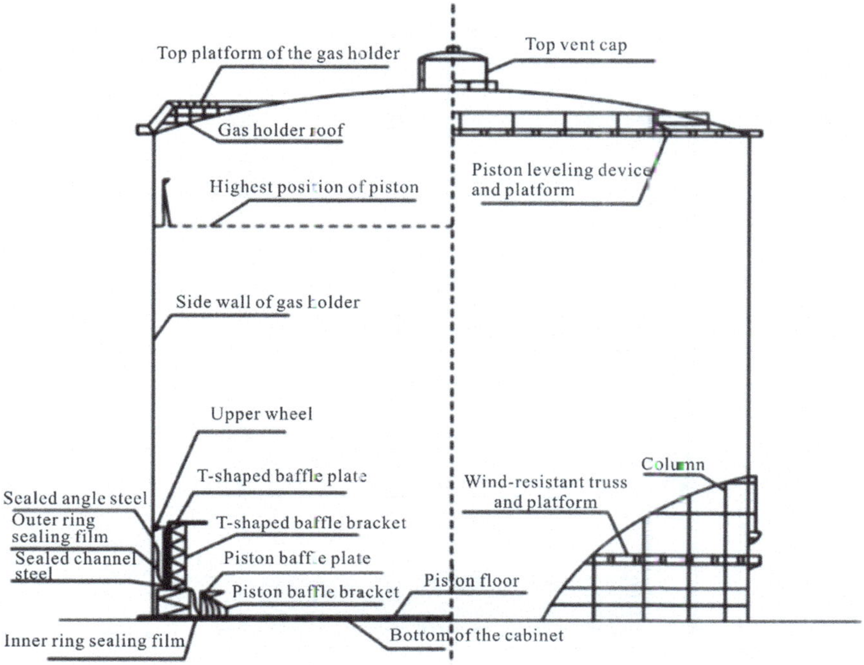

Fig. 7.7 Schematic diagram of a piston-type dry gasholder sealed with a rubber membrane (Tang and Wang 2013)

2. *The gas sac-type waterless gasholder*

The main components of a gas sac type waterless gasholder are a rigid casing, a flexible gas sac, pressure balance devices, etc. Auxiliary facilities mainly include a gas sac height display scale, platforms for gasholder operation and maintenance, gas sac limit devices and handrails, repair manholes, etc. (Fig. 7.8).

Biogas is stored in a flexible gas sac, and the pressure balance device is used to set the biogas storage pressure. The storage of biogas is realized by the flexible gas sac, so the gas storage pressure is not high. The gas sac is heat-pressed into tube bellows made by a special polyester fiber plastic film that allows regular folding of side wall when the gas sac contracts, and this special film improves the stability of the gas sac when the sac moves up and down. The rigid outer casing is not a part of the biogas storage space: its main function is to protect the internal flexible gas sac from external interference and damage, and at the same time it serves as an installation support for auxiliary facilities and parts. The rigid outer casing is usually cylindrical and can be made of welded steel plates, the Lipp tanks, enamel assembled tanks, or glass fiber reinforced plastic. The outer casing is fixed on the base of the gasholder and has a ventilation cap at the top.

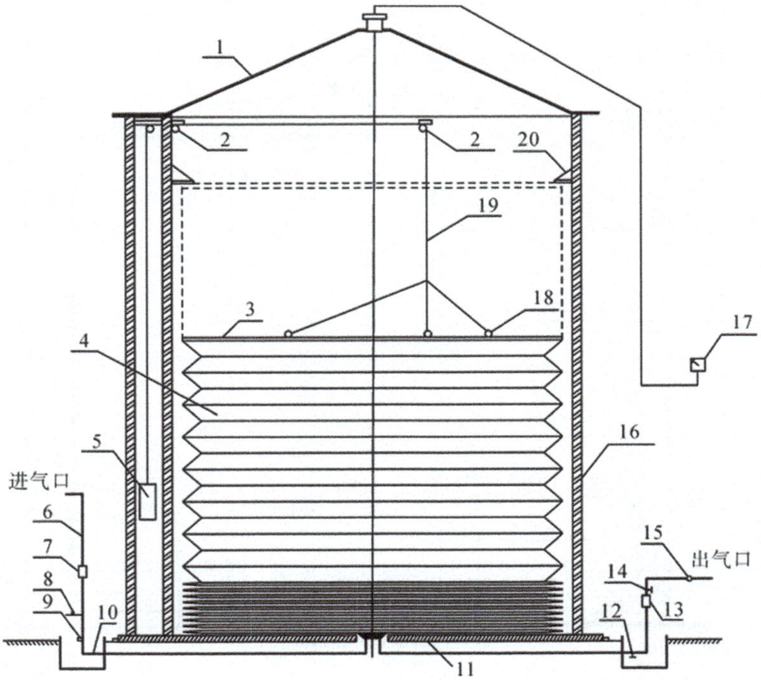

Fig. 7.8 Schematic diagram of the gas sac type waterless gasholder (Tang and Wang 2013). 1—Top cover; 2—pulley; 3—gas sac fixed ring frame; 4—gas sac; 5—balance weight; 6—inlet pipe; 7—pressure reducing valve; 8—valve; 9—valve; 10—water condenser; 11—basis; 12—water condenser; 13—fire arrester; 14—safety valve; 15—pressurizing pump; 16—protective shell sidewall; 17—sonic instrument; 18—flying ring; 19—sling; 20—limiter

7.4.2 Flexible-Structure Waterless Gasholders

In biogas plants, the most commonly used flexible-structure waterless gasholders are independent double membrane gasholders (commonly known as ground-mounted double membrane gasholders) and double membrane gasholders on the top of digesters. The latter is used for both gas production and gas storage (commonly known as tank rim-mounted double membrane gasholders). Unlike traditional gasholders, the double membrane gasholder is composed of polymers, which has better performances than the gasholder made of steel. The double-membrane gasholder can be manufactured in factory and the quality can thus be ensured. The membrane materials are soft and foldable, so it can be easily folded and transported. The utilization efficiency of ordinary water-sealed gasholders and waterless gasholders is only 60–70%, but the gas utilization efficiency for double membrane gasholders can reach 100%. In case of a negative pressure, a double-membrane gasholder will not be damaged due to its foldability; the steel gasholder can easily collapse under negative pressure. Furthermore, double-membrane gasholders contain fluoride which can effectively protect the gasholder against natural aging caused by ultraviolet rays, microorganisms, and weathering. Double-membrane gasholders also have good corrosion resistance when used to store corrosive gas, allowing a long lifespan for the holders. Since a double-membrane gasholder does not need a water sealing tank, the gasholder does not need a specific consideration of freezing protection, so this design is particularly suitable for cold regions, except that biogas inlet and outlet pipes as well as condensate drain pipes need to be protected against freezing. Compared with ordinary low-pressure gasholders, double-membrane gasholders have a simple structure and a better look. They can be industrially produced in factory, and they have the characteristics of high efficiency, low cost, and simple construction and maintenance. In addition, double-membrane gasholders have an important advantage: a lower cost than that of the water-sealed gasholders. When storage volume is large, the advantage becomes more obvious. Compared with water-sealed gasholders, double membrane gasholders have a disadvantage that biogas must be pressurized to meet the biogas pressure requirements for end uses, and they cannot work normally during power outages.

1. **Independent double membrane gasholders**

1) Gas storage principle and system composition

A double membrane gasholder is composed of a membrane body and auxiliary parts. Two spaces are formed by the bottom membrane, the inner membrane and the outer membrane. The space formed by the bottom membrane and the inner membrane is used for biogas containment, and the space formed by the inner membrane and the outer membrane is filled with air for regulating the biogas pressure in the inner membrane while supporting the outer membrane to resist external snow and wind pressure. When biogas storage in the inner space increases and the inner membrane rises, the air between the inner and outer membranes is squeezed out to accommodate the inner membrane volume increase, so that biogas can smoothly enter the gasholder. When the inner membrane rises to the point of maximum inflation, the excess biogas will be released through the safety protector on the inner membrane, so that the inner membrane will not be damaged by excessive pressure; when biogas stored in the inner membrane is reduced and the inner membrane deflates, the space between the inner and outer membrane spaces will be filled with air so that the biogas pressure in the inner membrane is adjusted, and the outer membrane rigidity is maintained and the stored gas can be smoothly discharged out of the gasholder.

The main components of the independent double membrane gasholder are shown in Fig. 7.9.

The shape of an independent double membrane gasholder used in the biogas plant is usually a 3/4 sphere (Fig. 7.10). The common dimensions are shown in Table 7.3.

2) Auxiliary facilities of double membrane gasholder

Due to limitation of the mechanical properties of the gasholder materials, a double-membrane gasholder usually takes a storage pressure of about 1 kPa to prolong the service life. This pressure cannot reach the inlet pressures of terminal biogas

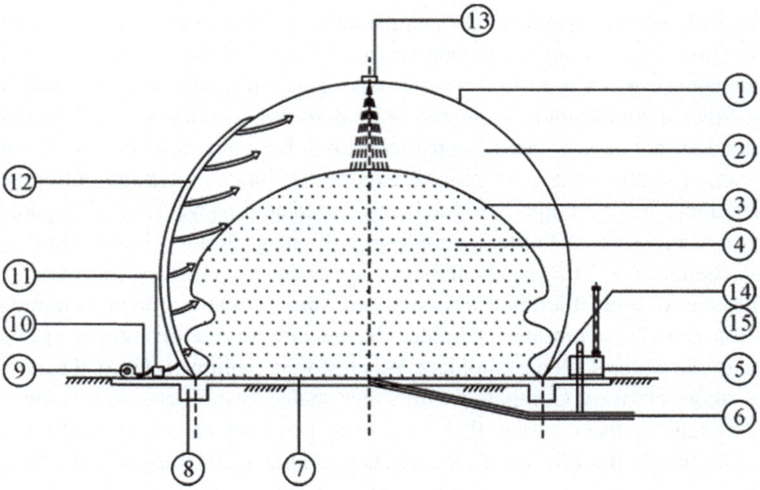

Fig. 7.9 Schematic diagram of the composition of the double membrane gasholder. 1—Outer membrane; 2—air chamber; 3—inner membrane; 4—biogas storage chamber; 5—pressure protector; 6—drainage pipe; 7—base slab; 8—embedded parts; 9—blower; 10—air tube; 11—check valve; 12—air supply passage;13—ultrasonic probe; 14—upper clamping rail; 15—compression bolts

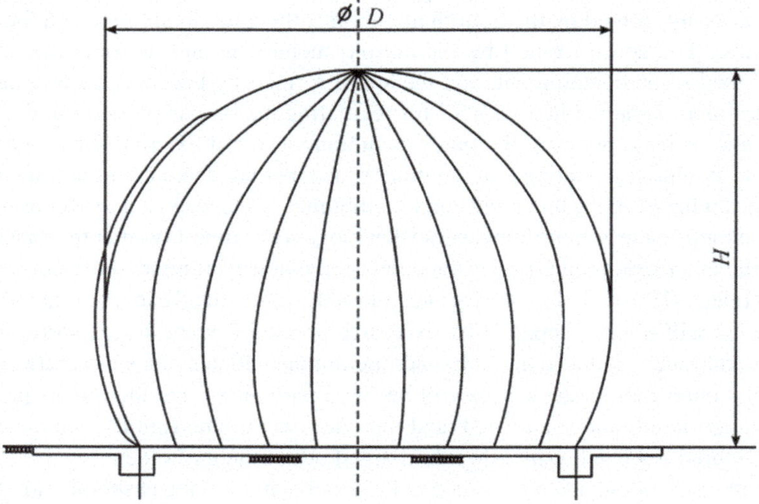

Fig. 7.10 Dimensions of the double membrane gasholder

Table 7.3 Common dimensions of independent double membrane gasholders

No.	Volume (m^3)	Ball diameter (m) D (米)	Base diameter (m)	Height (m)
1	100	6.58	5.4	4.45
2	200	7.79	6.95	5.65
3	300	9.11	8.49	6.21
4	500	10.83	9	7.8
5	800	12.48	10.66	9.1
6	1000	13.55	11.91	9.61
7	1500	15	13	11.28
8	2000	17	15.02	12.1
9	3000	19.8	18.69	13.16
10	5000	22.8	19.5	16.82
11	10,000	28.29	24.5	21.22
12	20,000	37.7	35.31	24.7

appliances. Therefore, corresponding supportive auxiliary facilities must be installed. The auxiliary facilities of double membrane gasholders usually include an outer membrane constant pressure control system, an inner and outer membrane pressure protection system, and a post-plant pressure boosting and regulating system.

(1) Constant pressure control system on the outer membrane

Some constant pressure control systems on outer membrane have a constant pressure control block at the exhaust port, while some employ a variety of sensors and a central controller to detect gasholder pressure and the volume of the inner membrane. Information such as membrane volume, concentration of leaked methane, and system flow can also be integrated if needed, and automatic air injection to adjusts pressure based on instantaneous pressure reading is also achievable.

(2) Inner and outer membrane pressure protection system

A membrane gasholder has a large diameter. In order to prevent the inner membrane from outlasted by the outer membrane, the inner membrane pressure is generally about 1 kPa higher than that of the outer membrane. Under normal conditions, the inner membrane is subjected to a pressure of 1 kPa when it is at full capacity, but when the pressure of the outer membrane cannot be maintained (such as power failure, loss of sealing water, or any other malfunction), the inner membrane may be subjected to a pressure as much as 4 kPa. Although a highly tensile membrane can be used, in this case, the inner membrane is likely to be deformed and infiltration might occur. The outer membrane suffers from natural aging, strong ultraviolet rays, and corrosion from gas in the inner membrane, so the life expectancy is greatly reduced. The aesthetic value will also be greatly reduced. When methane in the air layer between the inner membrane and the outer membrane reaches the explosion limit, a serious accident might happen. Therefore, the double membrane gasholder requires a pressure protection system with high precision and reliability. One simple protection system uses a safe water seal. The principle of a water seal is to stick a pipe down into water, in which one end of the pipe is immersed. The system is sealed as long as the pressure of biogas balances out the pressure from water. When the biogas pressure is greater than water pressure, water is squeezed by the biogas and the immersed end becomes exposed. The biogas can then be released from the pipeline and discharged to the atmosphere. The protection system of a double-membrane gasholder is classified into inner membrane protection and outer membrane protection. The inner membrane and outer membrane with water seals as safety protection can only be individually protected. It is also possible to use a protection control with both electromagnetic and physical pressure release with the following constituents.

① High-pass solenoid valve for active pressure relief.

A high-pass solenoid valve can achieve a long lifespan. It is an explosion-proof valve that can handle large flow rates. The valve is also able to start up with a low pressure and a low energy. It can be manually operated by a central controller or a touch screen, so that the pressure of the gasholder is always controlled within the safe range.

② Spare waterless physical safety valve.

Although high-pass solenoid valve has many advantages, it can not work when there is power blackout. Therefore, the gasholder needs an alternate waterless physical safety valve with an accuracy of 0.15 kPa and an adjustable pressure, which automatically starts to function when the solenoid valve stops working. Without water, there are no considerations regarding corrosion, freezing, or pressure changes, so this type of safety valve can reduce maintenance of the protection system.

③ An inner and outer membrane pressure balance system.

The pressure exerted on the inner and outer membranes can be kept at the same level by the pressure balance system. When the inner membrane pressure is 0.3 kPa greater than the outer membrane, the inner membrane pressure is released (the inner membrane is already saturated). That is to say, when the outer pressure is 1 kPag, the inner pressure more than 1.3 kPag is released. As long as the pressure difference between the inner and outer membranes is always kept at less than 0.3 kPa, the inner membrane will bear a pressure less than 0.3 kPa, by which the inner membrane is protected.

(3) Post-plant pressure boosting and sustaining system

The post-plant pressure boosting and sustaining system usually includes a booster fan and a pressure sustaining tank situated downstream of the gas plant. When the biogas pressure in the double membrane gasholder cannot meet the gas pressure requirement at the inlet end of biogas end users, this system needs to be installed.

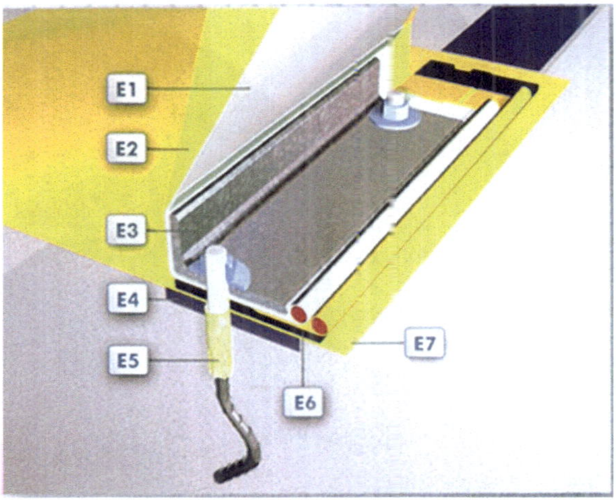

Fig. 7.11 Installation and fixation of the bottom of the double membrane gasholders. E1: outer membrane; E2: inner membrane; E3: clamping rail; E4: bottom rail; E5: anchor bolt; E6: sealing; E7: bottom membrane

3) Installation of double membrane gasholders

Double membrane gasholders are usually produced in factory and installed on site. Figure 7.11 can be referred to for installation. Attention should be paid to the following steps.

(1) First check whether the surface and installation location of the base slab are correct, firm, and flat, and whether the inlet pipes, outlet pipes, and the drain pipes are in correct configurations. Ensure the aforementioned installation requirements are met before installation.
(2) Place the clamping rail at a position corresponding to the bottom rail.
(3) Lay the bottom membrane on the foundation slab, place the inner membrane on top of the bottom membrane, place the outer membrane on the already laid inner membrane, and evenly lay the bottom rail on the edge of the inner membrane and outer membrane (need manual alignment), check the dimensional deviation, make certain adjustments (installers should not wear sharp objects when installing the membrane, need to take off shoes or wear rubber sole shoes to enter).
(4) Lay the rubber sealing strip on the bottom rail, and the sealing strip interface is pasted with special flexible glue.
(5) The bottom membrane, the inner membrane and the outer membrane are perforated together with a puncher. When punching, first hit the two ends of the rail, fix the membrane, finish all the holes, then fix them with bolts.
(6) Installation of auxiliary facilities.
(7) The system starts, the inner membrane is filled with air, and it is checked whether there is air leakage at the joint and whether the auxiliary facilities are operating normally.

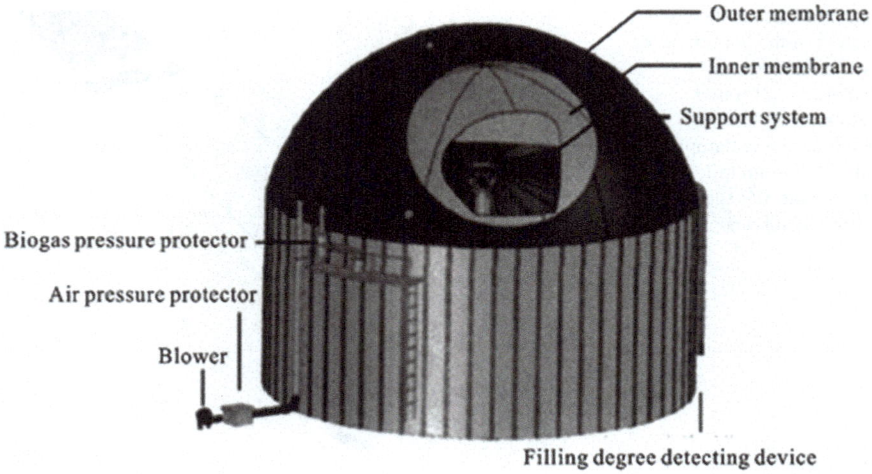

Fig. 7.12 Schematic diagram of integrated gas production and storage system

2. **Double membrane gasholder mounted at tank rim**

1) Gas storage principle and system composition

A tank rim-mounted double membrane gasholder is constructed by anchoring a double membrane gasholder on the top wall of the digester, thus integrating the structure and functions of the biogas generation unit and the gas storage unit (Fig. 7.12), and an integrated biogas production and storage system is formed by this configuration. Compared with the traditional split system, a digester roof and a gasholder base are eliminated. The utility model has the advantages of a compact structure, land saving, relatively low engineering cost, together with low operation and maintenance cost.

In the integrated biogas production and storage system, design and construction of the digester and the double membrane gasholder are not much different from the split system. The difference lies in the edge of the digester top wall in the integrated system. The package steel angle was replaced by a compression flange. The integrated gas production and storage system is mainly composed of the following parts.

(1) Digester

The digester is the lower part of the integrated gas production and storage system. The digester can be made of steel or reinforced concrete, and the volume is generally 500–3000 m^3, which can be constructed by assembly. The digester is usually 12, 14, 16, 18, 20, 22, or 24 m in diameter, and it normally has a height of 6–8 meters. Common ratios of diameter to height varies from 2:1 to 3:1. The digester is generally equipped with a stirring device which usually takes the form of a side stirrer, a rod stirrer (low-speed energy-saving stirring technology) or a combination of both. When a mesophilic or a thermophilic digestion is undertaken, the digester is equipped with a temperature increasing (circulating hot water pipe) and heat preserving device. The hot water circulating pipe of a steel structure digester is generally installed outside the tank wall, and the hot water circulating pipe of a reinforced concrete digester is generally installed at the lower inner wall of the digester.

(2) Gasholder

The gasholder is the upper part of the integrated gas production and storage system, and the system usually adopts a double membrane gasholder. The main configurations include a conical shape and a spherical crown shape. The double-membrane gasholder is composed of an inner membrane and an outer membrane. The outer membrane protects the inner membrane and maintains the structure, and the inner membrane stores the biogas. The double membrane gasholder adjusts air pressure in between the inner and outer membranes through a blower and a pressure regulator, which help maintain the shape and structure of the gasholder while feeding biogas in the inner membrane into gas pipelines.

(3) Support grid

The tank rim type double membrane gasholder has no bottom membrane. In order for the inner membrane of the double membrane to avoid contact with liquid in the digester, it is usually necessary to add pillars at the center of the digester. The pillars can be made of concrete, stainless steel, or wood. An umbrella ribs were installed between the apex and the tank rim, and a flexible web is usually used to ensure that the inner membrane cannot touch the mixed liquid when the digester is under negative pressure. The support grid can also work as a packing of biological desulfurization bacteria.

(4) Support blower

It is used to inflate the interlayer between the inner and outer membranes to maintain the design pressure in the interlayer, thereby ensuring that the outer membrane is always inflated and capable of sustaining forces from wind, rain and snow within the designed range.

(5) Biogas pressure control

The gas storage pressure is preset at a certain allowable range. When biogas pressure exceeds the set range, the pressure control will automatically release biogas or draw in air to maintain the biogas pressure within the set range, ensuring that the gasholder and the digester are not damaged.

(6) Inflation level detecting device

It is used to detect the degree of biogas inflation in the gasholder.

2) Installation

When a tank rim-mounted double membrane gasholder is installed, the gasholder needs to be lifted by a hoist and then unloaded on top of the target digester, following which the membranes are anchored to the digester rim. The compression flanges at digester top are pre-bent based on the designed shell diameter. Next, matching drilling of the flanges with the upper clamping rail of the double membrane gasholder is performed, and the upper clamping rail is welded to the top edge of the digester as steel angles that wrap along the digester top. The drilling side of the flanges are pointing outwards of the digester. Drill the upper clamping rail together with the bottom rail and stick in the sealing strip. After the double membrane gasholder is unloaded in place, install the upper clamping rail and press it until tight, then blow in air to check for leakages and whether the auxiliary facilities are working properly.

7.5 Medium Pressure or Sub-high Pressure Gasholders

Medium pressure or sub-high pressure gasholders are usually used for biogas supply over long distances. The storage containers of bio-methane produced from biogas upgrading are usually in the form of fixed volume spherical tanks (Fig. 7.13). The pressure of medium pressure storage is not more than 0.4 MPa (gauge pressure), the sub-high pressure is not more than 1.6 MPa (gauge pressure).

The spherical gasholder of steel consists of a body, pillars, and supportive parts. Amongst gasholders of all geometries, the spherical gasholders require a least amount of steel materials given the same storage volume because of the geometric properties of spheres. Moreover, given the same diameter, pressure, and material, the thickness of steel plate required for manufacturing a spherical gasholder is only half that of a cylindrical gasholder. The advantages of a spherical gasholder are smaller surface areas, simple foundations, and small wind loads.

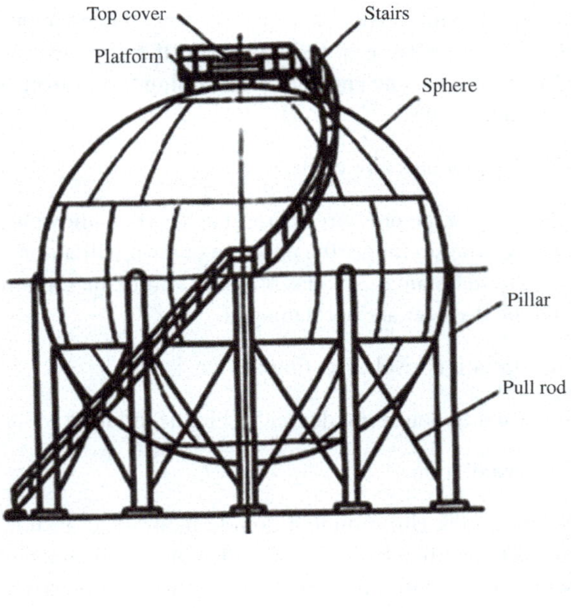

Fig. 7.13 Schematic diagram of a spherical gasholder

The spherical body is the main part of a spherical gasholder that is made of steel plates, and it is the part subjected to pressure. The pillars are the part that bears the weight of the spherical body and the weight of the stored biogas. The pillars should be as short as possible as long as operation and maintenance requirements are met to lower the center of gravity of the spherical body. The attachments of spherical gasholders in biogas plants typically include ladder platforms, pressure gauges, top safety valves, bottom drain valves, manholes, and biogas pipelines.

The effective volume of the medium or sub-high pressure gasholder depends on the geometry of the spherical body, tank pressure, and gas supply pressure from the biogas pipe network. The calculation formula is (7.4).

$$V_n = V(P - P_c)\frac{T_B}{P_B T} \tag{7.4}$$

where

- V_n Gasholder effective volume, m^3;
- V Gasholder geometric volume, m^3;
- P Gasholder top pressure, Pa;
- P_c Biogas pipe network pressure, Pa;
- P_B Standard pressure, 101,325 Pa;
- T Operating temperature, °C;
- T_B Standard temperature, °C.

Medium or sub-high pressure gasholders are pressure vessels, which are completely different from atmospheric pressure gasholders such as low pressure gasholders and digesters. It is necessary to strictly abide by the relevant regulations and technical specifications of pressure vessels. Legislation and technical specifications have both strictly regulated design, manufacturing, installation management, and safety accessories of pressure vessels. Attentions should be paid to the following points:

(1) Production of pressure vessel materials must be approved by the National Safety Supervision Agency. The production entity shall provide the user with quality certificates (original) in accordance with corresponding standards, and each produced material shall have a clear and irremovable mark of steel seal or another mark on a clearly observable place of the material. The mark shall include at least the material manufacturing standard code, material trademark and specifications, the furnace (batch) number, the approval sign of national safety inspection organization, name of fabrication entity, and inspection seal. The content of the material quality certificates must be complete and clear, and the certificates shall be stamped together with the quality inspection certificates of the material fabrication entity. In addition to mechanical properties and tensile properties, material compatibility should also be considered.

(2) Fixed pressure vessel manufacturing entity shall obtain a pressure vessel manufacturing license of AR or BR.

(3) The entity engaged in installation must be the entity that has obtained corresponding manufacturing qualification or the entity approved by the provincial safety supervision agency where the installation entity is located. Supervision engineers engaged in installation supervision shall have the professional knowledge of pressure vessel, and they shall be certified by trainings and assessments approved by the national safety supervision organization.

    The installation entity or the user entity shall report the pressure vessel name, quantity, manufacturing entity, user entity, installation entity, and installation location to the safety inspection organization where the pressure vessel shall be registered before applying for registration procedure. When the user entity is to purchase or carry out project bidding, the pressure vessel design, manufacturing (or group welding) unit with corresponding manufacturing qualifications should be selected. The qualifications, design category, and variety scope of the design entity shall comply with the "Regulations on the Administration and Supervision of Pressure Vessel Design entity".

(4) The responsible technicians of the manufacturing entity is responsible for safety management of the pressure vessel. The engineering and technical personnel who own the expertise of the pressure vessel and are familiar with relevant national laws and regulations are assigned for safety management of the pressure vessel. Before a pressure vessel is put into use, it shall go through the registration procedures one by one at the safety supervision organization or the authorized department according to the requirements of the "Regulations on the Registration of Pressure Vessel Use". During process operation and after operation, the safety operation requirements of pressure vessels should be clearly stated, and the operators should regularly participate in professional training and receive safety education. All operators should be certified in order to work.

(5) The user entity and their authorities must arrange timely and regular inspections while reporting the annual inspection plan to the local safety supervision organization and inspection entity. When making use of pressure vessels, in accordance with "inspection procedures for the pressure vessels in use" and "pressure vessel use registration management rules", the user entity and their authorities should conduct regular inspections, assess safety conditions, and apply for registration. The first internal and external inspection cycle generally takes 3 years, and the subsequent internal and external inspection cycle will be reported to the local safety supervision organization by inspection entity after consulting with the user entity based on results of previous internal and external inspections.

As can be seen from the above provisions, when a biogas plant involves a medium or sub-high pressure gasholder, procedures, requirements for installation, and uses are completely different from those of the low pressure gasholders, so close attention is required for such pressure vessels.

References

Cai, Zhuoning, Lei Cai, and Cai Changda. 2009. Integrated gas production and storage device—New-type large biogas plant. *Agriculture Engineering Technology (New Energy Industry)* 1: 31–33.

Department of Personnel and Labor of the Ministry of Agriculture. 2004. *Biogas production workers*, vol. II. Beijing: China Agriculture Press.

GB 50009-2001. *Loading code for the design of building structures*. National Standard of the People's Republic of China.

GB 50016-2006. *Code of design on building fire protection and prevention*. National Standard of the People's Republic of China.

HG/T20679-1990. *Anti-corrosion design regulations for chemical equipment and pipelines*. Industry Standard of People's Republic of China.

HG20517-92. Steel low pressure water-sealed gasholders. Industry Standard of People's Republic of China.

NY/T1220.2-2006. *Technical code biogas engineering, Part 2: Design of biogas supply*. Agricultural Industry Standards of the People's Republic of China.

Tang, Yanfen, and Yuxin Wang. 2013. *Design and application of large and medium biogas plants*. Beijing: Chemical Industry Press.

Chapter 8
Biogas Utilization

Accompanying the depletion of fossil fuels and rising concerns in ecological issues such as environmental pollution, more and more people have arrived in a consensus of using biogas as an important energy source. Biogas utilization has undergone role changes from a traditional household fuel to transportation fuels, energy sources of power generation, as well as raw materials of modern industry. Utilization of biogas becomes diversified, and the fields of biogas application is continuously expanding.

8.1 Biogas Combustion Characteristics

The combustion characteristics of biogas are determined by its composition. Methane, carbon monoxide, hydrogen, hydrogen sulfide, and carbohydrates in biogas are combustible gases, while carbon dioxide and nitrogen are inert gases. Methane has a relatively high auto-ignition temperature, but a high carbon dioxide ratio in biogas has a strong inhibitory effect on its combustion, and biogas combustion rate is slowed (Zhang 2013).

8.1.1 Combustion of Biogas

Methane, hydrogen, hydrogen sulfide and other components in biogas are all combustible gases. With sufficient oxygen, they will burn when exposed to an open flame, emitting light and heat (Zhou et al. 2009). When biogas is completely combusted, the flame is blue-white. The theoretical combustion temperature of biogas is about 1807.2–1943.5 °C. The chemical reaction equation for complete combustion of biogas is as follows.

$$CH_4 + 2O_2 \rightarrow CO_2 + 2H_2O + 35.91\text{MJ} \tag{8.1}$$

L. Deng et al., *Biogas Technology*,
https://doi.org/10.1007/978-981-15-4940-3_8

$$H_2 + 0.5O_2 \rightarrow H_2O + 10.8\text{MJ} \tag{8.2}$$

$$H_2S + 1.5O_2 \rightarrow SO_2 + H_2O + 23.38\text{MJ} \tag{8.3}$$

8.1.2 Theoretical Air Demand and Excess Air Coefficient for Biogas Combustion

1. *Theoretical air demand for biogas combustion*

Biogas combustion needs an appropriate amount of oxygen. Too much or too little oxygen is detrimental to its combustion. Oxygen required for biogas combustion is generally obtained directly from air. Since the volume fraction of oxygen in air is 20.9%, and the rest is nitrogen and carbon dioxide, the volume ratio of nitrogen to oxygen in dry air is about 3.76.

Theoretical air demand for biogas combustion refers to the volume of air required to completely burn a unit volume of biogas according to the stoichiometric equation of the combustion reaction. As the volume fractions of methane and carbon dioxide in biogas vary, the theoretical air requirement for biogas combustion will vary (Table 8.1).

The theoretical air demand for biogas combustion can be determined by Eq. 8.4.

$$V^0 = \sum_{i=1}^{n} r_i V_i^0 \tag{8.4}$$

where

V^0 theoretical air demand for biogas combustion, m^3/m^3 biogas;
n number of combustible components in the biogas;

Table 8.1 Main characteristic parameters of biogas

Item	CH_4 (50%) CO_2 (50%)	CH_4 (60%) CO_2 (40%)	CH_4 (70%) CO_2 (30%)
Density (kg/Nm^3)	1.347	1.221	1.095
Relative density	1.042	0.944	0.847
Theoretical air demand (m^3/m^3 biogas)	4.76	5.71	6.67
Theoretical flue gas amount (m^3/m^3 biogas)	6.763	7.914	9.067
Lower calorific value (kJ/m^3)	17,937	21,524	25,111
Explosive limit (%) upper limit/lower limit	26.10/9.52	24.44/8.80	20.13/7.00
Burning speed (m/s)	0.152	0.198	0.243

r_i volume fraction of the i-th combustible component in the biogas, %;
V_i^0 theoretical air demand for the i-th combustible component in the biogas, m^3/m^3 biogas.

2. *Excess air coefficient of biogas combustion*

The theoretical air demand for biogas combustion is the minimum volume of air required for biogas combustion. In real world conditions, due to heterogeneity of mixing biogas and air, if the theoretical air demand for the biogas combustion equipment is supplied in the exact amount, complete combustion of biogas cannot be achieved. Therefore, the actual air demand for biogas combustion is greater than the theoretical air demand for biogas combustion, and the ratio of the two amounts is called the excess air coefficient of biogas combustion. The excess air coefficient of biogas combustion can be determined by Eq. 8.5.

$$\alpha = V/V^0 \tag{8.5}$$

where

α excess air coefficient of biogas combustion;
V actual air demand for biogas combustion, m^3/m^3 biogas;
V^0 theoretical air demand for biogas combustion, m^3/m^3 biogas.

In actual conditions, the value α depends on the combustion products of biogas and the operation of the biogas combustion equipment. For civilian or public-use combustion equipment, α is generally 1.30–1.80. For industrial combustion equipment, α is generally 1.05–1.20. If α is too small, incomplete combustion occurs. However, too large α will increase the amount of flue gas, lower the combustion temperature, increase the heat loss of exhaust gas, and reduce the thermal efficiency.

The amount of air required for combustion and the resulting flue gas are essential parameters for designing and retrofitting various combustion devices, as well as for designing air supply devices, chimneys, flue, and other facilities. This amount is also important in calculating flue gas temperatures, basic data on thermal efficiency, and heat balance of gas utilization facilities, so α is especially important in obtaining and calculating reasonable parameters.

8.1.3 Combustion Products and Calorific Value of Biogas

1. *Combustion products of biogas*

Flue gas generated by biogas combustion is collectively referred to as biogas combustion products. Since the proportion of ashes and unburned solid particles carried in biogas combustion products are extremely small, the solid content in flue gas is generally not considered.

When theoretical air for biogas combustion is supplied, the amount of flue gas produced by complete combustion of biogas is called the amount of theoretical flue gas from complete biogas combustion. Components of theoretical flue gas are carbon dioxide, sulfur dioxide, nitrogen and water. Flue gas with moisture is called a wet flue gas, and flue gas containing no moisture is called dry flue gas. When the volume fractions of methane and carbon dioxide in biogas vary, the theoretical amount of flue gas will vary (Table 8.1).

The theoretical amount of flue gas from biogas combustion can be determined by Eq. 8.6.

$$V_f^0 = V_{CO_2} + V_{SO_2} + V_{N_2} + V_{H_2O} \tag{8.6}$$

where

V_f^0 theoretical amount of flue gas from unit-mass biogas combustion, m^3/kg;
V_{CO_2} volume of carbon dioxide in flue gas after a unit mass of biogas is completely burned, m^3/kg;
V_{SO_2} volume of sulfur dioxide in flue gas after a unit mass of biogas is completely burned, m^3/kg;
V_{N_2} volume of nitrogen in flue gas after a unit mass of biogas is completely burned, m^3/kg;
V_{H_2O} volume of vapor in flue gas after a unit mass of biogas is completely burned, m^3/kg.

When biogas is supplied with an actual amount of air, the amount of flue gas generated by complete combustion or incomplete combustion of biogas is called the actual amount of flue gas from biogas combustion. Under complete combustion conditions, components of actual flue gas are carbon dioxide, sulfur dioxide, nitrogen, oxygen and water. In case of incomplete combustion, components of the actual flue gas may contain combustible gases such as carbon monoxide, methane, hydrogen sulfide, and hydrogen in addition to carbon dioxide, sulfur dioxide, nitrogen, oxygen, and water. The actual amount of flue gas generated by biogas can be determined by Eq. 8.7.

$$V_f = V_f^0 + (\alpha - 1)V^0 + 0.00161d(\alpha - 1)V^0 \tag{8.7}$$

where

V_f actual amount of flue gas generated by unit-mass biogas combustion, m^3/kg;
V_f^0 theoretical amount of flue gas from unit-mass biogas combustion, m^3/kg;
α excess air coefficient of biogas combustion;
V^0 theoretical air demand for complete unit-volume biogas combustion, m^3/m^3;
d air moisture content, g/kg.

For biogas combustion, the theoretical flue gas volume corresponds to the theoretical air demand, and the actual flue gas volume corresponds to the actual air demand.

Therefore, the theoretical amount of flue gas and the actual amount of flue gas are also the basic parameters for designing and retrofitting various combustion devices, and these parameters play an important role in biogas utilization.

2. *Calorific value of biogas*

The calorific value of biogas is one of the physical and chemical properties of biogas. This value depends on the total amount of combustibles in biogas, and it refers to the maximum amount of heat that can be released when a unit mass or a unit volume of biogas is completely combusted. It is an important indicator of performances of biogas as an energy source.

The calorific value of biogas falls into the higher calorific value and the lower calorific value. The higher calorific value is the actual gross calorific value of biogas, which includes the energy released of water formation by the combustion of hydrogen and the latent heat of vaporization of water in biogas. As a result of actual combustion, the moisture produced by biogas combustion is discharged along with flue gas in the form of vapor. Therefore, this part of energy cannot be utilized. The actual calorific value of the biogas is a lower calorific value, a value obtained by deducting the latent heat of vaporization from the higher calorific value. When the volume fractions of methane and carbon dioxide in biogas vary, the lower calorific value of the biogas will be different (Table 8.1).

The lower calorific value of biogas can be calculated from the lower calorific value of each single combustible gas in biogas, and the calculation is as shown in formula 8.8.

$$Q_{\text{net}} = \sum_{i=1}^{n} r_i Q_{\text{net}\,i} \tag{8.8}$$

where

Q_{net} lower calorific value of biogas, about 22,154–24,244 kJ/m^3;
n number of combustible components in the biogas;
r_i volume fraction of the i-th combustible component in the biogas, %;
$Q_{\text{net}\,i}$ lower calorific value of the i-th combustible component in biogas, kJ/m^3.

8.1.4 Ignition Temperature, Ignition Concentration Limit, and Burning Rate of Biogas

1. *Ignition temperature of biogas*

Combustion of any combustible gas mixture must first enter an ignition stage in order to burn. The ignition phase is the pre-heating process of the combustion phase and is caused by a rise in temperature. When combustible gases in fuel gas are subjected to a rapid increase of temperature, the process of triggering a combustion caused by transition from stable oxidation to unstable oxidation is called ignition. The lowest temperature at which such transition occurs is called the ignition temperature.

Ignition temperature of a gas in the air is not a fixed value: the ignition temperature also depends on the concentration of the combustible gas in the air, the degree of mixing, pressure, and the shape and size of the combustion chamber. The ignition temperature of methane is about 538 °C. Because biogas contains a large amount of inert gas such as carbon dioxide, the ignition temperature of biogas is higher than the ignition temperature of methane, which is about 650–750 °C, meaning that biogas is not as easy to ignite as methane.

2. *Explosive limit of biogas ignition*

The concentration range of a mixture of biogas and air that generates a fire which leads to explosion is called the flammability limit of biogas or the explosive limit of biogas. The explosive limit is divided into an upper limit and a lower limit. When biogas is mixed with air to form a combustible mixture, the biogas-air mixture can hardly reach the ignition point if too much or too little biogas is present. If there is too much biogas in the combustible mixture, oxygen content is so small that biogas can only get partially oxidized, and thus less heat is generated. Moreover, due to the insufficient amount of heat generated, the combustible mixture fails to reach its ignition temperature. On the other hand, if there is too little biogas in the combustible gas mixture, the amount of heat generated is also small, and the small amount of heat does not allow the combustible mixture to reach its ignition temperature. In both extreme cases, combustion cannot be achieved. When the ignition temperature of biogas is reached, the maximum volume fraction of the biogas in the combustible mixture is called the upper explosive limit of the biogas, and the minimum volume fraction of biogas in the combustible mixture is called the lower explosive limit of the biogas.

It can be seen that the ignition concentration limit of biogas greatly influences combustion characteristics of biogas. The concentration limit of biogas is determined by factors such as temperature, pressure, and inert gas content of the combustible gas mixture.

1) Temperature of combustible gas mixture

When the initial temperature of combustible gas mixture is increased, the flammable range of biogas can be extended. Tests show that the influence of temperature on the ignition concentration limit of biogas is mainly found in the upper limit. The lower limit is subjected to a weaker influence.

2) Pressure of combustible gas mixture

When the pressure of combustible gas mixture is high, influence of pressure on the explosive limits of biogas is weak, but as the pressure of the combustible gas mixture gradually decreases, influence of pressure on the explosive limit of biogas starts to show up.

3) Content of inert gas in the combustible gas mixture

Generally, when inert gas content in the combustible gas mixture increases, both the upper and lower explosive limits of biogas increase. As the volume fractions of methane and carbon dioxide in biogas vary, the explosive limit of the biogas will be different (Table 8.1).

3. *Burning rate of biogas*

The burning rate of biogas is also called the flame propagation speed of biogas. It is one of the most important physical and chemical properties of biogas combustion. When a biogas-air mixture ignition starts at one point, a very thin combustion flame surface is formed at the spark, and the high-temperature combustion flame surface rapidly transfers heat to adjacent biogas-air layers whose temperature is raised up immediately. When the ignition temperature is reached, the layers begins to ignite and form new burning flame surfaces. The interface between the unburned gas and the combustion products is called a combustion flame surface. The combustion flame surface continuously expands and moves toward the unburned gas so that each layer of gas successively undergoes heating, ignition, and combustion processes, and this phenomenon is called flame propagation.

The speed at which the combustion flame moves forward is called the burning speed or the flame propagation speed, and its unit is m/s. The burning speed is related to composition of the combustible gas, temperature, the mixing speed, the mixed gas pressure, the mixing ratio of the combustible gas and the air, etc. The burning rate of biogas is very slow due to the composition of biogas. When the volume fractions of methane and carbon dioxide in biogas vary, the burning rate of biogas will be vary (Table 8.1).

Methane present in biogas has the lowest flame propagation speed among many pure combustible gases, and the presence of carbon dioxide further limits the flame propagation speed of biogas combustion. The maximum propagation rate can only reach about 0.2 m/s, which is less than 1/4 of the burning rate of liquefied petroleum gas and only about 1/8 of the burning rate of coke gas. Because the burning speed of biogas is low, when the flow rate of an unburned gas coming out from the chamber of a combustion device is greater than the burning speed, the unburned biogas can be easily blown off, thereby causing flame-out. Therefore, the stability of biogas combustion is poor.

8.2 Biogas for Civilian Use

Biogas is used as an energy source for households or public buildings. Civilian use of biogas is one of the most important ways to use biogas in China.

8.2.1 Biogas Transportation and Distribution Technology

1. *Consumption and supply pressure of biogas*

The consumption philosophy of biogas should be determined according to local gas supply principles and conditions, including taking into consideration residential gas consumption, gas consumption of boilers, gas consumptions of other purposes, and unpredictable gas consumptions. The gas consumption index of households should be determined by collecting daily consumption statistical data from local residents. In case of lacking actual statistics of gas consumption, the consumption can be estimated to be 0.8–1.6 m^3/d as derived from actual situation of each household. Gas consumption of boilers can be calculated based on actual fuel consumption. Other types of gas consumption can be determined according to the current national standard "Code for design of city gas engineering" GB 50028. The unpredictable gas consumption is margined 5–8% of the total gas consumption.

Biogas can only be transported and used at a certain supply pressure. Generally, the supply pressure of biogas from wet gas holders is within 3–5 kPa (0.003–0.005 MPa). Under conditions of a gas supply radius of 500 m and reasonable pipeline design, this pressure range can meet the requirements of various gas appliances without pressure boosting. If the transport distance is long or if there are numerous gas utilization facilities, it is necessary to pressurize and regulate the biogas pressure to ensure required pressures of the gas utilization facilities. Biogas supply is often at low pressure (not more than 0.01 MPa) or grade B medium pressure (more than 0.01 MPa and not more than 0.2 MPa), whereas grade A medium pressure (more than 0.2 MPa and not more than 0.4 MPa) is rarely applied and high pressure (more than 0.4 MPa) is barely taken (see standard GB/T 51063 and NY/T 1220.2).

2. *Material of biogas pipes*

According to their materials, biogas pipelines can be classified into steel pipes, polyethylene pipes, steel skeleton reinforced polyethylene composite pipes (hereinafter referred to as composite pipes), etc. The selected materials must comply with relevant standards. The main features of steel pipes, polyethylene pipes, and composite pipes are as follows.

1) Steel pipes

Steel pipes are the major pipes used in biogas transmission and distribution. They have advantages of high strength, good tightness, and mature welding technology, but they have poor corrosion resistance and need preservation. Steel pipes are classified into seamless steel pipes (see standard GB/T 8163) and welded steel pipes (see standard GB/T 3091) according to different manufacturing methods. In biogas transmission and distribution, straight seam welded steel pipes are commonly used. According to different surface treatments, steel pipes are classified into galvanized (white iron pipes) and non-galvanized (black iron pipes). According to different wall thicknesses, the steel pipes are classified into three types: ordinary steel pipes, thick steel pipes and thin-wall steel pipes.

Small-diameter seamless steel pipes are mainly made from galvanized pipes and are usually used indoors. If they are used for outdoor underground installation, they must also take an anti-corrosion treatment. Seamless steel pipes with a diameter larger than DN150 are non-galvanized black iron pipes. Biogas pipeline transportation pressure is not high, so standard seamless pipes or fuel gas transmission steel pipes made of soft carbon steel are used. However, for large-diameter biogas pipes, butt weld pipes an spiral welded steel pipes are usually adopted (Tang and Wang 2013).

Steel pipelines must be externally protected against corrosion (see standard CJJ 95 and SY 0007). The anti-corrosion layer of the steel pipes should adopt extruded polyethylene, sintered epoxy powder, double-layer epoxy, and so forth as the anti-corrosion coating materials. The basic structure of an ordinary grade and a reinforcement grade anti-corrosion layer should meet the requirements in Table 8.2.

According to the specific conditions of a project, underground steel pipes can select coatings such as petroleum bitumen, polyethylene anti-corrosion tape, epoxy coal bitumen, polyethylene thermoplastic coating, and chlorosulfonated polyethylene coating. Selection of these coatings should comply with the relevant national standards. The corrosion grade of soil and the grade of the anti-corrosion coating of underground steel pipes shall be determined according to the geological conditions and electrical resistivity of the soil where pipeline passes through (Table 8.3).

The steel pipes exposed to the atmosphere should be made of anti-corrosion coating with stable paints, strong surface adhesion, and good weather resistance, and the surface of the pipeline should be treated according to the coating requirements. The outer surface of the pipes shall be coated with anti-corrosion yellow paints. Coated steel trunk pipes should also be protected by cathodic protection at the same time.

Table 8.2 Basic structure of anti-corrosion layer of steel biogas pipelines

Anti-corrosion layer	Basic structure	
	Ordinary level	Enhanced level
Extruded polyethylene coating	≥120 μm epoxy powder	≥120 μm epoxy powder
	+≥170 μm adhesives	+≥170 μm adhesives
	+1.8–3.0 mm polyethylene	+2.5–3.7 mm polyethylene
Fusion bonded epoxy powder coating	≥300 μm epoxy powder	≥400 μm epoxy powder
Double epoxy coating	≥250 μm epoxy powder	≥300 μm epoxy powder
	+≥370 μm modified epoxy	+≥500 μm modified epoxy

Table 8.3 Soil corrosion grade and anti-corrosion coating grade

Soil corrosion grade	Low	Medium	Relatively high	High	Super high
Soil resistivity (Ω)	>100	100–20	20–10	10–5	<5
Anti-corrosion coating grade	Ordinary level		Enhanced level	Special reinforcement	

2) Polyethylene plastic pipes

Polyethylene pipes are a replacement product of traditional steel pipes (see standard GB 15558.1 and GB 15558.2). Biogas pipes must withstand a certain pressure, and usually polyethylene resin with a high molecular weight and good mechanical properties is used for this purpose. Low density polyethylene (LDPE) resins, especially linear low-density polyethylene (LLDPE) resins have become common materials for biogas pipes due to their good hygiene. LDPE resins and LLDPE resins have a small melt viscosity and good fluidity, and both are easy to process. Therefore, they have a wide range of melt flow index (MFI) values, and usually the MFI value is between 0.3 and 3.0 g/10 min.

Advantages of polyethylene pipes include low density (only 1/4 the density of steel pipe), convenient transport, easy processing and installation, good electrical insulation, low susceptibility to electrochemical corrosion, up to 50 years of service life (2–3 times longer than that of steel pipes), smooth inner wall of the pipe, strong wear resistance, small resistance along pipe pathway, no deposition of impurities in biogas, improved biogas transmission capacity, flexibility, strong earthquake resistance. Polyethylene pipes can also be easily pinched and repaired in an emergency and can be flexibly adjusted when the construction encounters obstacles. Simple construction process, no need to remove rust, and corrosion resistance are also significant advantages of such pipes. Furthermore, pipe connection is simple and reliable, and the pipeline maintenance is simple (Qi 2013).

Disadvantages of polyethylene pipes include lower strength than steel pipes, in general low-pressure application (high density polyethylene pipes with maximum pressure of 0.4 MPa), easy aging under oxygen and ultraviolet light (therefore, they should not be laid above ground), extreme sensitivity to temperature changes (when temperature increases, the elasticity of the material increases, the rigidity decreases, and the dimensional stability of the product is poor; when the temperature is too low, the material becomes hard, brittle, and easy to crack), poor rigidity (if the pipe foundation sinks or water accumulates in pipes, the pipes can easily deform and have partial blockage), non-polarity, causing static electricity, difficulty of finding buried pipelines, and inconvenience of on-ground marking (Qi 2013).

3) Steel skeleton reinforced polyethylene composite pipes

The composite pipe is a new type of improved pipe (see standard CJ/T 125 and CJ/T 126). The pipe is made of high-strength plastic-coated steel mesh frame and thermoplastic polyethylene. The wire mesh is used as the skeleton reinforcement of polyethylene pipes. The pipe is based on high-density polyethylene (HDPE) and uses high-performing HDPE modified adhesive resin to closely bond the steel wire skeleton to inner and outer high-density polyethylene to give it excellent composite properties. Because the high-strength steel wire reinforcement is coated onto continuous thermoplastics, the composite pipe has overcome the respective disadvantages of steel pipes and polyethylene pipes, while maintaining their respective advantages.

The outdoor high-pressure biogas piping systems should adopt steel pipes, whereas medium-pressure and low-pressure biogas piping systems should adopt polyethylene gas pipes, steel pipes, or steel skeleton reinforced polyethylene composite pipes. Biogas transmission from household digesters and digesters of domestic wastewater treatment plants usually adopt polyethylene pipes.

3. *Arrangement and installation of biogas pipelines and gas pressure regulation*

Biogas is flammable and explosive, so the safety of biogas pipelines must be guaranteed. Biogas pipelines should be equipped with safety protection devices, concentration alarms to prevent overpressure, and lightning protections, and the antistatic measures for biogas pipelines must also be reliable. At the entrances and exits of the biogas pipelines in buildings, on outdoor roofing pipes, on risers, on discharge pipes, on introduction pipes, and on biogas equipment shall have lightning protection and anti-static grounding implemented (see standard GB 50057 and HGJ 28). In addition, in order to ensure the safety of biogas pipelines, safety facilities such as ventilation pipes, emergency shut-off valves, and biogas concentration detection alarms must also be equipped.

1) Arrangement and installation of outdoor biogas pipelines

The outdoor biogas pipeline should be able to safely and reliably meet the gas consumption and supply pressure of various users. When arranging and installing outdoor biogas pipelines, the requirements for biogas use should be met first, and once the requirements are met, the pipeline length should be shortened as much as possible to save investment and reduce waste. The layout and installation of outdoor biogas pipelines should be based on the principles of comprehensive planning, integration of long and short-term plans with a focus on short-term plans, and phased construction.

When preparing for the installation of outdoor biogas pipelines, special attention should be paid to the following specific matters:

(1) The location of the outdoor biogas trunk lines should be close to large users.
(2) In order to ensure the reliability of biogas supply, the outdoor biogas trunk lines should be connected into a ring shape in order.
(3) Outdoor biogas pipelines should occupy as few fertile lands as possible and stay as close to roadways as possible. Avoid lands which have been planned.
(4) Outdoor biogas pipelines shall not be laid in the same trench with other pipelines or cables.
(5) It is strictly prohibited to lay outdoor biogas pipelines in high-voltage cable corridors, under warehouses, under storage spaces for flammable and explosive materials or corrosive liquids, or in traffic tunnels.
(6) Outdoor biogas pipelines shall not be set across the river bottom.
(7) Outdoor biogas pipelines should be installed underground. When it is difficult to do so, the steel pipes can be arranged aboveground.
(8) When an outdoor biogas pipeline is buried, it should be buried below the frost line.
(9) When an outdoor biogas pipeline is buried, traffic pathways should be avoided as much as possible to avoid crossing with railways and rivers. When outdoor biogas pipelines cross railways, highways, or main urban roadways,

the pipelines should comply with the latest national standards. When biogas pipelines pass through other roads in general, casings of water supply and drainage pipes (ditches), heat pipes (ditches), trenches, tunnels, and so on shall be equipped.

(10) When an outdoor biogas pipeline is buried, the foundation should be the original soil layer, and all the sections that may cause uneven settlement of the pipeline should be pre-treated.

(11) The horizontal distance between outdoor buried biogas pipelines and buildings, structures, or adjacent pipelines shall meet the requirements of Table 8.4.

(12) The vertical distance between the outdoor buried biogas pipelines and the adjacent structures or pipelines shall meet the requirements in Table 8.5.

(13) When an outdoor biogas pipeline is laid overhead, it can be fixed along the outer wall or pillar of residential or public buildings with a fire-resistance rating of not less than grade two.

(14) The vertical distance between outdoor overhead biogas pipelines and railways, roads and other pipelines shall meet the requirements in Table 8.6.

Table 8.4 Horizontal distance between outdoor buried biogas pipelines and buildings, structures, or adjacent pipelines (unit: m)

Item		Outdoor buried biogas pipeline pressure (MPa)		
		Low pressure $\leq$ 0.01	Medium pressure	
			$0.01 < B \leq 0.2$	$0.2 < A \leq 0.4$
Building foundation		0.7	1.0	1.5
Water supply pipe		0.5	0.5	0.5
Sewage, rainwater drain pipe		1.0	1.2	1.2
Power cable (including tram cable)	Directly buried	0.5	0.5	0.5
Communication Cable	Directly buried	0.5	0.5	0.5
	Inside the catheter	1.0	1.0	1.0
Other gas pipeline	DN $\leq$ 300 mm	0.4	0.4	0.4
	DN > 300 mm	0.5	0.5	0.5
Heat pipe	Directly buried	1.0	1.0	1.0
	In the trench (to the outer wall)	1.0	1.5	1.5
Foundation of pole (tower)	$\leq$35 kV	1.0	1.0	1.0
	>35 kV	2.0	2.0	2.0
Communication or lighting pole (to pole center)		1.0	1.0	1.0
Railway embankment		5.0	5.0	5.0
Tram rail		2.0	2.0	2.0
Street tree (to the center of the tree)		0.75	0.75	0.75

Table 8.5 Vertical distance between outdoor buried biogas pipelines and structures or adjacent pipelines (unit: m)

Item		Outdoor buried biogas pipeline (when casing is used, it is measured by casing)
Water supply pipe, drain pipe or other biogas pipeline		0.15
The tube bottom (or top) of the heat pipe		0.15
Cable	Directly buried	0.50
	Inside the catheter	0.15
Railway track bottom		1.20
Tram (track bottom)		1.00

Table 8.6 Vertical distance between outdoor overhead biogas pipelines and railways, roads, and other pipelines (unit: m)

Building and pipeline name		Minimum vertical distance	
		Under the biogas pipeline	Above biogas pipeline
Rail top		6.0	–
Urban road pavement		5.5	–
Factory road pavement		5.0	–
Pedestrian road surface		2.2	–
Overhead power line, voltage	Below 3 kV	–	1.5
	3–10 kV	–	3.0
	35–66 kV	–	4.0
Other pipe, pipe diameter	$\leq$300 mm	Same pipe diameter, but not less than 0.10	Same as left
	>300 mm	0.30	0.30

2) Arrangement and installation of indoor biogas pipelines

Indoor biogas pipelines includes an introduction biogas pipe and an indoor biogas pipe. The introduction pipe refers to a pipe between an outdoor gas distribution pipe and the main valve at an inlet pipe of an indoor biogas user. The indoor biogas pipe refers to a pipeline between the main valve at an inlet pipe of an indoor biogas user and a biogas utilization facility.

The types of biogas introduction pipes are not completely the same according to specific conditions. The introduction pipes can be classified into underground introduction and ground introduction pipes based on their different introduction methods. The introduction pipe for conveying wet gas in the heating area generally introduces the gas into the room underground. When anti-freezing measures are taken, the wet gas can also be introduced from the ground. In non-heating areas or when conveying dry gas, if pipe diameter is not more than 75 mm, the gas can be directly introduced into the room from the ground.

When arranging the installation of indoor biogas pipelines, special attention should be paid to the following specific points:

(1) Introduction biogas pipes should be directly introduced into kitchens or other places of biogas utilization from outdoor pipes. Indoor biogas pipes should not be installed in flammable or explosive warehouses, rooms with corrosive materials, power distribution rooms, transformer rooms, cable trenches, or flue or air inlets.
(2) Biogas pipelines are strictly prohibited from entering into bedrooms. When horizontal biogas pipelines pass through bedrooms, bathrooms, or basements, they must be welded and installed in casings. When biogas pipelines enter an gastight chamber, the gastight chamber must be modified with a proper ventilation, and the ventilation frequency should be no less than 3 times an hour.
(3) When biogas inlet pipes pass through foundations, walls or trenches of a building, they shall be encased. The space between the casings and enclosed biogas pipes shall be filled with bitumen and batched jute, and the casings can be sealed by heat-asphalt. Sizes and positions of holes of casing passing through the wall should take into consideration the extent of building settlement.
(4) The connection method of biogas introduction pipes and biogas indoor pipes varies depending on pipe materials. When biogas indoor pipes and biogas introduction pipes are steel pipes, welding or wire bonding is generally adopted. When biogas indoor pipes are plastic pipes while biogas introduction pipes are steel pipes, steel-plastic joints are generally used.
(5) Indoor biogas pipes should be exposed. When buildings or processes have special requirements, indoor biogas pipes can be concealed on condition that the following requirements are met: (a) The concealed biogas risers can be placed in pipe troughs on the wall or in pipe wells, and hidden horizontal biogas pipes can be placed in furred ceilings or pipe trenches. (b) Pipe channels of biogas pipelines shall have movable doors and ventilation holes. Trenches of biogas pipelines shall have movable cover plates filled with dry sand. (c) Biogas pipelines used for industrial or laboratory purposes may be set on concrete floors. The introduction and exit of biogas pipelines shall be equipped with casings that extend 50–100 mm from the ground, and the ends of the casings shall be sealed with flexible waterproof material. Biogas pipelines shall each have an anti-corrosion insulation layer. (d) Concealed biogas pipelines may be settled in pipeline wells, pipe trenches or together with air, inert gas, water supply pipelines, heat pipes, etc., only if the biogas pipelines are welded. Biogas pipelines shall not be placed in trenches that is subjected to penetration by corrosive materials. (e) When trenches of biogas pipelines intersect with other trenches, the biogas pipe trenches shall be sealed, and the biogas pipes shall be enclosed in steel casings. (f) Equipment platforms and pipe wells of biogas should be well ventilated. Each of the wells should be provided with

a fire-proof layer which has the same fire resistance as the floor ducts, and an easy access for maintenance should be set. (g) Gas pipes should be coated with recognizable yellow anti-corrosion paints.

(6) The minimum distance between biogas indoor pipes and electrical equipment. as well as adjacent pipelines, shall meet the requirements of Table 8.7.

(7) Indoor biogas pipes on walls, columns and floors shall be fixed by pipe clamps, brackets, or hangers.

(8) The height of indoor biogas pipes shall not be less than 2.2 m from the indoor floors, not less than 1.8 m from the kitchen floors, and not less than 0.15 m from ceilings.

(9) The slope of biogas indoor pipes should be no less than 0.003, and the pipes should be inclined to the riser and the cooker respectively.

(10) Indoor biogas pipes should each have control valves in upstream of the flow meters and the gas utilization facilities.

(11) Discharge pipes shall be set on the pipelines of large and medium gas utilization facilities. The discharge pipes should be more than 1 m above roof, and measures should be taken to prevent rain and snow from entering the pipes and to avoid purging discharges into the room.

(12) The connection between indoor biogas pipes and civilian municipal biogas stoves can be achieved by hoses. The design should meet the following requirements: (a) The length of the connecting hoses should not exceed 2 m, and there should be no joint in the middle. (b) Soft gas pipes should be oil-resistant rubber gas hoses. (c) The connection between hoses and biogas pipelines, biogas

Table 8.7 Minimum distance between indoor biogas pipes and electrical equipment or adjacent pipelines (unit: m)

Pipe and equipment	Clear distance from biogas pipeline	
	Parallel laying	Cross laying
Surface mounted insulated wire or cable	0.25	0.10 (see Note)
Concealed or interior insulated wire	0.05 (from the edge of the groove or tube made)	0.01
Exposed wire with a voltage less than 1000 V	1.00	1.00
Switchboard or distribution box, electricity meter	0.30	Not allowed
Electric socket, power switch	0.15	Not allowed
Adjacent pipe	Ensure the installation and maintenance of biogas pipeline and adjacent pipeline	0.02

Note When the exposed wire is insulated with the insulating sleeve and the two ends of the casing protrude from the biogas pipeline by 0.10 m, the cross distance of the casing and the biogas pipeline can be reduced to 0.01 m. When the arrangement is difficult, the distance can be appropriately reduced after effective measures are taken stoves, and other gas utilization facilities should be fixed by pressing nuts or pipe clamps. (d) Hoses must not pass through walls, windows or doors.

3) Gas pressure regulation of biogas pipelines

Gas pressure regulation of biogas pipelines is a technical measure to ensure that the biogas pressure can meet the requirements of biogas utilization facilities (Deng et al. 2015). In biogas supply, the commonly used pressure regulating devices include a pressure regulating box and a pressure regulator. The main technical parameters of a pressure regulating device include gas flow rates, inlet pressure, outlet pressure, working temperature, and pressure regulation accuracy.

Under normal circumstances, when climate conditions and surrounding environment is acceptable, the pressure regulating device can be set in open air, but protective fences, guardrails, or vehicle buffer stops should also be installed. When aboveground operations are restricted, the pressure regulating device can be installed in a separate building underground or in a separate underground box, but the installation must comply with the requirements of the current national standard "Code for design of gas transmission pipeline engineering" GB 50251. The ambient temperature of the non-heating pressure regulating device should ensure normal functioning of the moving parts of the pressure regulating device, and the pressure regulating devices without anti-freezing measures should only be employed in places where ambient temperature is above 0 °C.

(1) Pressure regulating box

A pressure regulating box is an important part of the medium-pressure biogas pipework. It is a device that reduces the pressure in the medium-pressure biogas pipework to a pressure suitable for gas uses in biogas utilization facilities. It can be used for gas supply pressure adjustment of residential quarters, public users, direct combustion equipment, biogas boilers, industrial furnaces and kilns. The pressure regulating box is widely used due to its compact structure, small land occupancy, investment savings, and easy installation and use. The basic components inside a pressure regulating box include an inlet valve, an outlet valve, filters, pressure regulators, and corresponding measuring instruments, and the box can also be equipped with accessory equipment such as corrugated compensators, pressure relief valves and overpressure shut-off valves. At the same time, according to the purposes of use and user requirements, a pressure regulating box can be assembled into a single (1 + 0), single plus bypass (1 + 1), dual (2 + 0), dual plus bypass (2 + 1), multi-channel parallel, and other structural form. The single-channel plus bypass (1 + 1) pressure regulating box and the two-way (2 + 0) pressure regulating box are shown in Figs. 8.1 and 8.2.

The single pressure regulating box is equipped with a pressure regulating system. Since gas must be shut off during maintenance, the single-channel pressure regulating box is generally used for users who can bear intermittent gas supply.

The two-way pressure regulating box is equipped with a two-way pressure regulating system, one of which is used to operate the pressure regulating system while

Fig. 8.1 Single plus bypass (1 + 1) pressure regulating box

Fig. 8.2 Dual (2 + 0) pressure regulating box

the other is used to ensure safety. This way, maintenance of the pressure regulating system can be realized without affecting the use of gas. Therefore, the two-way pressure regulating box is generally used for users who must have continuous gas supply. If the two-way pressure regulating system is equipped with an overpressure shut-off valve, automatic switching from one mode to the other can be realized by setting different cut-off pressures.

Installation of a pressure regulating box should meet the following requirements: (a) The bottom of a pressure regulating box should be 1.0–1.2 m high from the floor, and it can be installed on the outmost wall of the gas-using building or suspended on a special support bracket. When installed on the outmost wall of a gas-using building, the diameter of the inlet and outlet pipes of the regulator should be no greater than DN50. (b) The horizontal distance of a pressure regulating box to doors, windows, or other building access shall not be less than 1.5 m. A pressure regulating box shall not be installed under windows of a building or in its balcony. It should not be installed on the wall of an indoor ventilator. (c) The wall on which a pressure regulating box is to be installed shall be a permanent solid wall whose fire resistance rating shall not be less than Class II. (d) A bypass pipe shall be set between biogas inlet and outlet pipes of a pressure regulating box, and a bypass tank can be omitted for the user pressure regulating box (suspended type).

(2) Pressure regulator

Regulators are commonly known as pressure reducing valves or gas pressure regulating valves. They maintain a specified pressure of outlet gases by automatically changing the gas flow rate. Based on the operations, regulators can be classified into two types: direct acting regulator and indirect acting regulator. A direct acting regulator and an indirect acting regulator are shown in Figs. 8.3 and 8.4.

A direct acting regulator consists of a measuring element (film), a transmission part (valve) and an adjustment mechanism (valve). Assume a sudden fluctuation in gas flow that the amount of gas increases at the outlet or the inlet pressure decreases, the outlet pressure is then decreased. At this moment, due to the pressure reflected by the impulse pipe, the force of gas flow acting upward onto the bottom side of the membrane is smaller than the weight (or spring) of the membrane, and the membrane is therefore lowered. The valve disc also moves down with the valve stem, causing the valve to open so that the gas flow rate increases, and the outlet pressure returns to the original set value. Conversely, when the amount of gas at the outlet is suddenly reduced or the inlet pressure increases, the valve is open to a lesser extent and the flow rate is reduced. The outlet pressure is still restored. The outlet pressure value can be set by adjusting the weight or spring force of the membrane. The working principle of the direct acting regulator is shown in Fig. 8.5.

An indirect acting regulator consists of a main regulator, a pilot valve, and an exhaust valve. When the outlet pressure is lower than the given value, the film of the pilot valve is lowered, and the pilot valve is opened. The throttled gas is thereby replenished to the membrane space of the main regulator so that the main regulator valve is opened and the flow is increased. The outlet pressure returns to the set value. Conversely, when the outlet pressure exceeds a given value, the film of the pilot valve

Fig. 8.3 Direct-acting pressure regulator

Fig. 8.4 Indirect acting pressure regulator

rises, causing the valve to close. At the same time, due to the force acting on the underside of the exhaust valve film, the exhaust valve is opened, and a part of the gas is released into the atmosphere, so that the force on the underside of the membrane of the main regulator is offset. The valve opening is narrowed, and the outlet pressure is restored to the given value. The working principle of the indirect acting regulator is shown in Fig. 8.6.

Installation of pressure regulators shall comply with the following requirements: (a) At the inlet or outlet of a pressure regulator, safety protection devices shall be set

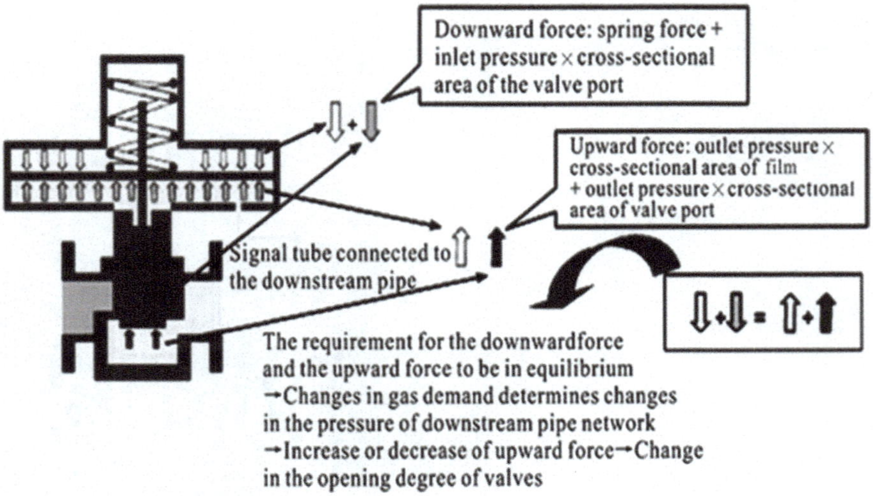

Fig. 8.5 Working principle of the direct acting regulator

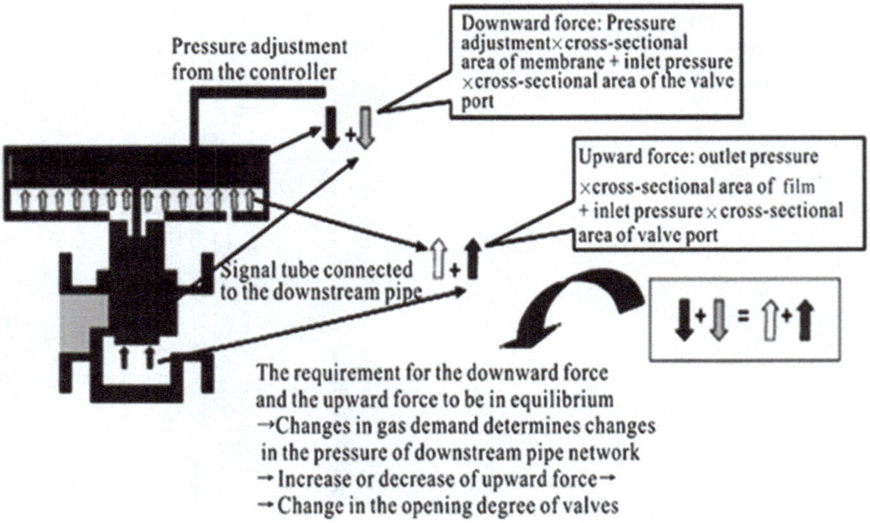

Fig. 8.6 Working principle of the indirect acting regulator

to prevent excessive pressure at the biogas outlet. When a pressure regulator itself has safety protection devices, the set can be omitted. (b) The safety protection device of a regulator should be manually reset. A starting pressure value must be set for the safety protection (distribution or shut-off) device. The starting pressure should not exceed 50% of the maximum working pressure of the outlet. Pressure of the gas utilization facilities directly connected to low pressure biogas pipelines should be within the safe working pressure limits.

4. *Operation management of biogas pipeline networks*

1) Basic tasks of operation management

(1) Safely supply biogas to all users without interruption.
(2) Regularly inspect and repair the biogas pipe network and its ancillary facilities to ensure that they are intact.
(3) Quickly eliminate leakage, damage and malfunctions in biogas pipeline network.
(4) Responsible for connecting new users to pipe network.
(5) Responsible for supervision of construction quality of biogas pipeline network. Participate in completion and acceptance check of the biogas pipeline network.
(6) Responsible for conflict management or cooperation between other construction units and the operating biogas pipeline network.
(7) Regularly discharge condensate from the biogas network.

2) Operation and safety technology of biogas pipe network

(1) Maintenance and repair work of biogas pipe network in operation is usually carried out with biogas. Fire is strictly prohibited on working sites, and workers should wear gas masks. When the biogas pipe network is in operation with biogas, the biogas pressure should be controlled within the range of 200–800 Pa, and there should be no fewer than two operators, together with at least one observer.
(2) At least two inspections of low-pressure biogas pipeline network shall be conducted every month, and regular preventive inspections of gates and underground facilities shall be carried out simultaneously. Integrity of gate wells, removal of condensate along pipelines, and the pollution level of underground facilities are the main items to be checked. When a gate is open, smoking, ignition, and use of non-explosion-proof lamps are prohibited.
(3) The main work of daily maintenance management of biogas pipe network is the leakage inspection. According to the intensity of biogas odor, the approximate leakage range is initially determined, and the following empirical methods are used to pinpoint the leak location: (a) Drilling and leak detection: check on the ground along the direction of the biogas pipelines drill holes along at a certain distance on the ground (2–6 m) and check with an olfactometer or a leak detector. (b) Excavation pit: dig a hole at the target pipe location or at the target joint to expose the pipe or joint and check for biogas leakage. (c) Well chamber inspection: under the ground where biogas pipelines are laid, various shields or manhole covers for water wells, drainage wells, rainwater wells, cable wells and other underground wells or other underground facilities can be used to check if there is a leak with an olfactometer or a leak detector. (d) Observing plant growth: The leakage of biogas from underground pipelines can penetrate into soil and cause the foliage of plants to turn yellow and dry. (e) Judging the leak by

using a condensate: If the condensate volume suddenly increases during regular biogas moisture condensation, a crack in the biogas pipeline could be the reason so that the groundwater seeps into the pipeline and flows into the condensate. By the aforementioned methods, the leakage of biogas can be predicted to some extent.

In addition, the period and number of leak inspections should be set based on a pipeline's operating pressure, materials, buried age, soil quality, groundwater level, road traffic, and previous leak inspection records. Leakage inspection work should be carried out by a professional personnel throughout the year. Apart from the usual leakage inspection, leakage inspection should also be carried out in a focused and thorough manner at regular intervals. The inspection method can be appropriately selected after taking consideration of the specific conditions of the biogas pipeline network. Some projects can also use safety monitoring systems to monitor the leakage of the biogas pipeline network in real time, so that the leakage inspection can be implemented smoothly.

(4) Pipeline blockage and its removal require attentions regarding the following aspects: (a) Biogas contains vapor, and if temperature decreases or pressure rises, water vapor will condense into liquid and flow into a condenser or a low point drain along the pipe. If the condensate reaches a certain amount and is not removed in time, the pipe will be blocked. In order to prevent condensed water from clogging the pipelines, gathered condensate should have a location and pumping history record to mark the pumping date and pumping quantity as an important basis for determining the removal cycle and for finding abnormal conditions such as groundwater infiltration as early as possible. (b) When the groundwater pressure is higher than the biogas pressure in a pipeline, the pipes that are in disrepair may cause penetration of unwanted substances into the biogas pipeline system via the pipe joints, corrosion holes, or cracks. When the amount of condensate increases sharply, water seepage might have happened. The suspected pipe section can be isolated or bypassed followed by a shutoff. Biogas with a pressure higher than the infiltration pressure can be pressed in, and the leakage site can be found by empirical method of leakage inspection. (c) Due to various reasons, sometimes biogas pipelines are unevenly settled, and the condensed water will be stored in the sinking part of the pipeline to form the so-called "liquid pockets". The method of finding "liquid pockets" is to drill holes in the biogas pipeline of interest and stuff a rubber bladder into the borehole, followed by checking the sound on the inflatable side of the drilled hole after inflation, and the "liquid pocket" is found after correction. The methods of altering pipe slope or adding another drain can be used to remove the "liquid pocket". (d) For steel pipes without inner wall coating nor good treatment of inner coating, corrosion is more serious, and more rust stains are generated, which not only reduces the effective flow cross section area of the pipeline, but also often causes blockage in the branching pipes. (e) A method to remove impurities is to carry out mechanical cleaning of the main pipes section by section, with each cleaning pipe section being about 50 m long. If pipe connection fittings, valves, and drainers are blocked, they can be removed for cleaning.

8.2.2 Biogas for Domestic Use

Residential biogas utilization facilities include biogas meters, biogas stoves, and biogas water heaters (see standard GB 16914). These utilization facilities use low-pressure biogas, and the biogas pressure at the inlet end of the utilization facilities should be controlled within the range of $0.75P_n$ to $1.5P_n$ (P_n is the rated pressure of the combustion appliance).

1. *Domestic biogas utilization facilities and their installation*

Since biogas, natural gas, and liquefied petroleum gas differ in composition, they cannot share the same combustion facilities except gas meters (see standard CJJ 12).

1) Biogas meter

Household biogas meters usually have membrane structures. Membrane gas meters belong to volumetric mechanical instrument. The driving force of the diaphragm movement inside the gas meter comes from a pressure difference between the inlet and outlet of the gas meter. When gas pressure fluctuates on either side, the diaphragms generates out-of-phase stroke motions, thus continuously directing gas into the measuring chamber and comparting gas into each single metering volume (circulating volume) before discharging it to the outlet. Since the diaphragm compartment is connected to the counter by a mechanical transmission mechanism, counting and calculation of the number of single metering volumes are performed, so that the total amount of circulating gas can be finally measured (metered) (Fig. 8.7).

The maximum flow rate required by standard technical specifications is called the capacity of a gas meter. Manufacturers produce gas meters of different capacity

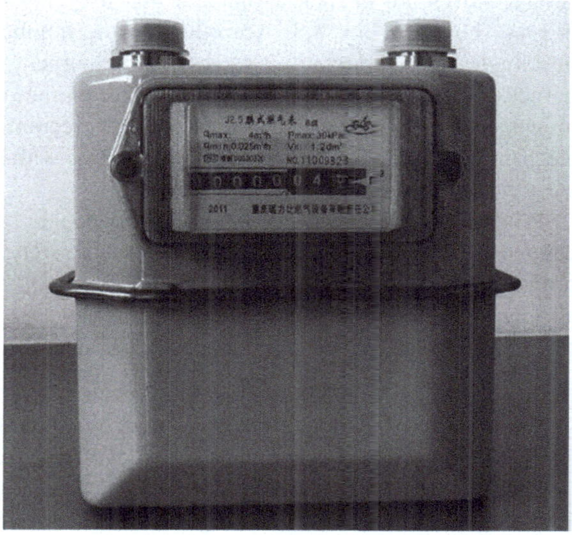

Fig. 8.7 Membrane gas meter

specifications according to different user requirements. In general, it is not possible to produce one single specified gas meter to satisfy various usage requirements. For example, a J1.6C gas meter has a nominal flow of 1.6 m^3/h and a maximum rated flow of 2.5 m^3/h. Such a gas meter cannot meet the requirements of a large enterprise or restaurant. When a gas meter is used beyond its range and allowable conditions, its function cannot be guaranteed. Capacity should be selected according to the maximum gas consumption. In real-world applications, the gas flow rate should be within the range of 20–70% of the maximum rated flow of its gas meter. When estimating the maximum capacity, one should not only examine the current situation but also consider future possible gas uses. For example, a gas meter should not only meet the gas need for a gas stove and a water heater at the moment, but also the possible needs for gas ovens, gas fireplaces, etc.

A gas meter should be installed in a non-combustible structured room with good ventilation and convenience to check and repair. It is strictly forbidden to install a meter in the following places: bedrooms, bathrooms, dressing rooms, toilet, pipeline wells with power supply, electrical switch and other electrical equipment, or hidden places where there may be leakages of biogas. installation of gas meters should also avoid places where the ambient temperature is higher than 45 °C, often wet places, and places stacked with flammable, corrosive, or radioactive materials.

A gas meter should be installed vertically without any obvious tilt. When a gas meter is installed at a high level, the bottom of the gas meter is not less than 1.4 m from the ground. When a gas meter is installed at a low position, the bottom of the gas meter shall not be less than 0.1 m from the ground. When a gas meter is installed above a gas cooker, the horizontal distance between the gas meter and the gas cooker shall not be less than 0.3 m.

2) Biogas stoves

Biogas stoves are kitchen appliances that use biogas as the fuel for direct heating, as shown in Fig. 8.8. According to the number of burners, it is classified into single-burner (single-eye) stoves, double-burner (double-eye) stoves, and multi-burner (multi-eye) stoves. According to ignition mechanisms and control methods, biogas stoves can be classified into piezoelectric ceramic biogas stoves, electric pulse electronic sparking biogas stoves, biogas stoves with a flameout protection device, etc.

When a biogas stove is used, it will produce an open flame. Once gas leakage happens, it may cause fire, explosion, and other safety accidents. Some biogas stove products use AC power as the energy source and may also cause electric shock accidents. Therefore, safety requirements of biogas stove are very high. In terms of safe gas stove use, selections of biogas stoves should consider the following aspects:

(1) Biogas tightness of the stoves. In the event of biogas stove leakage, an accident such as an explosion or fire may occur, resulting in injuries and property damages. Therefore, the national standard has strict requirements on the upper limit of biogas leakage. Leakage from the biogas inlet to the inlet valve should be no more than 0.07 L/h. Leakage from the automatic control valve should be no

Fig. 8.8 Biogas stoves

more than 0.55 L/h. There should be no gas leakage from the biogas inlet to the fire hole of a burner.

(2) Concentration of carbon monoxide in flue gas. Biogas combustion produces waste gases such as carbon monoxide and carbon dioxide, of which carbon monoxide is highly toxic and is not easily detected. Most of the exhaust gas from combustion in a biogas stove is directly discharged into the kitchen and cannot be discharged to the outside immediately. The excessive concentration of carbon monoxide in the flue gas poses a potential hazard. The national standard stipulates that the volume fraction of carbon monoxide in dry flue gas should be no more than 0.05%.

(3) Flameout protection device. Biogas stoves may have accidental flameouts caused by liquid spills, wind blows, etc. If there are no control measures, a decent amount of gas leakage would happen, resulting in serious consequences. In order to prevent leakage, biogas stoves must be equipped with a flameout protection device. The flameout protection device generally has two control modes: pyroelectric and ion sensing.

(4) Electrical safety performances. Some biogas cookers use AC power due to diversified functions and intelligent control. Biogas stoves are often used in hot and humid kitchen environments. Therefore, in addition to the safety requirements for gas in such stoves, there are also strict requirements for their electrical safety performances. It is necessary to have anti-shock protection measures, reliable grounding measures, large insulation resistance, small leakage current, and sufficient withstand voltage to ensure electrical safety of such stoves.

In addition to safety considerations, thermal performance indicators should also be an important criterion of the stove selection. (a) Thermal load, i.e., heating power, which is one of the most important thermal performance indicators for gas stoves. Heat load is determined by the stove structure and the gas combustion system. The heat load of stoves is generally 3–5 kW. The national standard stipulates that there should be a main burner in a double-burner or multi-burner stove. Calculation of the main burner heat load is converted as follows: the infrared type is not less than 3 kW, other types are not less than 3.5 kW, and the deviation between the actual measured converted heat load and the standard nominal value should not exceed 10%. (b) Thermal efficiency refers to the efficiency of thermal energy utilization by biogas stoves. It is the most important indicator for measuring thermal performances of biogas stoves. The overall thermal efficiency of embedded stoves is lower than that of bench top stoves. Thermal efficiency of embedded stoves is not less than 50%, while thermal efficiency of bench top stoves is not less than 55%.

Biogas stoves should be installed in kitchens with natural ventilation and natural light. When using an en suite bedroom or a hallway connected to the bedroom as a kitchen, the kitchen should be gated and separated from the bedroom. The height of the room where a biogas stove is installed should be no less than 2.2 m. The distance between the biogas stove and the wall should be no less than 10 cm. When the wall is flammable, fire insulation panels should be added. The distance between the biogas stove surface and wooden furniture shall not be less than 20 cm. When the distance is shorter than specified, a fireproof insulation board shall be added. Biogas stoves should be made of non-combustible material, and when using flammable materials, a fireproof insulation board should be added.

3) Kitchen facilities of biogas users

(1) Kitchens using biogas should have windows for ventilation and lighting, and the layout of the cook tops, cabinets and sinks should be reasonable.
(2) There should be fixed cooking countertops made of bricks or cooking cabinets in a kitchen. Tile countertops are desirable and hardened materials such as cement, and floor tiles should be used for flooring.
(3) The length of a countertop should be greater than 100 cm, with its width greater than 55 cm and a height of 65 cm. When biogas as a fuel is supplied from rural household digesters, the horizontal distance of a regulating purifier from the stove is 50 cm, and the vertical distance from the ground is 150–170 cm, as shown in Fig. 8.9.
(4) Above a countertop, one can use ventilation facilities such as automatic exhaust duct, exhaust hood or exhaust fan.

4) Biogas water heater

Biogas water heaters basically have the same structure as ordinary gas water heaters. The working principle is that cold water enters the biogas water heater and passes a water-gas linkage valve. The flow sensor inside can sense cold water coming in due to a pressure difference introduced by the incoming flow, and the linkage valve between the sensor and the gas valve forces biogas to flow in, pressing on the DC micro switch simultaneously. The micro switch is turned which starts the pulse igniter. At the same time, the biogas gas solenoid valve opens and the ignition is automatically triggered by the pulse igniter until the ignition succeeds, entering the normal working state. The process lasts for 5–10 s. An example of biogas water heaters is shown in Fig. 8.10.

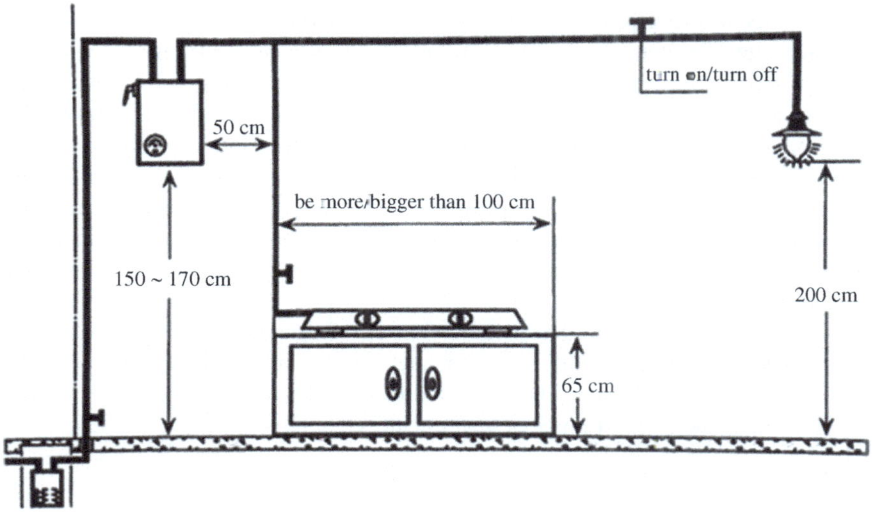

Fig. 8.9 Layout of the kitchen facilities in a biogas household

Fig. 8.10 Biogas water heaters

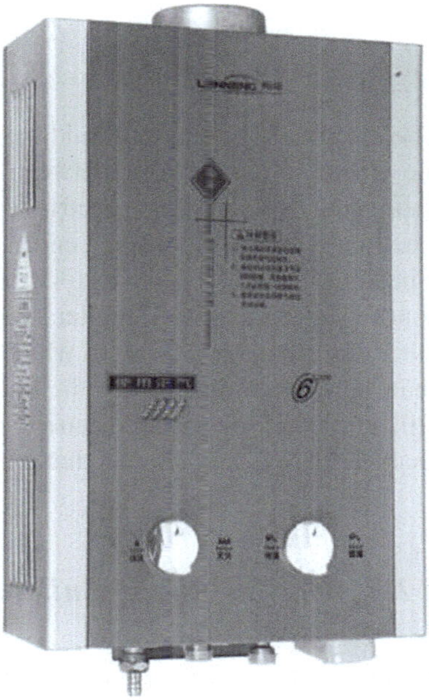

During its working process or ignition process, when the biogas water heater has a shortage of water, insufficient water pressure, lack of electricity, lack of biogas, high temperature of hot water, accidental blowout, etc., the pulse igniter will automatically detect the feedback signal from the induction needle. Then power is turned off and the electromagnetic valve of biogas supply immediately returns to the original normally closed state in the absence of electricity supply. That means that, the biogas passage is cut off to prevent biogas from continuing to flow out. The valve cannot be automatically turned on again unless the error above is manually eliminated, the biogas water heater is restarted, and normal functioning of the water heater is confirmed.

Biogas water heaters should be installed in well-ventilated non-residential rooms, aisles or balconies. When installing water heaters on combustible or flammable walls, effective fire-proof and heat insulation measures should be taken. Metal pipes should be employed as the exhaust pipes of water heaters.

4) Flue gas exhaust of domestic biogas appliances

The flue gas generated by biogas combustion must be discharged to the outside. When the indoor volume heat load index of the in-line type burning appliance exceeds 207 W/m^3, an effective exhaust device must be set to discharge the flue gas to the outside.

3. *Electrical system for household gas appliances*

The electrical system and building wires of biogas utilization facilities, including the electrical connections between ground wires, shall comply with the relevant national electrical codes. Electric ignition, burner controllers, and ventilation of the electrical system should not endanger biogas utilization facilities in the event of a power outage or power restoration. Circuits of automatically operated main gas control valves, automatic igniters, room temperature thermostats, limit controllers, or other electrical devices (all of which are used together with gas utilization facilities) shall be in accordance with the wiring diagram supplied with the equipment. All utilization facilities using electrical controllers should be connected to permanently charged circuits and circuits that are not controlled by the light switches.

8.2.3 Gas for Public Building Use

1. *Installation of biogas utilization facilities for public buildings*

1) Biogas utilization facilities for public buildings

There is a big difference between biogas utilization in public buildings and for residents. Because of the versatile functions of public buildings such as shopping malls, companies, hotels, restaurants, etc., the flow of people in public buildings is larger, so the biogas consumption is larger, and the use of biogas is mainly concentrated in restaurants for stove cooking, atmospheric pressure hot water boilers powered by biogas, and direct-fired hot and cold water units in hotels.

(1) Commercial gas meters

At present, there are two types of mainstream gas meters in market, one is a mechanical membrane gas meter, and the other is an electronic membrane gas meter. The mechanical membrane type gas meter adopts an internal mechanical structure that operates according to the amount of gas used. Each time one unit of gas is consumed, the roller count increases by one, by which the gas volume measurement is eventually realized. The technology of mechanical gas meters is mature, with reliable measurements and stable quality. However, the structure is complex, and the size is big, let alone the time and labor taken by manual meter reading. The electronic membrane type gas meter is an improvement from the mechanical membrane gas meter: electronic metering mode, display function, prepayment, and remote metering are added, and Semi-electronization is realized. Measurements by electronic metering are reliable, and issues of manual meter reading are effectively solved.

(2) Large pot stoves

A large pot stove is a kind of cooking stove in commercial kitchen equipment that can be generally classified into two types: large pot stoves using fuel gas and large induction pot stoves. The flame of large pot stoves using fuel gas is transmitted around the furnace and the heat transfer is uniform, making fuel gas pot stoves energy-saving and highly efficient. Usually, a large pot stove with a diameter of 80 cm or more is used, which is mainly used in large dining halls.

(3) Atmospheric pressure hot water boilers using biogas

The atmospheric pressure hot water boilers using biogas as the fuel heat water through the burner to supply heat and provide hot water for daily use and bathing. The boiler has high intelligence, fast heating speed, low noise, and no dust. An atmospheric-pressure hot water boiler is shown in Fig. 8.11.

The atmospheric-pressure hot water boiler using biogas as the fuel consists of three systems as discussed below.

a. Water system. Tap water enters from a cold-water pipe, which has a cold-water solenoid valve, a check valve, a gate valve, and a bypass gate valve. The function of the cold-water solenoid valve is to automatically control the cold-water shutdown. When the cold-water solenoid valve is turned on, cold water enters the boiler through the pipeline. When the cold-water solenoid valve is turned off, the pipeline is cut off and the cold water stops flowing. The purpose of the check valve is to prevent the mixing of hot water backflow and cold water. The bypass gate valve is an alternate temporary passage for cold water when the cold-water solenoid valve fails.
b. Biogas channel system. The biogas passage has a pressure gauge, a ball valve, a filter, a pressure reducing valve, a flow meter, a burner valve group, and so

Fig. 8.11 Atmospheric-pressure hot water boiler using biogas

on. As the fuel of the burner, biogas is burned in the boiler's furnace. The high-temperature flue gas enters the reburning combustion chamber along the furnace through the first flue pipe bundle, and it then enters the second flue pipe bundle by turning 180° after passing through the pressure-type front smokebox. Finally, after convective heat transfer, it enters the tail flue, and the flue gas is discharged into the atmosphere through the chimney. For the atmospheric boiler, a gas exhaust pipe that connects to the atmosphere is installed on the body of the boiler and does not bear any pressure, which is the major difference between atmospheric pressure boilers and high pressure boilers. When water is heated up to a desired temperature, the cold-water solenoid valve is turned on again. The hot water in the boiler is pressed into an insulated water tank through an upper circulation pipe by the pressure of the cold water. While the hot water in the biogas boiler flows to the insulated water tank, cold water is refilled from the cold-water pipe. The insulated water tank is connected with a hot water pipe to distribute hot water, and the pipe has a hot water solenoid valve for selectively supplying hot water to users at given time. The hot water bypass gate valve next to the solenoid valve can bypass the hot water solenoid valve and directly provide long-term continuous water supply. A liquid level gauge is installed in the insulated water tank, and when water in the tank is full, the boiler automatically stops boiling. The water tank is equipped with a temperature probe. When water temperature drops to the set point, hot water is pumped into the boiler through the circulating water pump of the lower circulation pipe to be reheated.

c. Electrical control system. At present, the atmospheric pressure hot water boiler generally adopts an automatic control system, and the automatic control system is the control center of the entire boiler system. Users adjust the operation of the boiler system through an automatic control system.

2) Installation of biogas utilization facilities for public buildings

Biogas utilization appliances should be installed in a well-ventilated room for the exclusive use of appliance installation. Appliances should not be installed in a guard room, duty room, civil air defense project spaces and any space that is also used as a bedroom. Biogas utilization appliances should not be installed in areas where flammable liquids are used, transported or delivered. When biogas utilization facilities are installed near access points of vehicles or equipment, appropriate guardrails or baffles that prevent damages to gas utilization facilities should be installed. The distance between gas utilization appliances and walls shall meet the requirements for operation and maintenance. Effective fire and heat insulation measures shall be taken to protect biogas utilization facilities from flammable or combustible walls, floors, and furniture.

When the specific room for installing biogas utilization facilities is a basement, a semi-basement or a closed space on the ground (including a closed space with no windows or with windows for lighting only), the following requirements shall be met: a manual quick shut-off valve such as a ball valve and an emergency automatic shut-off valve shall be installed at the biogas inlet pipe. The emergency automatic shut-off valve must turn off (normally turned on) when a power cut happens. The biogas utilization facilities should have flameout protection devices. The room using biogas should be equipped with a biogas concentration detection alarm, which is monitored and controlled in a management room. The biogas utilization room should be equipped with mechanical air supplies, exhaust systems, and independent mechanical ventilation facilities for accident prevention. The frequency air exchanges by mechanical ventilation facilities during normal accident prevention work should be no less than 6 times per hour, and the frequency of air exchanges when accidents happen should be no less than 10 times per hour.

When biogas utilization facilities are installed on rooftop, they should be able to withstand the influences of climatic conditions in the installation area. Their connecting parts, fasteners, and so forth should be made of corrosion-resistant materials. A 1.8 m wide operating distance, a 1.1 m high guardrail, a power disconnect device, and grounding devices should be provided for biogas utilization facilities. Safe maintenance should be guaranteed even if there is water on the roof.

2. *Flue gas exhaust of public buildings*

Biogas utilization facilities in public buildings cannot share the flue gas exhaust system with facilities that use solid fuels. A separate flue should be used by each set of biogas utilization facilities. When multiple sets of biogas utilization facilities share one flue, it should be ensured that the flue gases do not affect each other.

When the top of biogas utilization facilities or chimneys exhaust flue gas, the horizontal flue should be connected to a vertical flue with a minimal height of 0.3 m. Biogas utilization facilities with backwind-proof exhaust hoods shall not have flue dampers, while biogas utilization facilities without backwind-proof exhaust hoods shall have dampers on each branching pipe converging towards the main flue pipe. There should be holes larger than 15 mm in diameter on the dampers.

The chimney outlet shall have means to prevent entrance of rain and snow and chimney downdraft. The exhaust temperature of the chimney outlet shall be higher than the dew point of the flue gas by more than 15 °C. Metal flues installed in rooms below 0 °C shall be insulated.

The flue suction force of gas exhausts of biogas utilization facilities shall be no less than 3 Pa given a heat load of less than 30 kW from biogas utilization facilities. For biogas utilization facilities with a heat load of more than 30 kW, the flue suction force shall be no less than 10 Pa.

When a chimney of a biogas utilization facility is to be extended outward, the vertical distance from the chimney to the roof should be 0.6 m in any case. When such distance is less than 1.5 m from the roof (horizontal distance), the chimney should be 0.6 m above the rooftop. When the chimney is 1.5–3.0 m away from the roof (horizontal distance), the chimney can be as high as the rooftop. When the distance from the chimney to the rooftop is greater than 3.0 m (horizontal distance), the chimney top end should be on a straight line 10° below the horizontal line of the rooftop. When the chimney is located adjacent to a tall building, the chimney top end should be higher than the 45° shadow line boundary from the building.

When a flue gas contains corrosive or combustible fumes or gases, there should be safe disposal facilities. Biogas utilization facilities installed in cosmetics stores, barbershops, or other places that normally produce corrosive or combustible products (e.g., liquefied spray chemicals) should be placed in a separate space or an isolated equipment room. Air used for dilution (ventilation) should be taken from the outside, except for closed combustion appliances.

3. *Explosion-proof facilities for public buildings*

A discharge pipe shall be installed between the main valve and the burner valve of a public building biogas utilization facility. The blower and piping should be equipped with an electrostatic grounding device with a grounding resistance of no more than 100 Ω. Low-pressure and overpressure alarms and emergency automatic shutdown instruments should be installed on biogas pipelines. In places where it is easy to accumulate smoke and in closed furnaces, a venting device shall be installed, and the pressure relief port of the venting device shall be located at a safe place.

4. *Electrical system for biogas utilization facilities of public buildings*

For the electrical system of biogas utilization facilities of public buildings, refer to the electrical system requirements for residential biogas utilization facilities.

8.3 Power Generation from Biogas

Power generation uses biogas as a fuel to drive the engine through combustion, which in turn drives the generator to generate electricity. The generated electricity is delivered to users or integrated into electrical grids. The waste heat can be used for anaerobic digestion or as heating energy for surrounding users. Power generation is a biogas utilization method that has emerged along with continuous development of the construction and comprehensive utilization of biogas plants. Power generation by biogas has the advantages of synergy, energy saving, safety and environmental protection. It is a widely used technology for distributed energy.

8.3.1 Types of Power Generation from Biogas

Power is generated by biogas combustion, in which heat generated by biogas combustion is directly or indirectly converted into mechanical energy that drives the generator to generate electricity. Biogas is used as a fuel for a variety of power generation equipment, such as internal combustion engines, gas turbines, and boilers. In internal combustion engines and gas turbines, the heat released by fuel combustion is converted and reused by power generator sets and heat exchangers. Therefore, the overall thermal efficiency of internal combustion engines and gas turbines is higher than that of the boiler (steam turbine) unit that does not utilize waste heat. The structure of biogas engines is the simplest, and biogas engines have the advantages of low cost and easy operation (Zhang and Wang 2014). Generation of electricity in a fuel cell is the conversion of chemical energy from fuel into electrical energy, so a generator powered by biogas fuel cells is also called an electrochemical generator. It is a new type of biogas power generation technology. At present, internal combustion engine power generation is the most common way of generating electricity from biogas.

1. *Generation of electricity in an internal combustion engine*

The attempt to use biogas as a fuel for internal combustion engines began in the United Kingdom in the 1920s. In the 1930s, waste heat was recovered for the anaerobic digestion process, which is the original form of modern biogas power generation and cogeneration systems. In the early 1970s, in the processes of treating organic pollutants overseas, in order to rationally and efficiently utilize the biogas generated by anaerobic digestion, reciprocating internal combustion engines were commonly used for biogas power generation. By the 1980s, China's scientific research institutions and production enterprises had carried out extensive research and development on biogas gen-sets using internal combustion engines and created a series of products.

An internal combustion engine refers to a mechanical device in which fuel is burned in one or more cylinders, by which the working pistons are pushed to reciprocate, converting chemical energy of biogas into mechanical work and outputs of shaft power. An internal combustion engine is shown in Fig. 8.12. Biogas engines are generally classified into two types: the compression-ignition type and spark-ignition type.

The compression-ignition engine uses dual fuel of diesel and biogas (see standard NY/T 1223), and it burns a small amount of diesel to ignite biogas for combustion.

Fig. 8.12 Internal combustion engine

This kind of engine can adjust the fuel ratio between diesel and biogas. When biogas supply is normal and sufficient, the fuel intake of the engine can be kept basically constant, and biogas supply can be changed to adapt to external load changes. When biogas supply is insufficient or even stopped, the engine can automatically switch to the mode of diesel combustion. This mechanism is generally used in small-scale biogas power generation plants and does apply to grid-connected plants where the requirements for power supply reliability are demanding. One disadvantage is that the system is complicated, so large-scale biogas power generation and grid-connecting plants often use ignited biogas engines instead of such engines.

A spark-igniting engine adopts a single fuel of biogas, and it is characterized by a simple structure and convenient operations. Generally, a low compression ratio is adopted, and the spark plug is used to ignite and burn the mixture of biogas and air with no need of auxiliary fuel. This type of engine is suitable for medium and large-scale biogas plants. Biogas is burned by the internal combustion engine, and the exhaust gas generated can be recycled by a heat exchanger or a waste heat boiler. The system is slightly complicated, but it has good economic values, as well as environmental and social benefits.

An internal combustion engine power generation system using biogas as its fuel mainly consists of the following parts:

1) Biogas purification, pressure stabilizing, and anti-explosion devices. Biogas used for the engine needs to pass through a desulfurization device to reduce the corrosion of hydrogen sulfide to the engine. A pressure stabilizing device is installed on the biogas inlet pipe to adjust the biogas flow to achieve the best air-fuel ratio. In addition, in order to prevent tempering of the intake pipe, anti-tempering and anti-explosion devices should be placed onto the major biogas pipelines.
2) Biogas internal combustion engine. Like the general-purpose internal combustion engines, a biogas internal combustion engine also has four basic stages of gas intake, compression, combustion expansion, and exhaust. Since the calorific

value and characteristics of biogas are different from those of gasoline and diesel, a biogas internal combustion engine must be designed for combustion characteristics of methane. A biogas internal combustion engine has a high compression ratio in general. The ignition period is shorter than that of gasoline and diesel engines. Cylinders and pipes of this type of engine must possess corrosion resistance.

3) AC generator. AC generators are similar to general-purpose AC generators, and there is no other special requirements as long as the power of biogas internal combustion engine and other requirements are met.
4) CHP device. Waste heat discharged from the engine is recovered by a CHP device consisting of a water-gas heat exchanger, a cooling water-gas heat exchanger, and a preheating boiler to improve the total energy utilization efficiency of the unit. The recovered waste heat can be used for heating the mixed liquid in the digester.

Internal combustion engine power generation has the following characteristics:

1) High power generation efficiency. The electrical energy conversion efficiency of internal combustion engines is significantly higher than that of ordinary gas turbines and steam turbines. The power generation efficiency of gas internal combustion engines is usually between 30 and 45%.
2) Gas combustion with low heat value. Modern biogas engine groups use advanced electronic control technology and control devices to manage air-fuel mixture ratio. The range and types of fuel utilization are expanded to combustion of low-calorie gas fuels such as biogas and biomass gas.
3) Direct use of low-pressure gas sources. Gas internal combustion engines can use their own supercharged turbine to pressurize gases, so a low-pressure gas source can be directly utilized. Biogas storage and distribution use low-pressure systems, which can achieve a better match with this feature.
4) Wide power supply ranges and good adaptability. At present, the single-machine minimum power of gas internal combustion engine is less than 1 kW, and the maximum power has reached 4 MW. One type of such gas internal combustion engines is enough to serve various purposes and can achieve full load or partial load operation. These engines can be turned on and off quickly, and they have strong peak-shaving abilities, high machinery efficiencies, reliable operation, and easy maintenance.

2. *Generation of electricity in a micro gas turbine*

A micro gas turbine is a newly developed small heat engine with a single power range of 25–300 kW. The basic technical features are the use of radial flow impeller machinery (central turbine and centrifugal compressor) and regenerative cycle. The structure of the engine is similar to that of a small aero-engine. In order to improve efficiency, the heat cycle technology is widely used.

In addition to distributed power generation, applications of micro turbines also include its uses in backup power plants, combined heat and power, grid-connected power generation, peak load power generation, and etc. It is the best way of providing clean, reliable, high quality, multi-purpose, small distributed power generation, and combined heat and power. The method can be applied to both city centers and remote suburbs or areas.

At present, companies such as Capstone, Turbec S.p.A. (hereinafter referred to as Turbec), and Ingersoll Rand have developed a variety of models for power generation and CHP. The Capstone micro biogas turbine power generation plant is shown in Fig. 8.13. The main components of Capstone's micro gas turbines include: generators, centrifugal compressors, turbines, heat regenerators, combustion chambers, air bearings, and digital power controllers (converting high frequency electrical energy to parallel grid frequency 50/60 Hz and providing control, protection, and communication). The unique design of this micro turbine is that its compressor and generator are mounted on a rotating shaft supported by an air bearing that rotates at a speed of 96,000 r/min on a thin film of air. This is the only rotary part of the entire unit. It does not require a gearbox, an oil pump, a radiator or other ancillary equipment at all.

Micro turbine power generation has the following characteristics:

1) The structure is simple and compact. The micro gas turbine makes the high-speed alternator coaxial with the internal combustion engine to form a compact high-speed turbine alternator.
2) Easy operation and maintenance, low operating cost and long service life. With air bearing and air cooling, there is no need to change the lubricant and cooling medium. The planned annual maintenance simply needs to replace the air filter and check the fuel injector and sensor probe after continuous operation at full capacity throughout the year. The unit's first maintenance is after 8000 h of

Fig. 8.13 Capstone micro gas turbine biogas power generation plant

operation, which reduces maintenance costs. The life of micro gas turbines is above 40,000 h.

3) Low noise and low emissions. Micro turbine vibration is small, so the noise is small. For example, T100 of Turbec has a noise level of 70 dBA at 1 m, and C200 of Capstone has a noise level of 65 dBA at 10 m. In addition, micro gas turbines emit less exhaust gas.
4) Power generation efficiency is lower than that of internal combustion engines. At present, power generation efficiency of micro gas turbines is still lower than that of gas engines. The power generation efficiency of a regenerative micro gas turbine can reach 20–33%. However, efficiency of a cogeneration system consisting of micro turbines can exceed 80%.

3. *Generation of electricity in a fuel cell*

Fuel Cell, a device that uses fuel to generate electricity directly by electrochemical reactions, was first invented in 1839 by Grove in England. Proton-exchange membrane fuel cells are most commonly fueled by hydrogen and oxygen and are therefore the cheaper type of fuel cells. They have expose no chemical hazard to human body and are harmless to the environment, for this type of cell produce pure water and heat as products of power generation. They were applied to the US military in the 1960s. In 1965, it was applied to the Gemini 5 spacecraft of NASA's Project Gemini. Nowadays, there are also some laptops companies that are starting researches on the use of fuel cells. However, since the generated electricity is too small and cannot provide a large amount of instantaneous electric energy, it can only be used for steady power supply.

A fuel cell is a dynamic mechanism in which a battery body and a fuel tank are combined. There is a wide selection of fuels, including hydrogen, methanol, ethanol, natural gas, and biogas. Even the most widely used gasoline can be used as a fuel for fuel cells, which is not possible for all other power generators at present.

A gasoline fuel cell uses a specific catalyst to speed up the reaction of the fuel with oxygen to produce carbon dioxide and water. Since mechanical work from impelling such as an internal combustion engine or a turbine is not necessary, and heating water to vapor which is then converted into water by heat dissipation is not needed, the energy conversion efficiency by fuel cell power generation is as high as about 70%, which is about 40% higher than general power generation methods. In addition, carbon dioxide emissions are much lower than those from general power generation methods, and the generated water is nontoxic and safe.

Biogas fuel cell is a biogas power generation technology that attracts great attention (Wellinger et al. 2013). Under certain conditions, the highly purified biogas is subjected to hydrocarbon cracking reaction to produce a hydrogen-based mixed gas. The mixed gas cell system undertakes an electrochemical energy conversion to realize biogas power generation. A biogas fuel cell system generally consists of three units: a fuel processing unit, a power generation unit, and an inverter unit. The main component of the fuel processing unit is a reformer, which uses nickel as a catalyst to convert methane into hydrogen. The power generation unit is a fuel cell, the basic components of which consist of two electrodes and an electrolyte. Hydrogen and oxygen undergo electrochemical reaction on the two electrodes. The electrolyte constitutes the inner loop of the battery. The function of the inverter unit is to convert direct current into alternating current. The working principle of the fuel cell is shown in Fig. 8.14.

In cases where biogas is used as a fuel, hydrogen gas is produced by methane in the reformer in a preceding stage. In a reformer that maintains a high temperature, water becomes water vapor which reacts with methane to form hydrogen and carbon dioxide or hydrogen and carbon monoxide. The carbon monoxide then reacts again with water vapor to form hydrogen and carbon dioxide. In general, 1 mol of methane can produce 4 mol of hydrogen and 1 mol of carbon dioxide, and there must be an external energy supply in the process. The chemical reaction equation in the biogas fuel cell reformer is as follows:

$$CH_4 + 2H_2O \rightarrow 4H_2 + CO_2 \tag{8.9}$$

$$CH_4 + H_2O \rightarrow 3H_2 + CO \tag{8.10}$$

$$CO + H_2O \rightarrow H_2 + CO_2 \tag{8.11}$$

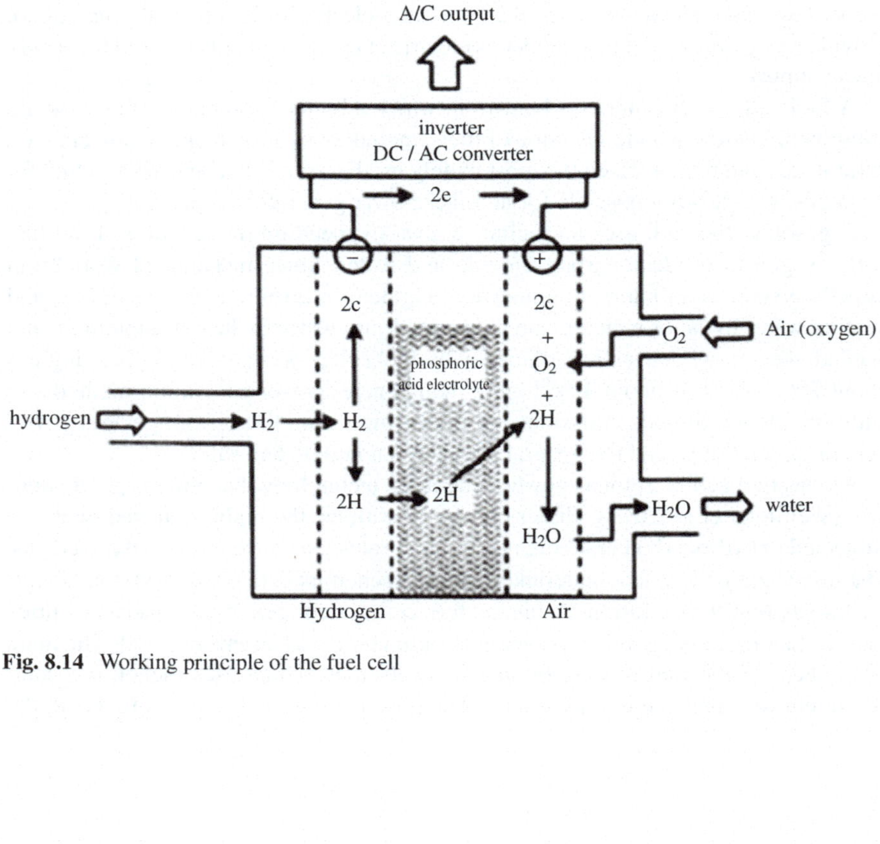

Fig. 8.14 Working principle of the fuel cell

Compared with other biogas power generation technologies, biogas fuel cell technology has the following advantages: First, the energy conversion efficiency is high. The actual energy conversion efficiency can reach more than 40%, and the total energy utilization ratio of the system from waste heat recovery is above 70%. Secondly, this technology is eco-friendly. Biogas fuel cells produce no or very few pollutants, and there is almost no noise during operation. On the other hand, however, biogas fuel cells have high requirements for biogas quality. Methane content needs to be above 85%, and hydrogen sulfide concentration needs to be below 5.5 mL/m^3. Biogas upgrading requirements for fuel cell power generation are more stringent than other biogas power generation technologies.

8.3.2 *Power Generation Plants Using Biogas Fuel*

1. *Composition of power generation plants using biogas fuel*

Power generation with internal combustion engines is still the most commonly used method of power generation by biogas fuels. An internal combustion engine power plant system consists of a biogas supply unit, a biogas generator set, a cooling unit, a transmission and distribution unit, an equipment control unit, and a waste heat utilization unit. Biogas produced by anaerobic digesters is dehydrated and desulfurized before stored in a gasholder. The biogas is discharged from the gasholder under the pressure of gasholder or through a biogas transmission device. Biogas is then supplied to a biogas engine after dehydration and stabilization, driving the generator connected to the biogas engine to generate electricity. The heat in cooling water and flue gas discharged from the biogas engine is recovered by a waste heat recovery device as a heat source for the digester or other heat-using facilities (Fig. 8.15). According to different operating modes, an internal combustion engine biogas power plant can be classified into an island-mode biogas power plant and a grid-connected biogas power plant.

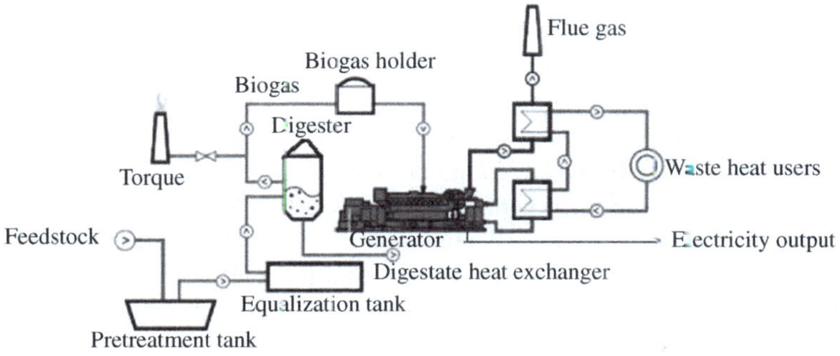

Fig. 8.15 Composition of power generation plants using biogas fuel

2. *Design of the installed capacity of biogas gen-sets*

The equipment selection of biogas power gen-sets is an important part of the biogas power generation plant design. The installed capacity of biogas gen-sets is the major parameter of equipment selection. It should be determined according to Eq. 8.12 on the basis of biogas flow rate and the lower calorific value of the biogas (see standard NY/T 1704).

$$P = k(V \times Q_{\text{net}}/g_{\text{h}}) \tag{8.12}$$

where

- P installed capacity of biogas power gen-sets, kW;
- k comprehensive scale factor of installed capacity and gen-set efficiency, about 1.08–1.20;
- V volumetric flow rate under standard conditions converted from the maximum flow of biogas per hour, m^3/h;
- Q_{net} lower calorific value of biogas, about 22,154–24,244 kJ/m^3;
- g_{h} thermal efficiency of biogas power generator set, kJ/(kW h) (see standard GB/T 29488).

In addition, attention should be paid to the following matters:

a. When biogas power gen-sets operate in parallel, the influence of distribution difference of active power and reactive power on power of biogas generator set should be considered.
b. When starting the motor with the largest capacity, the total bus voltage should not be lower than 80% of the rated value.
c. For power plants with a total installed capacity greater than or equal to 200 kW and with uninterrupted power generation, a standby unit shall be provided. The number of standby units shall be 1 when 3 units are in operation.
d. When the actual working conditions of biogas power generator sets cannot satisfy the technical requirements, the output power shall be first converted into engine power based on experimental results and then converted into electric power according to relevant regulations. The electric power shall not exceed the rated power of the generator set.

3. *Waste heat recovery of biogas generator sets*

A biogas generator generates a large amount of heat while generating electricity and the temperature of the flue gas is generally around 550 °C. By utilizing heat recovery technology, the heat in lubricating oil, intercoolers, liner water, and exhaust gas in the internal combustion engine can be fully recovered for winter heating and domestic hot water. The heat recovery system can be connected to a lithium bromide absorption chiller in summer to be used as an air conditioner. Generally, the heat absorbed from the heat recovery system of the internal combustion engine is supplied to a heat exchanger in the form of hot water at 90 °C. The normal temperature of the return water of the internal combustion engine is 70 °C. In biogas plants, this heat can also be used to heat the mixed liquid in digesters.

Biogas engine is cooled the same way as a normal gasoline and diesel engine. It is typically cooled with water. To prevent formation of scale deposits, soft water should be uses as cooling water and sometimes antifreeze liquid should be added. For such purposes, the modulated water is usually used as the primary cooling water circulated inside the engine. The indirect cooling method of transferring heat to the secondary cooling water by a heat exchanger is used to recover the residual heat in liner water. This method is used in a liner water cooling cycle. In addition, the heat absorbed by lubricating oil can also be transferred to cooling water through a lubricating oil cooler.

Biogas contains trace impurities and corrosive substances, so if the temperature of the flue gas after combustion and heat exchange is too low, some impurities will be generated. Therefore, the exhaust gas temperature of biogas engines should be dozens of degrees higher than the exhaust gas temperature of other gas engines, and the return water temperature should be slightly higher for the same reason.

At present, some overseas power generation equipment integrates the waste heat recovery equipment with generator sets, which means that, heat exchange devices are placed inside the power generation unit instead of being separately positioned. The output water of the integrated set is hot water. The device requires only three interfaces: connection points of inlet water, return water, and biogas. The equipment can be connected in parallel with the hot water boilers. Integration does not only simplify the system and reduce equipment and floor space, but also facilitates operation and maintenance while reducing the total investment in systems and engineering. This internal combustion engine unit is also equipped with a fully automated control system that can realize remote control of the cooling system, lubricating oil automatic replenishment system, exhaust muffler, and so on.

4. *Power grid connection*

Biogas power plants are distributed power sources. Distributed power supply means that power generation plants are located near users, and the power generated can be utilized locally and can be connected to the power grid at a voltage level of 10 kV and below, and the total installed capacity of a single grid connection point does not exceed 6 MW.

Figure 8.16 shows the connections of a distributed power source to distribution network systems. Connecting joints in the figure include grid points, access points, and public connection points.

1) Grid connection points

For a distributed power supply with a booster station, a grid connection point is a high-voltage side bus or node of the distributed power booster station. For a distributed power supply without a booster station, the grid-connected point is an output summary point of the distributed power source. As shown in Fig. 8.16, A1 and B1 are the grid connection points of distributed power supplies A and B respectively, and C1 is the grid connection point of conventional power supply C.

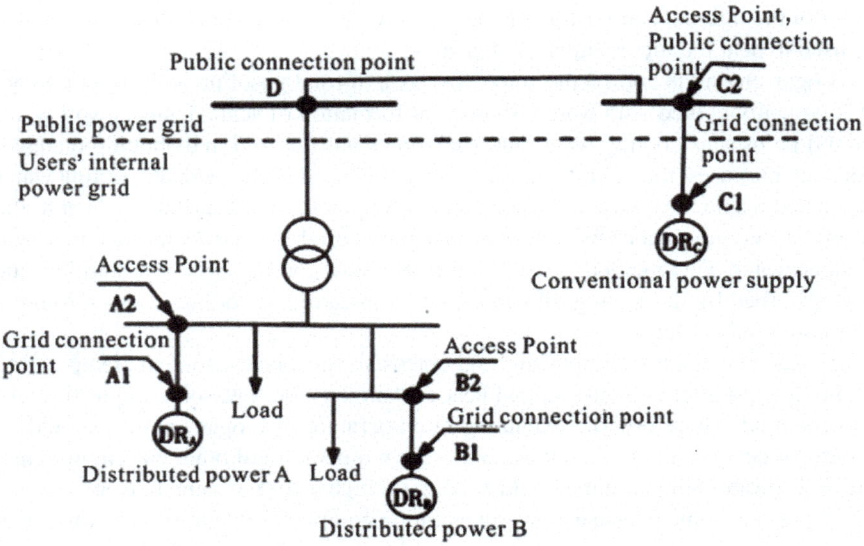

Fig. 8.16 Connection of distributed power source to distribution network system

2) Access points

An access point refers to a connection where a power source is connected to a power grid which may be either a public power grid or a user power grid. As shown in Fig. 8.16, A2 and B2 are the access points of distributed power supplies A and B respectively, and C2 is the access point of conventional power supply C.

3) Public connection points

A public connection point refers to the connection of a user system (power generation or utilization) to a public power grid. As shown in Fig. 8.16, C2 and D are common connection points, and A2 and B2 are not common connection points.

The grid connection design of biogas power generation should meet the "Design code for connection to distribution network" Q/GDW 11147. For a single grid connection point, the voltage level of distributed power supply which connects to the power grid should be determined based on the principles of safety, flexibility and economy. Consideration also should be given to the power generation capacity and conductor current capacity of the distributed power source, the upper transformer, line acceptability, and comprehensive comparison with the regional distribution network situation. The reference standard for the voltage level of grid-connected distributed power supply is based on the installed capacity: 8 kW or less can be connected to single-phase 220 V grids; 8–400 kW can be connected to three-phase 380 V grids; 0.4–6 MW can be connected to 10 kV grids. The final grid-connected voltage level should inclusively refer to all relevant standards and actual grid conditions, and the level can be determined after technical and economic comparisons.

Grid connection of biogas generator sets should meet three requirements, the first of which is that the voltage of the grid-connected generator set is equal to the voltage of the grid system. Second, the frequency of the grid-connected generator set is equal to the frequency of the grid system. Lastly, the phase angle of the grid-connected generator set is consistent with the phase angle of the grid system.

In grid connection operation, the phase sequence must first be adjusted so that the grid-connected generator set is in phase with the grid, and the generator set can then start operation. Adjust the excitation current of the generator set so that the gen-set voltage is as close as possible to the grid voltage. Adjust the engine speed so that the power generation frequency matches with that of the grid, creating conditions for grid connection. When the phase sequence reaches the allowable range, grid connection can be readily performed. The operations above must be manually performed by a skilled engineering technician under the direction of a dedicated grid connection device, or by a dedicated automatic device to complete the grid connection (Wang et al. 2016).

8.4 Biogas as Vehicle Fuel

The biomethane produced by biogas cleaning and upgrading can be used as a substitute for vehicle fuel, which can alleviate energy shortage and effectively reduce environmental pollution. At present, biomethane has been widely used as a fuel for motor vehicles in many countries, such as Europe and the United States and has broad prospects for development.

8.4.1 Gas Quality Requirements for Biogas as Vehicle Fuel

Biomethane produced by biogas upgrading for vehicle fuel should have corresponding gas quality standards. At present, there is no quality standard for biomethane in vehicle gas applications. If biomethane is used as vehicle gas, it must meet the existing standard "Compressed natural gas as vehicle fuel" GB 18047, and the main characteristic parameters should meet the requirements in Table 6.4.

8.4.2 Transmission and Distribution of Biomethane

For distribution of gas, pipe network transportation is undoubtedly one of the most efficient and environmentally friendly methods. In many countries, there is a wide range of natural gas pipeline networks. For example, more than 90% of people in Netherlands and 48% of residents in Ireland are connected to natural gas pipeline networks. The integration of biomethane into the gas pipeline network of motor vehicles is undoubtedly the best way for biomethane to be distributed as motor vehicle gases.

In addition to incorporating biogas and natural gas pipelines, there are several other methods of distributing natural biogas. Some instances include using mobile storage equipment for road transport or building local biogas pipelines. Research by the Swiss Gas Association shows that for local short-distance and large-scale transportation, local gas network transmission and distribution are optimal. For in-vehicle transportation with a transportation distance of more than 200 km, vehicle transportation for compressed biomethane (CBG) is more suitable than liquefied biomethane vehicle transportation (LBG). Another benefit of a local gas network is that biogas can be injected into pipe networks anywhere in the pipe network. From the perspective of economic interests, current commercial upgrading of biogas needs to reach a certain scale to be profitable. Therefore, this method can be used to construct a biogas pipeline network parallel to local gas network for collecting and transporting gas from small biogas plants, for centralized biogas upgrading is able to reduce the cost.

From the perspectives of investment cost and energy loss, pipe network transportation has great advantages, and large-scale land transportation of biomethane should be avoided as much as possible. Gradual expansion of local gas network and easier access to natural gas pipeline network is an inevitable trend in the development of the biogas market. Biogas and natural gas should be used in combination. Therefore, pipe network, land transportation of compressed natural gas, and LNG will coexist for a long time to meet different market needs. Currently in China, biomethane is mainly transported by vehicles.

8.4.3 Gas Filling Stations for Vehicles

1. *Structures of the gas filling stations*

Structures of the filling stations can be classified into two types: the open structure type and the skid-mounted structure type (Song 2003).

1) Open structure: all equipment of a gas filling station is installed in a factory building, and each equipment is assembled into a complete set through high and low-pressure pipelines and valves. The open structure has the advantages of large equipment space and convenient maintenance. However, the degree of automation is low, but due to its low investment and easy maintenance, it has been widely used in China.
2) Skid-mounted structure: a natural gas buffer tank, filters, the main unit of compressors, air coolers, driving machines, gas cylinders, electric control panels and priority discs, and instrument wind system before the compressor of the gas station are integrated on a sturdy base to form a comprehensive close loop-controlled equipment system. This type of structure has the advantages of convenient transportation, unconfined space installation, high degree of automation,

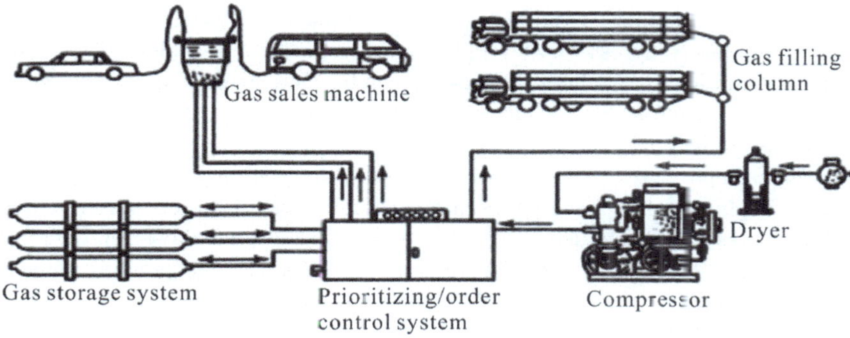

Fig. 8.17 Process system of gas stations

great safety and reliability, and reduced workload of on-site installation and adjustment, although the one-time capital investment is large.

2. *Process system of the gas filling stations*

As shown in Fig. 8.17, the process system of biomethane filling station consists of multiple subsystems such as pretreatment, compression, gas storage, gas sales, and control (Yang 2015).

1) Pretreatment subsystem

Biomethane entering a gas station must be purified and dried before or after compression. That means that it has to go through the processes of desulfurization and dehydration. According to the water and sulfur content of the biomethane, temperature and humidity of surroundings, the process and specifications of each device as well as the detailed structure of the dehydration and desulfurization can be determined. The process design of domestic compressor dehydration usually adopts high-pressure post-dehydration, which has the advantages of small equipment size and low dew point of dried gas. After upgrading and dehydration, purity of the biomethane can be guaranteed, and the upgrading and dehydration processes can secure normal operations of the compressor while ensuring desirable biomethane engine combustion that does not cause any harm to automobile engines. The process of a pretreatment subsystem mainly includes dust removal, equipment dehydration, desulfurization, and de-oiling of equipment. The purification equipment used for dust removal and desulfurization is mainly installed upstream of the compressor to ensure that the compressor is protected from damages and corrosions.

2) Compression subsystem

A compression subsystem is mainly composed of a main unit of compressor, a buffer tank for gas entrance, a cooling facility, and a lubrication facility. The main engine of the compressor is the most important part of all, since it is the heart of the gas filling station. Its performance directly affects the reliability and economy of the gas filling station. Gas filling stations of compressed biomethane generally use reciprocating compressors with high exhaust pressure and small displacement. Based upon lubricating methods, lubrication can be categorized into oil lubrication and oil-free lubrication. For oil lubrication, a de-oiling device should be installed downstream of the final exhaust port.

The exhaust pressure of a compressor for a gas filling station is generally 25 MPa. Some are slightly higher, reaching 28 MPa, and a few manufacturers have reported an exhaust pressure of as much as 32 MPa. Analysis shows that, however, the most economical, reliable, and safe exhaust pressure is 25 MPa. The inlet pressure range is 0.035–9 MPa and the compressor displacement can be selected according to different scales of filling stations. The range of exhaust volume varies from 16 to 2000 m^3/h, and a common exhaust volume range is 200–300 m^3/h.

3) Gas storage subsystem

A gas storage system mainly plays a buffering role and needs to balance intermittent productions and fillings of a station. The law of automobile refueling is determined according to the law of transportation operation. The basic rule of operation of a refueling station is determined by the flexible refueling time and gas volume, and the compressors are to work non-stop. In order to match with the working mechanism of a gas filling station, a gas storage system must be set up so that the number of startup and shutdown of compressors be reduced. The gas storage capacity should be 1/3–1/2 of the daily gas filling amount, and this conclusion comes from considerations of whether the refueling time of a gas station is concentrated and other relevant experience.

Gas storage equipment includes gas storage tanks, gas storage cylinder groups and gas storage wells.

A gas storage tank is a type 3 pressure vessel produced in accordance with the GB 150 "Pressure vessels" standard. The advantages of this type of gas storage are a small amount of joints, small gas transmission resistance, and few leakage points in the gas storage system. The disadvantages include relatively small volume of the storage containers and a large amount of steel consumption, so this gas storage tank is usually used as an alternative gas storage device.

A gas cylinder group is composed of a group of compressed gas cylinders as its primary structure, and several groups are connected in parallel to form gas cylinder groups. Based on the process requirements, gas cylinder groups are classified into high, medium and low-pressure storage systems. The advantages of this type of device is that it is flexible, but there are also problems such as difficulty in treating leaking joints, low corrosion resistance of cylinder surfaces, inconvenience in gas detection, and potential safety risks.

A gas storage well is an underground gas storage as opposed to a gas cylinder group and a gas storage tank. Pipes of gas storage wells adopt oil casings that meet specific requirements and should have high corrosion resistance. Leaving no solder joints by petroleum geological drilling, pipes of gas storage wells are buried deep underground to form gas-storage process equipment for gas storage wells. Gas storage wells have the advantages of safety and reliability, few hidden dangers, small land occupancy, simple operation, convenient maintenance and management, long service life, and short fireproof distance. Therefore, in current designs of CNG filling stations, using gas storage wells is a mainstream gas storage method.

4) Gas sales subsystem

A gas sales system consists of a gas filling stand and a computer management system. The compressed biomethane is piped to the gas filling machine through a pipeline, in which the biomethane passes an inlet ball valve, a filter, a check valve, a flow meter, a high-pressure hose, a gun valve, a gas filling gun and finally refill the target car. A flow meter in the pipeline measures the density, flow rate, and other parameters of the gas flowing through the gas filling machine. The corresponding analog signal is converted into electric pulse signals and transmitted to the computer control. The volume and price of the refilled gas are calculated and displayed on a display screen to users.

A gas storage facility and a gas sales facility can achieve efficient gas inflation and rapid refueling by the order of a priority control panel. In most cases, a gas filling station adopts storage method by pressure grading so that the gas storage bottles are divided into high-pressure, medium-pressure and low-pressure bottle groups, and the priority control panel automatically controls the inflation and gas withdrawal processes. For gas inflation, high-pressure group starts first. When the pressure of the high-pressure group rises to a certain level, the medium-pressure group starts to inflate. When the pressure of the medium-pressure group rises to a certain value, the low-pressure group begins to inflate. Next, the three groups of cylinders are inflated together. Inflation stops after the gas storage pressure rises to the maximum. For gas withdrawal, the low-pressure group starts first. When the pressure of the low-pressure group drops to a certain level, gas taking initiates for the medium-pressure group. When the pressure of the medium pressure group drops to a certain value, gas taking starts for the high-pressure group. Gas from all three groups are taken together. When the pressure in all three gas cylinder groups drop to the same level of highest gas storage pressure of the vehicle gas cylinder, gas filling process terminates. If there is another vehicle that needs to be refilled, take gas directly from the compressor exhaust line. When the car is refilled, the compressor inflates the three gas cylinder groups according to the inflating sequence described above before coming to a rest. The advantage of this operation mode is that it can ensure that the gas cylinder groups are inflated to the maximum. Gas utilization rate is improved, and cars can be refilled in a rather fast and efficient fashion.

5) Control subsystem

The functions of a control system are to control the normal operations of gas filling station equipment, to monitor the operating parameters of equipment, and to automatically alarm or stop when equipment fails. The control system of a gas filling station includes a power supply control, an operation control, a gas storage pressure control, a purification and drying control, a system safety control, and a gas sales control. Too much use of automatic control equipment will increase investment, but the number of operation and maintenance personnel and daily operation costs are greatly reduced, and the operational safety and efficiency of gas filling stations are also ensured. Moreover, the operation reliability is improved.

3. *Design of gas filling stations*

In China, gas filling stations of compressed biomethane must comply with the provisions of "Code for design of city gas engineering" GB 50028. Compressed biomethane can be transported by gas cylinder groups on vehicles, by gas cylinder trucks, or by ships.

Gas filling stations should be close to gas sources and should have appropriate transportation, power supply, water supply and drainage, communication, and engineering geological conditions. Fixed parking lots of gas cylinders trucks should be set in the station. The width of the fixed parking lot for each gas cylinder truck should be at least 4.5 m. The length should be the length of a gas cylinder truck. The boundary lines of each parking spot should be marked. Each parking spot should have one gas filling nozzle. There should be enough vehicle returning space in front of each fixed parking space. Gas cylinder trucks should be parked at a fixed parking spot and should be fixed since the trucks are strictly forbidden to move during gas inflation. The maximum gas storage capacity of a gas cylinder truck in a fixed parking spot shall not exceed 30,000 m^3. The gas-filling columns should be located near each fixed parking spot, at a distance of 2–3 m from the fixed parking spot. The distance from the gas-filling columns to gas storage tanks in the station shall be at least 12 m; the distance to surrounding walls shall be at least 6 m; the distance to the compressor chamber, pressure regulating chamber, and metering chamber shall be at least 6 m, and the distance to rooms with hot water boiler shall be at least 12 m. The design size of a gas filling station should be determined according to user demands and steady gas supply capacity of the biomethane source.

The design pressure of a compressed biomethane system should be determined according to process conditions and should be at least 1.1 times the maximum working pressure of the system. Gas cylinder trucks and gas cylinder groups that transport compressed biomethane to storage stations and gas supply stations shall have a filling pressure of not more than 20 MPa (gauge pressure) at a filling temperature of 20 °C. Models of biomethane compressors should be selected according to the pressure, dewatering process, and design scales. The models should be selected consistently and should incorporate backup units. The discharge pressure of the compressor should not exceed 25 MPa (gauge pressure). The single displacement of multiple parallel compressors should be equal to 80–85% of the nominal gas volumetric flow. The driver of biomethane compressors should be either electric motors or gas engines.

Compressors should be set in an unconfined space or in a single-storey building depending on site environment and climatic conditions. Skid-mounted equipment is also acceptable. Compressors should be arranged in a single row, and the width of the main passage of the compressor chamber should be no less than 1.5 m. The gas flow rate in the main pipe upstream of the compressor should be no greater than 15 m/s. Manual and electric (or pneumatic) control valves should be installed on the compressor inlet pipe. Safety valves, check valves and manual shut-off valves shall be equipped on the outlet pipe of the compressor. The discharge capacity of the outlet safety valve shall be no less than the safe discharge rate of the compressor. The discharge pipe outlet of the safety valve shall be more than 2 m above the building and should be at least 5 m from the ground.

Compressed biomethane pipelines should be made of high-pressure seamless steel tubes. The design pressure or pressure level of pipes, fittings, equipment, and valves for compressed biomethane systems should be no less than system design pressure, and all pipework materials should be compatible with biomethane. Gas-filling and degassing hoses on gas-filling and gas-removing columns should be made of corrosion-resistant pressure-bearing hose. The length of the hose shall not exceed 6 m, and the effective radius shall be no less than 2.5 m. Outdoor pipelines for compressed biomethane should be laid underground. The buried depth of the pipe top from the ground should be no less than 0.6 m, and the freezer zone of the station should be located below the frost line. Indoor compressed biomethane pipelines should be laid within pipe trenches. The clear distance between the bottom of the pipe and the bottom of the pipe trench should be no less than 0.2 m. The trenches are filled with dry sand, and a movable door and ventilation should be set.

References

CJJ 12—2013. *Specification for installation and acceptance of domestic gas burning appliances*. Industry Standards of Urban Construction Engineering of the People's Republic of China (in Chinese).

CJJ 95—2013. *Technical specification for external corrosion control of buried steel pipeline for city gas*. Industry Standards for Urban Construction Engineering of the People's Republic of China (in Chinese).

CJ/T 125—2014. *Steel framed polyethylene plastic pipes and fittings for supply of gaseous fuels*. Urban Construction Industry Standards of the People's Republic of China (in Chinese).

CJ/T 126—2000. *Fittings of steel framed polyethylene plastic pipes for supply of gaseous fuel*. Urban Construction Industry Standards of the People's Republic of China (in Chinese).

Deng, et al. 2015. *Biogas engineering*. Beijing: Science Press (in Chinese).

GB 150—2011. *Pressure vessels* National Standards of the People's Republic of China (in Chinese).

GB/T 3091—2015. *Welded steel pipes for low pressure liquid delivery*. National Standards of the People's Republic of China (in Chinese).

GB/T 8163—2008. *Seamless steel tubes for liquid service*. National Standards of the People's Republic of China (in Chinese).

GB 15558.1—2015. *Buried polyethylene (PE) piping systems for the supply of gaseous fuels—Part 1: Pipes*. National Standards of the People's Republic of China (in Chinese).

GB 15558.2—2016. *Buried polyethylene (PE) piping systems for the supply of gaseous fuels—Part 2: Fittings*. National standards of the People's Republic of China (in Chinese).

GB 16914—2012. *General safety technique conditions of gas burning appliances*. National Standards of the People's Republic of China (in Chinese).

GB 18047—2017. *Compressed natural gas as vehicle fuel*. National Standards of the People's Republic of China (in Chinese).

GB/T 29488—2013. *Medium/high-power biogas generating set*. National Standards of the People's Republic of China (in Chinese).

GB 50028—2006. *Code for design of city gas engineering*. National Standards of the People's Republic of China (in Chinese).
GB 50057—2016. *Code for design protection of structures against lightning*. National Standards of the People's Republic of China (in Chinese).
GB/T 51063—2014. *Technical code for large and medium-scale biogas engineering*. National Standards of the People's Republic of China (in Chinese).
GB 50251—2015. *Code for design of gas transmission pipeline engineering*. National Standards of the People's Republic of China (in Chinese).
HGJ 28—1990. *Design rules for electrostatic grounding in chemical enterprises*. Design standards of the Ministry of Chemical Industry of the People's Republic of China (in Chinese).
NY/T 1220.2—2006. T*echnical code for biogas engineering Part 2: Design of biogas supply*. Agricultural industry standards of the People's Republic of China (in Chinese).
NY/T 1223—2006. *Biogas-powered generating sets*. Agricultural Industry Standards of the People's Republic of China (in Chinese).
NY/T 1704—2009. *Biogas power generation technology criterion*. Agricultural Industry Standards of the People's Republic of China (in Chinese).
Q/GDW 11147—2013. *Design code for connecting to distribution network for distributed generation*. Enterprise Standards of the State Grid Corporation of the People's Republic of China (in Chinese).
Qi, Y. 2013. *Manual on biogas plant construction*. Beijing: Chemical Industry Press (in Chinese).
SY 0007—1999. *Design standard of corrosion control for steel pipeline and storage tank*. Oil and Gas Industry Standards of the People's Republic of China (in Chinese).
Song, X.Q. 2003. The development trend of CNG filling stations at home and abroad. *Oil & Gas Storage and Transportation* 8: 1–3. (in Chinese).
Tang, Y.F., and Y.X. Wang. 2013. *Design and application of large and medium biogas plants*. Beijing: Chemical Industry Press (in Chinese).
Wang, J.C., S.G. Yang, and X.C. Wan. 2016. *Manual of management on safe production of biogas plants*. Beijing: China Agriculture Press (in Chinese).
Wellinger, A., J. Murphy, and D. Baxter. 2013. *The biogas handbook: Science, production and applications*. UK: Woodhead Publishing Limited.
Yang, J. 2015. Process design and equipment configuration of CNG filling stations. *Chemical Enterprise Management* 8: 193–194. (in Chinese).
Zhang, Q.G. 2013. *Biogas technology and its application*, 3rd ed. Beijing: Chemical Industry Press (in Chinese).
Zhang, B., and L.N. Wang. 2014. Summary of biogas power generation system. *Science and Technology Vision* 6: 272–273. (in Chinese).
Zhou, M.J., R.L. Zhang, and J.Y. Lin. 2009. *Practical biogas technology*, 2nd ed. Beijing: Chemical Industry Press (in Chinese).

Chapter 9
Utilization of Digestate

The residue of animal manure, straw and other agricultural organic waste after anaerobic digestion is known as digestate. Digestate mainly consists of three parts: undigested feedstock, microbial organisms, and microbial metabolites. The solid part of digestate after mechanical solid-liquid separation or natural sedimentation is called solid digestate, while the liquid part is called liquid digestate. Digestate is rich in nutrients and bioactive substances required for plant growth, so digestate can be used as fertilizer. There are some differences in composition and utilization purposes between solid and liquid digestate, and this chapter will explain each in detail. Solid digestate is the solid part after solid-liquid separation, while liquid digestate contains the liquid part after solid-liquid separation and digestate that is difficult to separate.

9.1 Composition of Digestate

Although solid digestate and liquid digestate are both from anaerobic digestion, they are different in composition due to their difference in physical states. After solid-liquid separation, about one-third of dry matter, 13% of nitrogen, 28% of phosphorus, and 10% of potassium are separated into solid digestate (Table 9.1), and the rest remains in liquid digestate. In general, solid digestate is rich in organic matter, humic acid, and many other components that are suitable as base fertilizers, whereas liquid digestate is rich in soluble nutrients that are suitable as topdressing.

Table 9.1 Distribution of organics and nutrients in solid digestate and liquid digestate after solid-liquid separation (Tambone et al. 2017; Deng 2015)

Component	Solid digestate	Liquid digestate
Dry matter (%)	32.46	67.54
TKN (%)	13.13	86.87
P_2O_5 (%)	28.36	71.64
K_2O (%)	10	90

L. Deng et al., *Biogas Technology*,
https://doi.org/10.1007/978-981-15-4940-3_9

9.1.1 Physicochemical Properties of Solid Digestate

Solid digestate is brown or black, solid and semi-solid materials at the bottom of a digester. Table 9.2 shows some basic physical and chemical properties of solid digestate. Solid digestate is generally neutral or slightly basic, with a pH between 6.6 and 8.7. Due to different solid-liquid separation efficiencies and degrees of dehydration, water content of solid digestate varies. The content of dry matter is generally above 10%. Solid digestate produced from anaerobic digestion with different feedstock has different nutrient contents. Generally, solid digestate contains 10–54% of organic matter, 10–29.4% of humic acid, 25–34% of hemicellulose, 13–17% of cellulose, 11–15% of lignin, 1.6–8.3% of total nitrogen, 0.4–1.2% of total phosphorus, and 0.3–1.8% of total potassium. Total nutrient contents account for 2.4–24.6% of total digestate mass. It also contains crude protein, crude fiber, and a variety of mineral elements, amino acids, and many other components (Table 9.3).

Table 9.2 Basic physical and chemical properties of solid digestate

Component	Unit	Range	References
pH		6.6–8.7	Ge et al. (2014)
Dry matter	%	13–86	Chen et al. (2014), Zhang et al. (2008), Ge et al. (2014)
Organic matter	%	10–54	Chen et al. (2014), Nie et al. (2017)
Humic acid	%	10–29.4	Ge et al. (2014)
Hemicellulose	%	25–34	Tao et al. (2003)
Cellulose	%	13–17	Tao et al. (2003)
Lignin	%	11–15	Tao et al. (2003)
Total nitrogen (TN)	%	0.8–8.3	Tao et al. (2003), Nie et al. (2017)
Total phosphorus (P_2O_5)	%	0.4–1.2	Tao et al. (2003)
Total potassium (K_2O)	%	0.3–2.0	Tao et al. (2003), Nie et al. (2017)
Total nutrients (TN + TP + TK)	%	2.4–24.6	Zheng et al. (2014)

Table 9.3 Contents of amino acids in solid digestate (Wu et al. 2007)

Amino acids	Content (mg/g)	Amino acids	Content (mg/g)
Aspartic acid	4.09	Isoleucine	2.57
Threonine	2.46	Leucine	3.59
Serine	2.33	Tyrosine	1.67
Glutamate	5.51	Phenylalanine	2.32
Glycine	3.23	Lysine	2.75
Alanine	3.18	Histidine	0.77
Valine	2.72	Arginine	2.34
Methionine	0.2	Proline	1.987

Table 9.4 Heavy metal content in solid digestate produced by some biogas plants in China

Elements	Solid digestate from pig manure (mg/kg DM)	Solid digestate from cow manure (mg/kg DM)	Solid digestate from chicken manure (mg/kg DM)	References	Standard *Anaerobic digested fertilizer* (mg/kg)
Zn	40–525	86.5–129	15.13	Chen and Cui (2012), Zhang et al. (2016) Zhang et al. (2008) Sun et al. (2017), Wang et al. (2010)	–
Cu	45.8–605	2.29–46.4	1.96		–
As	0.13–36.4	0.11–4.33	1.09		≤15
Cr	10.3–27.6	0.447–1.89	11.35		≤50
Cd	0.09–8.6	0.009–0.020	–		≤3
Pb	3.22–17.8	0.033–5.143	–		≤50
Hg	0.002–0.01	0.0048–0.264	–		≤2

After anaerobic digestion, organic waste and heavy metals will have different distributions in liquid and solid digestate. In general, solid digestate will accumulate more heavy metals. Currently, Cu and Zn are the major heavy metals in solid digestate (Table 9.4). The content of heavy metals in most solid digestate must be within the limit required by *Anaerobic digested fertilizer* (*NY/T 2596-2014*) in order to be used as fertilizer. However, in some solid digestate, the content of heavy metal exceeds the maximum requirement. This kind of solid digestate should go through further harmless treatment and cannot be directly used for fertilizer.

9.1.2 Physicochemical Characteristics of Liquid Digestate

Liquid digestate is generally black or black brown, as shown in the physical properties listed in Table 9.5. Its chroma can reach above 2500, and its turbidity can reach more than 4000 NTU, especially the turbidity of liquid digestate deriving from chicken manure, which can be even higher. The density of liquid digestate is slightly greater than water density, and the liquid digestate could be weakly acidic, neutral, or weakly basic. Dry matter content in liquid digestate is generally less than 10%, and the dry matter is mainly organic matter, which takes up as much as over 3%. The COD content varies with different feedstock, generally within 20,000 mg/L. BOD_5 is relatively low, generally within 500 mg/L. Total nutrients are usually 0.1–0.5%. The salt content in liquid digestate is relatively high, and the total salt content is generally more than 1000 mg/L or as high as 4670 mg/L.

The main components of liquid digestate differ due to feedstock type, feedstock concentration, and digestion conditions, and the components can be roughly sorted out into nutrient salts, amino acids, organic acids, plant hormones, antibiotics and other small molecular compounds.

Table 9.5 Basic physicochemical properties of liquid digestate

Item	Unit	Range	References
Chroma	Degree	417–2838	Li et al. (2014b), Wang et al. (2017)
Turbidity	NTU	111–4189	Li et al. (2014b), Song et al. 2011, Han et al. (2014), Duan et al. (2015), Dong (2015), Wu (2015a, b)
pH		6.15–8.6	Jin et al. (2011), Wan et al. (2010), Li et al. (2014c), Wang et al. (2017), Wang (2017), Qin et al. (2013), Wu (2015a, b), Qiao (2015)
Dry matter	%	0.42–7.3	Wang et al. (2012a) , Chen e al. (2014), Gu et al. (2016), Deng (2015)
Organic matter	%	0.1–3.45	Wei et al. (2014), Chen et al. (2014), Li et al. (2013), Wang et al. (2012a, b), Qu et al. (2013), Xu et al. (2012), Liu et al. (2017)
Total nutrient	%	0.117–0.549	Dong et al. (2017), Deng (2015)
Total salt	mg/L	1420–4670	Lin et al. (2015), Li et al. (2014a, c)

(1) Nutrient salts

Carbon (C), hydrogen (H), oxygen (O), nitrogen (N), phosphorus (P), sulfur (S), potassium (K), calcium (Ca), magnesium (Mg), iron (Fe), manganese (Mn), zinc (Zn), copper (Cu), molybdenum (Mo), boron (B) and chloride (Cl) are the 16 essential elements for plants growth. Besides C, H, and O, the rest are taken from dissolved salts in surrounding moisture. During the process of anaerobic digestion, nutrients needed by most plants are retained in liquid digestate except for a significant loss of carbon, and the structures of nitrogen and phosphorous are optimized.

N, P and K are nutrient elements highly needed by plants. Generally, the nutrient content in liquid digestate is in the order of N > K > P, and mainly in a form of aqueous solution. The nutrient composition of liquid digestate produced from different feedstock is different to some extent. Among liquid digestate produced from pig manure, cow manure and chicken manure, liquid digestate from the last contains the highest amount of total nutrients and available nutrients (Table 9.6). Table 9.7 shows the content of secondary elements and micronutrients for plant growth in liquid digestate. Ca, Mg and S are secondary elements necessary for plant growth. Ca content in liquid digestate is generally high, at above 100 mg/L

Table 9.6 Nutrient content of liquid digestate produced from different feedstock (Qin et al., 2015)

Source	TN (%)	TP (%)	TK (%)	Available nitrogen (g/kg)	Available phosphorous (mg/kg)	Available potassium (mg/kg)
Swine wastewater	0.09–0.13	0.02–0.03	0.03–0.04	0.42–0.49	87–160	252–386
Cow wastewater	0.09–0.14	0.02–0.08	0.06–0.09	0.56–0.58	102–151	639–700
Chicken wastewater	0.13–0.20	0.01–0.03	0.09–0.14	1.06–1.09	289–293	1124–1226

Table 9.7 Content of secondary elements and micronutrients in liquid digestate

Element	Unit	Range	References
S	mg/L	68–115	Singh et al. (2011), Zhou et al. (2013)
Ca	mg/L	78.6–1163	Wu (2015a, b), Deng et al. (2017)
Mg	mg/L	5.31–89.46	Wu (2015a, b), Deng et al. (2017)
Fe	mg/L	1.58–57.4	Zhao et al. (2014), Wu (2015a, b)
Zn	mg/L	0.9–33.0	Zhao et al. (2014), Xia and Murphy (2016)
Cu	mg/L	0.09–27.4	Zhao et al. (2014), Xia and Murphy (2016)
Mn	mg/L	0.1–17	Zhao et al. (2014), Xia and Murphy (2016)
Mo	mg/L	0.031–0.034	Duan et al. (2015), Deng et al. (2017)
B	mg/L	0.51–4	Shen et al. (2014), Xia and Murphy (2016)
Cl	mg/L	160–438	Xia and Murphy (2016)

or even as high as 1000 mg/L. Mg content in liquid digestate is lower than Ca content and is generally within 100 mg/L. S content in liquid digestate is usually in the form of sulfate and is about 100 mg/L (Singh et al. 2011).

Fe, Mn, Zn, Cu, Mo, B and Cl are essential micronutrients for plant growth. The content of Cl in liquid digestate is relatively high, by and large 150 mg/L or above. The content of Cu and Zn varies greatly, and their content in liquid digestate from pig manure is normally large. Fe and Mn are also micronutrients in liquid digestate. B and Mo are generally low in content, but they are sufficient to meet the needs of plants.

(2) Amino acids

Anaerobic digestion breaks down protein into free amino acids, which are further converted into ammonia nitrogen. The content of ammonia nitrogen in liquid digestate is high, and liquid digestate contains a certain amount of amino acid. Amino acid content in liquid digestate is heavily influenced by digestion time and temperature. A temperature above 24 °C or over 14 days of digestion time facilitate accumulation of free amino acid (Shang et al. 2009). Amino acid content in liquid digestate varies with feedstock and is generally higher in liquid digestate having chicken manure as feedstock (Meng et al. 2000). In liquid digestate from pig and chicken manure listed in Table 9.8, the content of aspartic

Table 9.8 Composition of amino acids in liquid digestate (Wang 2012; Wu 2015a, b)

Amino acids	Content (mg/L)	Amino acids	Content (mg/L)	Amino acids	Content (mg/L)	Amino acids	Content (mg/L)
Lysine	37.8	Alanine	27.7–53.2	Valine	37.4	Cystine	4.8
Methionine	30.8	Isoleucine	32.9	Aspartic acid	81.6	Arginine	14.1–26.7
Leucine	14.5–48	Glutamate	35.2	Phenylalanine	36	Proline	26
Histidine	14.1	Tyrosine	10.9–42.6	Serine	11.6–36.2		
Threonine	11.2–41.4	Tryptophan	22.5	Glycine	24.9–42.7		

acid is as high as 81.9 mg/L, and the content of alanine is more than 50 mg/L. The total amino acid content in liquid digestate from pig manure can reach 651 mg/L, accounting for 9.5% of the total organic matter in liquid digestate (Wang et al. 2012a, b).

The presence of amino acids in liquid digestate enables it to be used as fertilizer and animal feed. In terms of utilization as fertilizer, amino acid is a supplementary source of organic nitrogen, which can improve fertilizer efficiency. Moreover, amino acid has the role of promoting complexation reactions with metal ions so that the necessary secondary elements and micronutrients can be easily carried into plants to improve the utilization efficiency of various nutrients. Amino acids are promoters and catalysts for synthesis of various enzymes in plants. Amino acids contained in fertilizers can strengthen seedlings, facilitate photosynthesis of leaves, enhance stress resistance of crops, and play an important role in plant metabolism.

(3) Plant hormones

Liquid digestate contains four major plant hormones: auxin (mainly indoleacetic acid, IAA), gibberellin (GAs), cytokinin and abscisic acid (ABA) (Table 9.9). Huo et al. (2011) detected plant hormones in liquid digestate from a biogas plant treating pig manure that had been in operation for more than one year, and they found that contents of IAA, gibberellin GA_4, gibberellin GA_{19}, gibberellin GA_{53}, and cytokinin isopentenyl adenosine (iPR) were 332 μg/L, 0.857 μg/L, 1.47 μg/L, 0.271 μg/L, 0.00194 μg/L, respectively. However, the content of IAA in liquid digestate found by Li (2016) was as high as 17.38–36.84 mg/L, while concentrations of GA_3 was found to be 16.37–44.83 mg/L and ABA concentration to be 13.23–35.39 mg/L.

According to the study of Li (2016), IAA in liquid digestate is mainly produced by anaerobic microorganisms metabolizing tryptophan. In anaerobic digestion, IAA content generally shows an upward trend with time and accumulates in liquid digestate. ABA content continues to increase throughout the whole digestion process, and its growth rate in the Methanogenic stage was significantly higher than that in the acidifying phase. GA_3 is produced in the hydrolysis phase of anaerobic digestion of animal manure, but the content of GA_3 decreased due to degradation at a later stage of anaerobic digestion.

Table 9.9 Content of organic acids in liquid digestate (Shi et al. 2013; Tatsuya 2009)

Feedstock	Organic acid in liquid digestate			
	Acetic acid	Propionic acid	Butyric acid	Pentatonic acid
Wheat straw	5.51–15.16 mg/L	0.61–2.67 mg/L	–	–
Corn straw	6.5 mg/L	1.95–2.17 mg/L	–	–
Cow manure	5.67 mg/L	–	–	–
Chicken manure	0.35–26.25 mg/L	9.22 mg/L	0.18–0.88 mg/L	0.03 mg/L
Pig manure	402 mg/kg	23 mg/kg	Trace amount	–

These plant hormones play important roles in plant growth and development. For example, IAA can promote burgeoning and germination of the plant tips. Abscisic acid (ABA) can control embryonic development and seed dormancy while enhancing the resilience of plants. Gibberellin can promote growth, germination, flowering and bearing of plants while stimulating fruit growth and improving seed set rate. Moreover, gibberellin can even significantly increase the yield of grain crops, cotton, vegetables, fruits and so on.

(4) Organic acids

Liquid digestate contains volatile organic acid such as acetic acid, propionic acid, butyric acid, and pentatonic acid, etc. In anaerobic digestion, bacteria decompose soluble sugars, peptides, amino acids and fatty acids to produce acetic acid, propionic acid and butyric acid. Acetogen further converts propionic acid and butyric acid into acetic acid. Methanogens eventually convert acetic acid into methane. Organic acid is the intermediate product of anaerobic digestion. Liquid digestate contains acetic acid, propionic acid, butyric acid, etc. Generally, acetic acid content and propionic acid content are higher compared with the content of butyric acid and valeric acid (Table 9.9). Studies have shown that secondary metabolites such as acetic acid, propionic acid and butyric acid have bacteriostatic effects. The higher the concentration of organic acid is, the more prominent the inhibitory effect is.

(5) B Vitamins

B vitamins in digestate can promote plant and animal growth and development while improving their resistance to diseases and pests. Liquid digestate contains B vitamins such as B_1, B_2, B_5, B_6, B_{11}, and B_{12}. The content of vitamin B_2, B_5 and B_{12} will increase in the process of anaerobic digestion. Studies have shown that vitamin B_{12} is produced by microbes, and metabolism of some methanogens such as *Methanobacterium omelianski* also produces such vitamin. The content of vitamin B_{12} in liquid digestate from pig manure can reach 150 μg/L (Min 1990).

(6) Antibiotics

Antibiotics in liquid digestate mainly derive from veterinary drugs and feed. Although anaerobic digestion is conducive to the degradation of some antibiotics, due to heavy use of antibiotics, liquid digestate deriving from animal manure still contains high concentration of antibiotics. Antibiotics in liquid digestate mainly include tetracycline, sulfonamide, macrolides, and quinolones. Liquid digestate from different feedstock contains different antibiotics, He et al (2017) used solid phase extraction-high performance liquid chromatography method to determine 3 kinds of tetracycline and 6 kinds of sulfonamides and found the existence of sulfadiazine, sulfamethoxazole, and sulfamerazine in liquid digestate from pig manure, with content ranging from 34.9 to 118.5 μg/L. Oxytetracycline hydrochloride, aureomycin hydrochloride, sulfadiazine and sulfamethoxazole were found in liquid digestate from cow manure, with the content ranged from 21.7 to 51.9 μg/L. In liquid digestate from chicken manure, there were 6 kinds of antibiotics, including oxytetracycline hydrochloride, sulfadiazine, sulfamethoxazole, sulfamethazine and sulfamisoxazole, with their content ranging from 28.5 to 125.5 μg/L. Analysis of quinolone antibiotics by HPLC and fluorescence detection showed that the content of antibiotics in liquid digestate from pig manure was the highest, in which norfloxacin was up to 204 μg/L (Table 9.10).

Wei et al. (2014) tested the content of 10 kinds of antibiotics in liquid digestate from10 pig farms in Jiaxing city (Table 9.11). In different seasons, 10 kinds of antibiotics were all detected in liquid digestate from each pig farm. However, the content of antibiotics in liquid digestate varied drastically with pig farms. The highest total concentration of antibiotics was about 24 times (spring group), 35 times (autumn group) and 25 times (winter group) of the lowest concentration.

Table 9.10 Content of quinolones in liquid digestate (μg/L) (He et al. 2016a, b)

Feedstock	Ofloxacin	Ciprofloxacin	Enoxacin	Norfloxacin
Pig manure	103.0	17.0	151.0	204.0
Cow manure	16.8	16.0	89.0	67.0
Chicken manure	76.0	5.0	67.0	56.0

Table 9.11 Seasonal change of antibiotic content in liquid digestate (Wei et al. 2014) unit: mg/L

Antibiotics	Spring		Autumn		Winter	
	Range	Average value	Range	Average value	Range	Average value
Tetracycline	0.75–43	11.2	0.4–8.98	1.84	0.3–14.2	3.07
Terramycin	1.6–994	269	4.58–332	77.3	21.2–672	161
Aureomycin	2.9–228	60.3	0.53–53.8	15.4	1.04–93.7	19.5
Sulfadimidine	0–30.2	7.37	0.008–3.47	0.636	0.008–1.66	0.277
Sulfamethoxazole	0–56.9	5.94	0–1.61	0.292	0.001–0.6	0.065
Enrofloxacin	0.1–2.85	1.3	0–1.98	0.587	0.03–5.88	1
Ciprofloxacin	0.65–4.6	1.87	0.2–5.92	1.39	0.44–5.91	1.55
Norfloxacin	0.65–4.4	1.28	0.08–1.97	0.59	0.04–0.4	0.191
Tylosin	0.4–22	8.62	0.01–1.22	0.274	0.012–0.38	0.199
Roxithromycin	0–3.4	0.34	0–0.26	0.099	0.007–6.86	0.728

Table 9.12 Heavy metal content in liquid digestate (Liu et al. 2014; Wu 2015a, b; Zhang 2014)

Element	Liquid digestate from pig farm (mg/L)	Liquid digestate from cow farm (mg/L)	Liquid digestate from chicken farm (mg/L)	Standard of anaerobic digested fertilizer (mg/L)
As	0.0018–3.82	0.0028	5.21	≤10
Cr	0.0025–15.32	0.1241	10.18	≤50
Cd	0.000095–7.51	0.0006	<0.025–4.30	≤10
Pb	0.0053–9.54	0.0235	<0.05–2.43	≤50
Hg	0.00015–0.32	0.0128	<0.005–0.022	≤5

(7) Heavy metals

Heavy metals in liquid digestate come from feedstock, while heavy metals in animal waste come from feed and veterinary medicine. For example, pig feed is usually added with heavy metals such as Zn, Cu and As, the digestion and absorption rate of which is extremely low in livestock and poultry body. The majority of Cu and Zn are discharged through feces, with a small amount through urine. The total excretion rates of Cu and Zn in feces and urine are 88%–96% and 87%–98% respectively (Zhu 2014).

Liquid digestate may contain heavy metals including As, Cd, Cr, Hg, Pb, Cu, Zn, Ni, etc. The content of Cu and Zn is relatively high. Although these two elements are classified as heavy metals, they are also indispensable micronutrients for livestock, plants and microorganism. These micronutrients are often added to feed, so there is no restriction on Cu and Zn in fertilizer standards in China and many other countries. Anaerobic digestion has a good removal effect on organic matter of animal manure, but the concentration of heavy metals will have a "relative concentration effect", and the content of heavy metals in solid digestate is much higher than that in liquid digestate. Table 9.12 shows the heavy metal content in liquid digestate produced by the manure from a pig farm, a cattle farm and a chicken farm. Although content varies, it does not exceed the standard of digestate fertilizer.

(8) Other metabolites

In addition to the aforementioned categories of components, liquid digestate also contains some alkanes and esters metabolites. Song et al. (2011) extracted some volatile components from liquid digestate concentrate by petroleum ether and found by GC/MS analysis that the volatile components were mainly alkyl compounds and a small amount of esters. These include 1, 2, 3-dicarboxyl propyl ester, heptadecane, 2, 6, 10-trimethyltetradecane, 2, 6, 10, 14-tetramethylhexadecane, 1, 2-phenyldicarboxylic acid-2-methyl propanes, 1, 1′-[1, 3-propylene (dioxygen)]-octadecane. Huo et al. (2011) found that the content of some quinoline such as 8-hydroxy-34-dihydroquinoline-2-ketone and 34-dihydroquinoline-2-ketone from liquid digestate of pig farm was 737.5 and

177.5 μg/L. These may have certain sterilizing effect. Some terpenoids separated from chicken manure liquid digestate has anti-cancer and anti-bacterial effects (Liu 2013; Wu 2015a, b).

9.1.3 Economic Value of Digestate

Digestate is rich in nutrients and thus have certain economic values. But at present most solid and liquid digestate produced by biogas plants are difficult to be sold commercially. The sales of solid digestate is relatively higher than that of liquid digestate. The price of solid digestate is comparable to the price of residue after solid-liquid separation, which is about 300–500 yuan/ton. In a few regions, the price of liquid digestate is 5–40 yuan/ton, but most local solid digestate is free, and some large livestock farms even subsidy users for the use of liquid digestate in order to deal with it.

At present, there are two methods to evaluate the commercial values of solid and liquid digestate. One is to determine the nutrient content of digestate by detecting the main nutrient element in it and determine the price according to that of the same element in fertilizers sold in local market. The second is to determine the price according to application effect, meaning comparing the actual crop yield by applying different amounts of chemical fertilizers and digestate to figure out the digestate equivalence of the fertilizer, and then the price of digestate is determined in accordance with the price of the chemical fertilizer. Zhang et al. (2011a, b) determined the price of liquid digestate from cattle wastewater as 78.12 yuan/m^3 by the nutrient content pricing, and 111.4 yuan/m^3 by liquid digestate application effect pricing. Later, the nutrient content pricing was revised considering the increased transportation and application cost of liquid digestate compared with other fertilizers as well as the popularity of it. The value of liquid digestate form the biogas plant treating cattle wastewater was finally determined as 38.50 yuan/10^3 kg (Zhang et al. 2012).

The formula for calculating the price of liquid digestate is as follows:

$$\mathrm{P} = [(\rho_{\mathrm{OM}} \times \eta_{\mathrm{OM}} + \rho_{\mathrm{N}} \times \eta_{\mathrm{N}} + \rho_{\mathrm{p}} \times \eta_{\mathrm{P}} + \rho_{\mathrm{k}} \times \eta_{\mathrm{k}}) - P_{\mathrm{M}} - P_{\mathrm{u}}] \times \delta \quad (8.1)$$

where,

P: Price of digestate, yuan/m^3;

ρ_{OM}: the price of organic matter, yuan/10^3 kg;

η_{OM}: is the content of organic matter content;

ρ_{N}: the price of nitrogen, yuan/10^3 kg;

η_{N}: nitrogen content;

ρ_{P}: the price of phosphorus, yuan/10^3 kg;

η_{P}: phosphorus content;

ρ_{K}: the price of potassium, yuan/10^3 kg;

η_{K}: potassium content;

P_M: additional transportation cost of liquid digestate compared with other fertilizers, yuan/m^3;

P_U: additional application cost of liquid digestate compared with other fertilizers, yuan/m^3;

δ: the popularity of liquid digestate compared to other fertilizers, expressed in terms of percentage.

Digestate has a dual identity of waste and a fertilizer. Therefore, in assessing its value, both its practical value as resources and environmental value in pollution abatement should be taken into account.

9.2 Function of Digestate in Agricultural Production

Digestate, rich in nutrients and active ingredients, is a good fertilizer, and its application is desirable in agricultural production due to the following advantages: (1) anaerobic digestion of volatile organic compounds can reduce odor emissions; (2) short-chain organic acids are significantly degraded, reducing the risk of seedling burning; (3) the main nitrogen is ammonia nitrogen, which improves the utilization efficiency of nitrogen; (4) weed seeds and some pathogens are killed in the anaerobic digestion.

9.2.1 Soil Improvement

Nutrients, organic matters and humic acid contained in digestate are beneficial to soil improvement. Therefore, the application of digestate has prominent improvement effect on soil. Humic acid in digestate plays an important role in the formation of soil aggregate structure. Coarse fiber organic components can help loosen soil and increase organic matter in soil, while nutrient salts can increase total nitrogen content and available N, P and K. Long-term application of digestate does not only improve soil fertility and structure, but it also improves its moisture and fertilizer retention capacity, as well as enhancing soil microbial activity and reducing heavy metal toxicity.

Solid digestate can improve soil structure. Long-term test results (5 years or 10 years) show that solid digestate can improve soil colloid, increase the number of complex colloid and microstructure while reducing soil bulk density, increase total porosity, decrease solid phase, and increase gas phase (Jin 1987; Li et al. 1992). Solid digestate can also increase rhizosphere microorganisms improve the activity of soil enzymes, promote conversion and absorption rate of nitrogen, phosphorus and potassium in soil, reduce accumulation of excessive nutrients in soil, and improve micronutrients content in soil (Xu et al. 2016a, b; Yin et al. 2017). Compared with soil applying chemical fertilizers and soil without fertilizers, the soil food web after solid digestate application was more structured, and soil nutrients were more enriched (Li et al. 2017).

Liquid digestate can change the physical and chemical properties of soil, loosening it and enhancing its color. After applying liquid digestate, soil particles become smaller, and soil moisture and fertilizer retention capability can be improved, making cultivation easier. The application of liquid digestate can also improve soil fertility and soil buffering capacity, promote balanced growth of soil microorganisms, and increase TN, available N, available P and available K in soil. The results of Cordovil et al. (2011) showed that after liquid digestate was applied, the content of ammonia nitrogen and nitrate nitrogen in soil increased, which demonstrated that liquid digestate has a good effect on soil improvement and remediation. The study of Pan (2009) showed that after using liquid digestate for more than two years, the organic matter, total nitrogen, alkali-hydrolyzed nitrogen, total phosphorus and available phosphorus in soil that plants alfalfa were increased by 181.6%, 59.6%, 93.4%, 153.3% and 100.0% respectively, compared with the group without liquid digestate application. According to the researches in different regions on different soil types (Table 9.13), liquid digestate can improve soil organic matter that can absorb more cations, so that the soil becomes fertile and has higher capability to buffer. The soil permeability and aggregate structure were improved as well. In conclusion, long-term application of liquid digestate has a positive impact on soil fertility.

The pH of digestate generally ranges from neutral to basic, which can increase the pH of acidic soil, modify acidified soil, increase the number of soil exchangeable bases, and boost soil base saturation, soil cation exchange capacity, soil organic matter content and so on. Digestate can also be used to improve saline-alkali soil, increase soil nutrient content, improve soil physical properties, and enhance crop output (Table 9.13).

9.2.2 Pest Control

Digestate has the function of disease prevention, bacteriostasis and pest control, and there are four main applications of disease and pest prevention and control: (1) application of digestate directly on roots to inhibit diseases or to improve the resistance to pests and diseases of crops (2) soaking seed in liquid digestate to improve the resistance of seedling or even adult plant; (3) spraying liquid digestate for crop pest control; (4) using liquid digestate in combination with pesticide. The above four methods can be used individually or in combination. In general, liquid digestate spraying is more commonly used in prevention and control of crop pests and diseases.

A field experiment showed that solid digestate as a base fertilizer could improve the disease resistance of corn (Zhang 2013). A Pot experiment on solid digestate application showed that phytophagous nematodes were significantly inhibited (Li et al. 2017). Application of liquid digestate from different biogas plants on crop roots has different inhibitory effects on hyphae growth of strawberry fusarium wilt, and the antagonistic microorganism in liquid digestate is the main factor for liquid digestate

Table 9.13 Improvement of digestate on soil in different regions

Soil type	Experiment site	Improvement	References
Cinnamon soil	Daxing County, Beijing, China	The total nitrogen content of surface soil increased by 65%, and the pH of soil decreased by 0.4	Chen et al. (2011a, b)
	Haining, Zhejiang, China	It has little effect on pH and available components, and the organic matter and total N content can increase by 23% and 32% respectively compared with the control group	Wang et al. (2011)
Red loam	Cixi, Zhejiang, China	Compared with chemical fertilizer, maximum soil pH was increased to 0.5, and the pH was close to neutral, which was not as effective as that of chemical fertilizer for N, P and K	Ni et al. (2008)
	Fuqing, Fujian, China	There is accumulation of total N in 0–30 cm soil. No effect was imposed on alkali-hydrolyzed N, total P and available P	Wu et al. (2011)
Sandy loam	Nanning, Guangxi, China	Compared with the control group, combined application of liquid digestate and chemical fertilizer leads to 50% of increase in mass fraction of organic matter, and significant accumulation of available N, P and K, but the pH showed a downward trend	Gan et al. (2011)

(continued)

Table 9.13 (continued)

Soil type	Experiment site	Improvement	References
Sandy loamy alkaline soil	China	Digestate can improve the nutrient content of alkaline soil, improve the physical properties of soil and increase the yield of crops	Cai et al. (1999)
Acid soil	China	Increase soil pH and the quantity of soil exchangeable base, improve soil salt-base saturation, soil cation exchange capacity, soil organic matter content, and improve crop yield	Shi et al. (2017), Xie et al. (2013)

to inhibit bacteria and prevent diseases (Ma et al. 2011). Integrated application of soaking seed in liquid digestate, spraying, and root application can prevent or reduce corn exscrohilum turcicum, bipolaris maydis, brown blotch, leaf sheath purpura and stem rot, etc., and the integrated application can improve disease resistance of corn and reduce the disease level of corn (Zhang 2014).

Liquid digestate spraying has obvious inhibitory effect on pathogens of some plants such as neck blast, rice bacterial leaf blight, leaf spot, leaf rust, powdery mildew, downy mildew, and gray mold, etc. Liquid digestate spraying can replace pesticide to some extent. The inhibition effect of liquid digestate on bacteria blight and sheath blight of rice was almost the same as that of validamycin, reducing the morbidity ratio by 65%. Spraying liquid digestate during budding and early flowering period of cotton can obviously prevent and control fusarium wilt. Spraying liquid digestate before flowering period of cotton can significantly prevent and control verticillium wilt. Compared with carbendazim, spraying liquid digestate twice has a higher control effect on scab up to 44.88% (Zhang 1994). Spraying liquid digestate can also inhibit aphids, cabbage caterpillar, spider mite, whitefly and other 10 types of pests. Another study also shows that the mix of pesticide with 5 times of 2% liquid digestate can improve the insecticidal effect by more than 30% compared with the applying pesticides alone. Compared with red Fuji apple trees without liquid digestate spray, the number of aphids and spider mite in red Fuji apple trees with liquid digestate spray was significantly reduced, and the efficiency of killing insects reached to over 80% (Zhang et al. 2001).

The mechanism of disease and pest control by digestate is as follows: (1) anaerobic digestion kills pathogenic bacteria (such as salmonella, Pasteurella, etc.), insect eggs, and some plant viruses. (2) Liquid digestate has many reducing substances with low Redox reaction potential, which readily takes away oxygen and removes fat when reducing substances come into contact with pests. (3) High concentration of NH_4^+

in liquid digestate has certain bactericidal effect; gibberellin, indoleacetic acid, and vitamin B contained in liquid digestate can inhibit growth of harmful microbes. (4) Bacteria in liquid digestate secret special substances that can inhibit the growth of harmful bacteria and drive away pests, and a part of the microbial species can inhibit the growth of other bacteria through competition, antagonism and hyper parasitism. (5) Nutrients and bioactive substances in digestate can also improve the resistance of crops to diseases and pests (Cao et al. 2015).

9.2.3 *Crop Yield Increase*

Because digestate can provide nutrients, improve soil structure, and prevent and control pests and diseases, it has obvious effect on increasing output of crops, vegetables and fruits (Table 9.14). Whether solid digestate or liquid digestate is applied alone, or whether the two are combined, or liquid digestate and chemical fertilizer are combined, liquid digestate irrigation and foliar spraying have a certain degree of yield increase effect on crops. The yield increase of grains and fruit was generally less than 20%, while that of vegetable was much higher, reaching 98%.

Although solid digestate, liquid digestate or mixed application can improve yield and replace chemical fertilizers, a large number of studies have shown that solid digestate and liquid digestate combined with chemical fertilizers have higher yield increase. The reason may be that by combined application, nutrients are more balanced. A particularly significant yield increase can be observed when adequate nitrogen is supplied.

9.2.4 *Quality Improvement of Agricultural Products*

The application of digestate does not only increase crop yield, but it also improves the quality of agricultural products. The main parameters to evaluate crop quality include vitamins, sugars, proteins, nitrates and minerals. For grain crops, the application of digestate can improve the content of protein and minerals in grains, such as Fe, Mn, Ca and Mg in rice and corn (Table 9.16). The main function of digestate on fruit crop is to improve soluble solids, sugar content, fruit sugar-acid ratio, flavor, and the content of vitamin C. For vegetable crops, liquid digestate does not only increase the content of vitamin C and soluble sugar, but it also reduces the content of nitrate in vegetables. In addition, liquid digestate can also improve the chemical quality, sensory quality and processing quality of tea, such as aroma, color of tea, and taste (Table 9.15).

Table 9.14 Effect of liquid digestate on crop yield increase

Crop	Location	Application method and effect	References
Rice	Jiaxing, Zhejiang, China	The yield of rice increases significantly with the increase of liquid digestate dosage, up to 14% increase in production	Li et al. (2011)
	Dengta city, Liaoning, China	The yield is the highest when digestate is applied at 2000 kg/666 m^2, which is higher than that of a single chemical fertilizer application, but the yield of mixed fertilization (digestate + compound fertilizer) is higher than that of a single digestate application	Chen (2016)
Corn	Ziyang, Sichuan, China	When the applied amount of liquid digestate per hectare is between 75000 kg ~ 105000 kg, the yield can be similar to that of chemical fertilizer treatment	Wu et al. (2014)
Wheat	Feidong county, Anhui, China	The yield of wheat applied with 60 000 kg/hm^2 of liquid digestate and 150 kg/hm^2 urea is the highest, reaching 6525 kg/hm^2, which is 6.5% higher than control group applied with 150 kg/hm^2	Li et al. (2014d)
Soybean	China	Soybean yield increases by 1.02%, 10.48% and 0. 16% respectively when applied with 100% solid digestate, 30% solid digestate and chemical fertilizer, 20% solid digestate and chemical fertilizer	Zhang et al. 2011a, b

(continued)

Table 9.14 (continued)

Crop	Location	Application method and effect	References
Chinese Cabbage	Shanghai, China	Yield increases to different degrees in all groups that applied liquid digestate, with the highest increase reaching 98%	Wang et al. (2008)
Apple	Baoding, Hebei, China	Individual output is the highest when applied with both liquid digestate irrigation and foliage spraying, reaching 400 kg, an increase of 13–20%	Qian et al. (2005)
Orange	Ninghai, Zhejiang, China	Replacing chemical fertilizer by 300 000/hm^2 of liquid digestate can increase output by 8.59%	Wang et al. (2011)
Grass	Nanping, Fujian, China	Liquid digestate irrigation (52,500/hm^2) increases yield (52,500/hm^2) by 37.95–58.74% compared with chemical fertilizers	Xie et al. (2010)

9.3 Utilization of Solid Digestate

Solid digestate can be directly used as a fertilizer or be further processed to commercial fertilizer and other products. Solid digestate used as fertilizers should be completely digested, and the fertilizer should meet the standards of anaerobic digested fertilizer regulating a moisture content of 60–80%, a pH of 5.8–8.0, and a total nutrient content of dry solid digestate base sample of ≥3.0%, and an organic matter content ≥30%.

9.3.1 Solid Digestate Used as Fertilizers

(1) Solid digestate as base fertilizers
Solid digestate contain high nutrient content, rich organic matter, and humic acid. It is a kind of organic fertilizer with high values and can replace chemical fertilizers to enhance soil fertility. At present, the most common utilization of

Table 9.15 Effect of digestate on enhancement of crop quality

Crop	Application method and dosage	Application effect	References
Rice	Base fertilizer and top fertilizer, 11,250–18,750 kg/hm^2	The protein content and mineral content of Fe, Mn, Ca and Mg in rice increased	Tang et al. (2010)
Rice	Conventional fertilization and solid digestate 30,000 kg/hm^{-2}	Improved polished rice ratio, head rice ratio, and gel consistency	Wang et al. (2012a)
Corn	Base fertilizer and top fertilizer, 75,000–105,000 kg/hm^2	The contents of Fe, Mn, Zn, Cu, Ca and Mg in corn were higher	Wu et al. (2014)
Sweet corn	Combined application of liquid digestate, solid digestate and chemical fertilizer	Increased the content of reducing sugar in seeds	Liu et al. (2009)
Orange	Watering, 300,000 kg/hm^2	Increased the content of soluble solid and sugar-acid ratio, and improved some quality indicators	Wang et al. (2011)
Apple	Fertigation and foliage spraying	The content of Vitamin C was 12 mg/kg higher than the control group. Fruit was bigger, crispier, juicier with finer color	Qian et al. (2005)
Grape	Topdressing and spraying liquid digestate and solid digestate	Improved the content of soluble solid in fruit, and increased sweetness by 1.75°	Yan et al. (2012)
Chinese cabbage	Topdressing, 70% of liquid digestate	Improved the content of Vitamin C and soluble reducing sugar while reducing nitrate	Wang et al. (2008)
Chinese cabbage	Compound fertilizer of solid digestate, organic and inorganic fertilizer	The content of nitrate decreased by 7.8–56.7%, water soluble sugar increased by 7.8–56.7%, reducing Vitamin C increased by 18–38.2%	Dong et al. (2014)
Tea	The ratio of liquid digestate and clean water is 1.5, spraying	Improved chemical quality and sensory quality of tea	Xiang et al. (2006)
Tea	5000 kg/hm^2 of solid digestate or 110,000 kg/hm^2 liquid digestate	Improved quality of tea products, such as aroma, color and taste	Lin et al. (2010)

Table 9.16 Reference amount of annual application of solid digestate in several main crops

Crops	Annual application amount of solid digestate (kg/hm^2)
Rice	22,500–37,500
Wheat	27,000
Corn	27,000
Cotton	15,000–45,000
Oilseed rape	30,000–45,000
Apple	30,000–45,000
Tomato	48,000
Cucumber	33,000

solid digestate is to directly use it as a base fertilizer, the application ways of which include cave application, banding application and surface spreading application. After application, soil should be immediately ploughed so that solid digestate and soil can fully mix, and soil should be immediately covered in order to be sowed or planted after a week. The amount depends on the different crop needs. The specific annual application amount is shown in Table 9.16.

When solid digestate is used as a base fertilizer, chemical fertilizer can be used as topdressing in jointing stage and booting stage. For dry land that lacks phosphorus and potassium, a suitable amount of phosphate fertilizer and potassium fertilizers can be supplemented (Table 9.17).

(2) Solid digestate used as topdressing

Solid digestate can also be used as topdressings, which is mainly applied to fruit trees. Each apple tree can be applied with 20–25 kg of solid digestate as top dressing (Wu et al. 2009), and each citrus can be applied with 50–100 kg of solid digestate as topdressing (Wu 2010). It can also be used as topdressings for crops and vegetables. The dosage per hectare is about 1000–1500 kg. The solid

Table 9.17 Reference amount of annual combined application of solid digestate and chemical fertilizer in several crops (one of the nitrogen fertilizers is selected)

Crop	Amount of solid digestate (kg/hm^2)	Amount of urea (kg/hm^2)	Amount of ammonium bicarbonate (kg/hm^2)
Rice	11,250–18,750	120–210	345–585
Wheat	13,500	150	420
Corn	13,500	150	420
Cotton	7500–22,500	75–240	240–705
Oilseed rape	15,000–22,500	165–240	465–705
Apple	15,000–30,000	165–330	465–945
Tomato	24,000	255	750
Cucumber	16,500	180	510

digestate can be directly applied around the roots once ditching and caving are done, and soil should cover the applied solid digestate to improve the fertilizer efficiency.

(3) Solid digestate used as fertilizers through further processing
Solid digestate can also be reused after aerobic composting. High-temperature aerobic composting does not only stabilize and improve solid digestate, but it also enhances the efficiency of utilization of organic compounds within. Solid digestate composting process requires sufficient oxygen supply. If the separated solid digestate is too wet and too thick, organic fibers (such as wood chips and straw) should be added to improve the composting environment and adjust the C/N ratio. Because of the fact that most organic matters are decomposed in the anaerobic digestion process, the temperature of composting can only reach up to 55 °C, which is lower than that of direct composting of animal manure. Composting can degrade lignin, cellulose and other organic macromolecules and remove harmful substances. At the same time, a large amount of water evaporates, increasing the nutrient concentration of the solid portion but also causing nitrogen loss.
Compost from solid digestate can be dried by drum dryers, belt dryers and feed dryers. In the drying process, however, ammonia nitrogen in solid digestate is converted to ammonia and transferred to dryer exhaust. Therefore, the exhaust gas should be treated to prevent ammonia emission. Through drying, solid digestate can achieve 70% or even 80% of dry matter content, which facilitates digestate storage and transportation. Solid digestate can be directly used after aerobic composting and can also be processed into commercial organic fertilizer through granulation and some other processes. The process of fertilizer production by solid digestate is shown in Fig. 9.1.

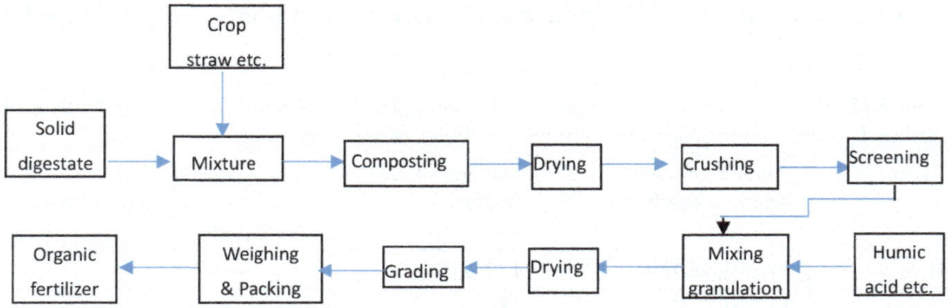

Fig. 9.1 Process flow of producing organic fertilizer through solid digestate

9.3.2 *Utilization of Solid Digestate as a Substrate*

Since most water-soluble available nutrients in digestate are separated into liquid digestate, the nutrients in solid digestate are mainly slow-release nutrients. In addition to its function as a base fertilizer, solid digestate can also work as a substrate. The pH and electrical conductivity (EC) are two of the most important indices to evaluate a substrate, for both parameters have a great influence on crop growth and quality. Compared with commonly used organic matter peat, solid digestate has a slightly higher pH that is still within the ideal range. The Electric conductivity (EC) of solid digestate also conform to the requirements of an ideal substrate, which is beneficial to the growth of vegetables (Table 9.18). However, due to the high nutrient content of solid digestate, it is inappropriate to directly use it as a cultivation substrate. Solid digestate is generally used for seedling cultivation, soilless cultivation and edible mushroom cultivation after mixing with materials such as vermiculite, perlite, and peat.

The main function of nursery substrates is to fix and support seedlings, hold water and nutrition, and provide an environment for sound growth and development of roots. Solid digestate can be compounded with other materials and directly used in seedling raising plates, or it can be mechanically pressed into a round shaped seedling nutrition block. The addition of solid digestate in substrate can promote growth and quality of seedlings, but the ratio of solid digestate should not be too high, and the ratio should depend on the content of solid digestate nutrients and types of breeding crops. The ratio of solid digestate generally should not exceed 50%. The ratio of solid digestate from chicken manure is usually low, while that from straw is high.

As is shown in Table 9.19, solid digestate is often used in vegetable seedling breeding and soilless cultivation, mainly compounded with vermiculite, perlite and plant ash. Solid digestate can be used as the substrate for tree seedling cultivation, turf cultivation and rice seedling cultivation, and solid digestate can be compounded with coconut bran and soil. However, when cultivating edible fungi, it needs to be compounded with more components such as wheat straws or other straws, cottonseed hulls, gypsum, lime and so on. When cultivating mushrooms, the mixture of cultivation ingredients should follow a 1000:300:3:12:5 ratio of solid digestate: wheat straws or other straw: cottonseed hulls: gypsum: lime. When cultivating oyster mushroom, the ratio of solid digestate to cottonseed hulls should be 6:4 or 7:3. In

Table 9.18 Comparison of pH and electrical conductivity of solid digestate and other organic matters (Zhao et al. 2005)

Sample	pH	Electric conductivity (EC) (mS/cm)
Imported peat	5.74	0.11
Huamei peat	5.1	0.24
Solid digestate	7.22	1.48
Ideal substrate	6–7.5	<2.5

Table 9.19 Ratio of solid digestate used as organic substrate

Crop	Type	Combination ratio	References
Tomato	Seedling breeding	Compound with peat, vermiculite, and perlite. Application dosage of solid digestate should not be more than 30%, and 10% of solid digestate is the best ratio	Chang et al. (2010)
Eggplant	Seedling breeding	The 1:5 ratio of substrate to solid digestate can promote e growth of eggplant seedlings and improve the quality of seedlings	Li et al. (2012)
Tomato, pepper	Soilless culture	Volume ratio of 2:3:1 for solid digestate:vermiculite:perlite can promote plant height and leaf growth of crops	Wang (2006)
Tea-oil tree	Seedling breeding	Mixed with coconut bran and yellow heart soil, the new shoot length of oil tea cuttings seeding with solid digestate content of 15% is higher than content from other formulas.	Li et al. (2012)
Birch	Seedling breeding	Ratio of solid digestate:coconut bran:vermiculite is 4:4:2	Su (2017)
Tall fescue	Soilless sod culture	Volume ratio of solid digestate: soil is 6:4	Song et al. (2015)
Rice	Seedling breeding	The ratio of solid digestate to bed soil was 1:4	Cui et al. (2015)
Enokitake	Substrate for culture	40% solid digestate, 30% bran, 29% corn cob, 1% gypsum, water content 63–65%	Yu et al. (2012)

cultivation of Ganoderma lucidum, 50% of cottonseed hulls, a small amount of corn powder, and sugar should be added to solid digestate.

9.3.3 Other Utilization of Solid Digestate

Solid digestate is rich in nutrients and amino acids, and thus it can be used as animal feed for fishkeeping, pig raising and earthworm feeding, etc. For solid digestate that cannot be used as fertilizers because of excessive heavy metal or other reasons, it can be used as industrial raw materials and an energy source.

(1) Utilization as feed
Solid digestate contains a certain amount of crude protein, crude fiber and amino acid, and thus it can be used as animal feed. Solid digestate can be used for breeding finless eel, and the nutrients in the solid digestate can be directly absorbed by finless eel, promoting the growth of plankton and reducing the cost of breeding. Solid digestate can also be used for breeding earthworm, which requires leached solid digestate to be mixed with 20% of crushed rice straw, wheat straw, and leaves.
(2) Utilization as bedding
Solid digestate contains crude fiber. Dry solid digestate, which is loose in texture, can replace sand, straw used as livestock bedding. For example, in raising cows, the sand in bedding of cows is often flushed into the cow manure, which affects the manure composting or anaerobic digestion. It is difficult to remove sand in cow manure with current technology. The adoption of solid digestate as a substitute for sand as the bedding of cows can avoid such problems.
(3) Utilization as raw materials
Some studies show that solid digestate can be used as low-grade construction products and composite materials to produce medium-density fiberboard or wood-plastic composite board (WinAndy and Cai 2008). Wuhan University of Science and Technology used solid digestate as pore forming materials, and added it to specific raw materials, achieving good plastic forming capacity. When the dosage of solid digestate was controlled at 10% and sintering temperature controlled at 1000 °C with heat preservation, a sintering wall material that meet performance index of MU10 (GB 5101-2003) and environmental protection can be made (Jian et al. 2015).
Solid digestate is rich in nutrients, suitable for the growth and proliferation of bacillus thuringiensis. Solid-state fermentation of 50% solid digestate, 35% of brewery mash, 10% of corn powder, and 5% of bean cake powder can produce Bt biological pesticide. Compared with the traditional medium, the production cost was reduced by 36.3%, and the fermentation performance was excellent (Zhang et al. 2013).
(4) Utilization as an energy source
Solid digestate that cannot be used as a fertilizer can be used as an energy source. Kratzeisen et al. (2010) conducted a combustion test on dry solid digestate from two different raw materials and found that their calorific value was equivalent to that of wood, and the generated flue gas could meet the discharge standard, proving the feasibility of using solid digestate as a fuel. Utilization of solid digestate as a fuel requires additional processing such as fiber separation, drying, and even granulation.

9.4 Utilization of Liquid Digestate

Liquid digestate has the characteristics of large volume, low value density, and high transportation and treatment cost. Application of liquid digestate on nearby fields is still the main way of utilization.

9.4.1 Direct Application of Liquid Digestate on Fields

Direct application of liquid digestate refers to direct spreading of liquid digestate on crops after simple storage. The process of applying digestate onto fields includes liquid digestate storage, transportation, application method, machines and tools, etc. Land loading capacity and the cost efficiency of storage, transportation and application process should be taken into account.

(1) Land loading capacity
Appropriate application of liquid digestate on fields is beneficial to land fertilization and soil structure improvement, but excessive application of liquid digestate in farmland will lead to soil and water pollution. Therefore, it is necessary to evaluate the loading capacity of target farmland near a biogas plant before application. To protect underground and surface water from nitrate pollution, *Nitrate Directive (91/676/EEC)* of the European Union limits nitrogen inputs to farmland with a maximum of 170 kg/(hm^2 year) and bans winter application. Requirements in some European countries on nutrient load of farmland are shown in Table 9.20.
In China, the *Technology Code for Land Application Rates of Livestock and Poultry Manure (GB/T 25246-2010)* and *Regulation Regarding the Safety Using of Livestock and Poultry Manure (NY/T1334-2007)* stipulate that the limit of returning livestock and poultry manure to field shall be based on the production needs. The yield of crops depends on land, and the amount of manure to be applied depends on yield of crops. Based on soil fertility, the expected crop yield is determined and the amount of nutrients absorbed per unit yield should be calculated. Along with the nutrient content in livestock and poultry manure

Table 9.20 Requirements of nutrient load on farmland in some countries

Country	Maximum nutrient load	Required storage capacity	Mandatory application season
Austria	170 kg/(hm^2 year)	6 months	February 28–October 25
Denmark	140 kg/(hm^2 year)	9 months	February 1–harvest
Italy	170–500 kg/(hm^2 year)	90–180 days	February 1–December 1
Sweden	Based on the breeding scale	6–10 months	February 1–December 1

Table 9.21 Limit of pig manure per stubble for wheat and rice (in terms of dry matter) unit: 10^3 kg/hm^2

Local fertility level of farmland	I	II	III
Limit in wheat and corn fields	19	16	14
Limitin rice fields	22	18	16

Table 9.22 Annual limit of pig manure in orchards (in terms of dry matter) unit: 10^3 kg/hm^2

Type of fruit tree	Apple	Pear	Orange
Limit	20	23	29

Table 9.23 Limit of pig manure per stubble in vegetable field (in terms of dry matter) unit: 10^3 kg/hm^2

Type of vegetable	Cucumber	Tomato	Eggplant	Green pepper	Cabbage
Limit	23	35	30	30	16

and the utilization rate of crops in the current year or season, the amount of livestock and poultry manure needed for base or topdressing can then be calculated. In the absence of fertilizer application, the limits (in terms of dry matter) of pig manure in wheat, rice, corn and vegetable fields are shown in Tables 9.21, 9.22 and 9.23. Digestate application amount should be calculated by converting nutrient content of liquid digestate into that of dry manure. The reference nitrogen content in pig manure given by *Regulation Regarding the Safety Using of Livestock and Poultry Manure (NY/T1334-2007)* is 1.0% (in terms of dry matter). According to this manure, nitrogen application limit for wheat and rice is 140–220 kg/(hm^2 crop), whereas for orchards it is 200–290 kg/(hm^2 year) and for vegetable fields it is 160–350 kg/(hm^2 crop). The calculated nitrogen limits are close to nutrient loads of farmlands in European countries. According to the estimation of nitrogen emission from pigs, every hectare (hm^2) of crops (wheat, rice and corn) can load manure from 2–3 pigs or their liquid digestate, and every hectare(hm^2) of orchard and vegetable field can load manure or from 2–5 pigs or their liquid digestate.

Sheng et al. (2015a, b) conducted a research on pig farms with ten thousand pigs. According to the proportion of the pig population structure, waste output, content of nitrogen and phosphorus, and nutrient loss rate in the process of waste treatment and utilization, the minimum land area for safe digestate absorption under complete anaerobic digestion was 272.5–285.4 hm^2 for grains and oil crops, 149.4–188.2 hm^2 for solanaceous vegetable, or 599.4–1 248.8 hm^2 for fruit tree seedlings. In other words, under this model, each hectare of grains and oil crop field, solanaceous vegetable field and fruit tree seedling field can load digestate equivalent to manure from 35–37, 53–67, and 8–17 pigs respectively. This amount also equals to manure from 5–6, 8–10, and 1–3 dairy cattle, or 10–12, 16–20 and 2–5 beef cattle, or 200–1700 poultry.

(2) Liquid digestate storage
Liquid digestate is produced continuously, while tillage and fertilization are carried out intermittently. Proper storage does not only solve this contradiction, but it also maintains the quality of liquid digestate as a fertilizer, prevent ammonia volatilization, methane emissions, nutrient leakage and loss, odor and aerosol emission, and reduce the environmental influence of liquid digestate.

1) Storage method of liquid digestate
In order to prevent ammonia loss and methane emission from open storage pools, liquid digestate storage in Germany must be covered to reduce nutrient loss, ammonia volatilization, methane pollution from residue, and odor release, as well as to reduce dilution of digestate from rainwater. Liquid digestate storage pool is commonly covered with sealed air bags (membrane top cover) that are made of polymer membrane materials, fixed around the storage pool, and supported by a central column. If a membrane top cover is not available, the storage pool should have at least one layer of scum or crust made of crushed straw, clay or plastic. The crust must be produced manually because liquid digestate does not form a crust layer like fresh manure. Before liquid digestate is ready for transportation or mixing, the crust layer must be kept in its original state.
Liquid digestate can be stored in biogas plants or in places adjacent to users. Liquid digestate storage pools are usually built on ground in Europe. Lagoon and storage bags are also used for liquid digestate storage. On a contrary, liquid digestate storage pools are usually built underground in China and as of now there is no mandatory requirement for cover.
2) Storage time of liquid digestate
Production of liquid digestate is a continuous process. For example, biogas plants treating livestock and poultry manure produce liquid digestate every day, but fertilization is seasonal. Therefore, liquid digestate should be stored until it can be applied. The storage capacity and storage time depend on the location of the fields, soil type, winter precipitation amount, and crop rotation systems. Some European countries restrict the season of liquid digestate application, which means that liquid digestate must be stored for 4 to 9 months (Table 9.23). In warm climates where crops grow throughout the year, the storage time can be shortened. *Technical Specifications for Pollution Treatment Projects of Livestock and Poultry Farms(HJ 497—2009)* requires that the storage period of manure should not be shorter than the maximum interval of fertilizing and the maximum freezing period in winter or the maximum rainfall period in rainy season, and generally the total storage time should not be less than 30 days.
3) Substance changes and environmental risks in liquid digestate storage process
Liquid digestate still contains a certain amount of organic matter and a large amount of soluble nutrients, which continue to produce methane and carbon dioxide in the storage process. The escape of carbon dioxide will increase the pH of liquid digestate, thus increasing NH_3 volatilization. In the process

Table 9.24 Changes in basic physical and chemical properties of liquid digestate before and after storage (Huang et al. 2012)

Time	Storage volume (L)	TN (mg/L)	NH_4^+–N (mg/L)	TP (mg/L)	COD (mg/L)	TS (%)	VS (%)	Ash content (%)
1st day	75	697	564	63.9	945	–	–	0.0955
46th day	74.5	477	398	22.8	465	0.164	0.069	0.0952

of liquid digestate storage, a certain amount of greenhouse gases CH_4 and N_2O will be produced. Volatile NH_3 will form ammonium sulfate in the atmosphere and become the core of PM (particulate matter) 2.5 particles, increasing the probability of haze. In addition, ammonia volatilization can also enhance nitrogen deposition or formation of acid rain, which affects the ecosystem.

The study of Huang et al. (2012) showed that CH_4 and NH_3 emissions was mainly found in an early stage of storage, in which the concentration of CH_4 continuously increased and reached the peak value in the first 12 days of storage, followed by a downward trend. Table 9.24 shows the changes of basic physical and chemical properties of liquid digestate after 46 days of storage compared with properties in the first day. The contents of nitrogen and phosphorus were reduced to some extent, especially ammonia nitrogen that had a significant decrease. Ammonia volatilization was the main cause. Acidifying liquid digestate can effectively reduce NH_3 emission, and it is beneficial to retaining nitrogen nutrients in liquid digestate at the same time, effectively ensuring nitrogen content in organic fertilizer while increasing the emission of CH_4 and N_2O.

(3) Transportation of liquid digestate

1) Modes of transportation

Transportation modes of liquid digestate relate closely to the yield of liquid digestate. The amount of liquid digestate produced from household digester is small; thus, it can be transported by manpower to nearby places. For some biogas plants, the large amount of liquid digestate requires large surrounding lands for using. Therefore, liquid digestate in this case is generally transported by ditches and pipelines. Pipelines are mainly used for short distance transportation, while liquid digestate truck are used for long-distance transportation.

2) Transportation distance and cost

Transportation distance affects the cost of liquid digestate application and its economic efficiency. Zeng et al. (2004) studied the economic transportation distance of manure in Fujian province and believed that the economic transportation distance of pig manure was 13.3 km. European biogas plants

(TS concentration of feedstock is above 8%) recommend a digestate transport radius of 15 km. In large-scale pig farms, washing water is generally more than 10 times of pig manure, with a TS ratio of 1%. Therefore, the economic transportation distance of liquid digestate of large-scale pig farms should not exceed 5 km.

(4) Application method of liquid digestate
According to the characteristics of liquid digestate, in practical application, it can be used as base fertilizers for food crops, vegetables, fruits, grasses, top fertilizer, and leaf surface fertilizers. In recent years, the technology of water-fertilizer integration of liquid digestate has developed.

1) Utilization of liquid digestate as base fertilizers and top dressing
Liquid digestate can be used as base fertilizers, which can be irrigated before sowing of grain crops and vegetables. Liquid digestate as top dressing can be used alone or with other fertilizers.
Liquid digestate can be applied by means of irrigation by flooding, hole fertilization, row replacement and broadcasting, and so forth. In China, foliar spraying is commonly used when liquid digestate is used as a liquid fertilizer, although this method is forbidden in many countries because of air pollution and nutrient loss (Fig. 9.2a). European countries require that in digestate application, surface air exposure should be reduced, and the digestate should

Fig. 9.2 Liquid digestate applicator (Nkoa 2014)

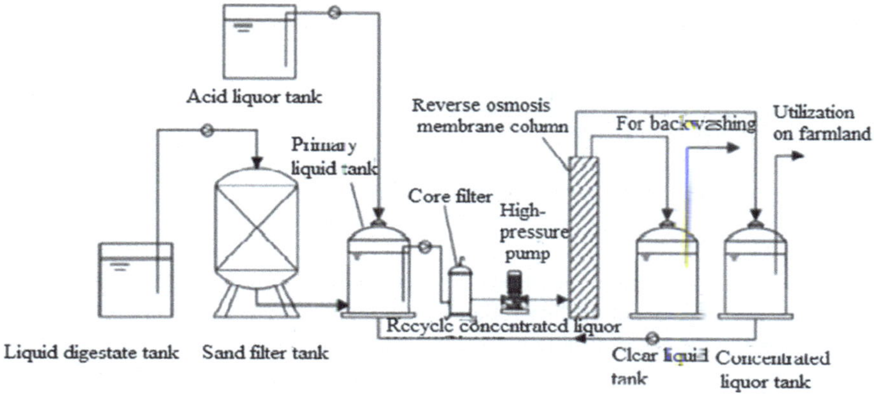

Fig. 9.3 Flow chart of membrane separation of liquid digestate (Liang et al. 2011)

permeate into soil as soon as possible. Because of these reasons, the application of liquid digestate should be conducted through trailing hoses, trailing shoes, or manure injectors. Figure 9.2b–d shows some digestate applicators which works properly and has low fertilizer losses. Once liquid digestate is applied, it should be fully mixed with soil and immediately covered with soil. After aging for one week, sowing and planting can be performed. In a combined application of solid digestate and liquid digestate, the former is applied as a one-time base fertilizer, while the liquid digestate is applied in furrow between the booting stage and the heading stage for wheats and oil crops, covered with a 10 cm-thick soil layer. When conditions permit, liquid digestate and soil can be mixed and sealed in a pit and kept for 7–10 days before application.

2) Foliar application of liquid digestate fertilizers

Liquid digestate is a kind of quick acting water-soluble fertilizer and can be used as a foliar fertilizer. Using liquid digestate for foliar fertilization has the following advantages: first, it is always on demand and easy to use; second, it has quick responses and a high utilization rate, and the blades can absorb 80% of the spray in 24 h. In addition, liquid digestate spraying does not only promote the development of plant roots and the growth of fruit and grains or increases the production of fruits, but it also reduces morbidity and the gnawing from pests (Xia et al. 2014).

Foliar application of liquid digestate usually takes the form of spraying with sprayers as the major tools, so liquid digestate should be cleared and filtered before spraying. The liquid digestate to be used should be taken from biogas plants that have digested for more than one month under normal temperature. After clearing, liquid digestate can be sprayed directly, diluted appropriately, or used after adding an appropriate amount of chemical fertilizers.

The spraying amount of liquid digestate as a foliar fertilizer should be determined according to crop and fruit tree types (Table 9.25), growth period,

Table 9.25 Application dosage of liquid digestate as foliar fertilizer

Crop	Dosage of liquid digestate	References
Spring tea	The volumetric ratio of liquid digestate to clean water is 1.5	Xiang et al. (2006)
Lettuce	10% liquid digestate	Yuan (2010)
Orange	Add a large amount of nutrients in liquid digestate to make the ratio of N:P:K to 2.5:1:2; microelements are not added	Yuan (2010)

and environmental conditions. Spraying should be carried out in mornings or evenings of sunny days rather than at noon when temperature is high, and spraying is not desirable before precipitation. When temperature is high or crops are in the seedling stage, spraying should be performed with dilution. Liquid digestate can be directly applied when temperature is low or when crops are in the middle or later growing stages. As to foliar fertilizer for fruit trees, it is advised to spray once every 7–10 days, and application should be paused one month before fruit harvesting. When spraying, ensure liquid digestate is sprayed on the back of the leaves, which facilitates absorption by crops.

3) Soaking seeds in liquid digestate
Soaking seeds in liquid digestate refers to soaking crop seeds in liquid digestate before sowing. This treatment is convenient, safe, involving no increase of investment with good effect. Therefore, it is widely applied and promoted in rural areas of China. Seeds soaked by liquid digestate can improve germination potential and germination rate while promoting seedling growth and disease resistance. The main reason is that the liquid digestate is rich in nutrients such as N, P, K and some resistance and physiological active substances, which can infiltrate seed cells in the process of soaking. Cell division and growth is therefore promoted, and nutrients required for seed germination and seedling growth are provided by liquid digestate. At the same time, this method can also eliminate pathogens, bacteria and so on carried by seeds. Therefore, after soaked in liquid digestate, seeds obtain a high germination rate, even budding, healthy seedling and root system, prosperous growth, strong stress resistance, and resistance to diseases and insects.
Liquid digestate used for seed soaking should be from anaerobic digesters that have produced biogas for more than two months. Liquid digestate should have a temperature between 10 and 35 °C and a pH value between 7.2 and 7.6. Before soaking, the seeds should be screened to remove impurities and blighted seeds. The seeds should be sun-bathed for at least 24 h. Put the seeds in a bag that can filter out water, and hang the bag in liquid digestate, followed by determination of soaking time according to liquid digestate concentration and crop seed condition. After soaking, rinse the seeds with clean water.

The recommended methods in *Technical Code for Application of Anaerobic Digestate Fertilizer (NY/T 2065-2011)* are as follows: conventional rice seeds should adopt one-time soaking, and the total length should be 48 h for early rice seeds, 36 h for intermediate rice seeds, 36 h for late rice seeds, and 6 h for japonica and glutinous rice seeds. Conventional rice seeds with poor stress resistance should be soaked in a two-fold aqueous liquid digestate dilution for 36–48 h. Hybrid rice seeds should be soaked in liquid digestate by batch method with three soaking batches with three drying, and the soaking time in liquid digestate is as follows: 42 h for early hybrid rice seeds in three 14-hour rounds, each followed by a 6-hour drying. Intermediate hybrid rice should be soaked for 36 h that consists of three 12-hour soakings, each followed by a 6-hour drying. Late hybrid rice should be soaked for 24 h that consists of three 8-hour soakings, each followed by a 6-hour drying. The soaking time of wheat seeds and corn seeds is 12 h and 4 h–6 h respectively, while that of cotton seeds is 36–48 h. After soaking, the seeds are to be rinsed off with water, and rice seed hulls should be broken to accelerate germination.

In addition to applications in grain crops, seed soaking with liquid digestate is also applied in other crops such as fruits, vegetables, grass, Chinese medicine herbs, etc. As liquid digestate comes from different sources and the biological characteristics of crop seeds differ, it is necessary to determine the appropriate soaking concentration and time.

4) Technology of water-fertilizer integration

The technology of water-fertilizer integration is a technology that combines micro irrigation and fertilization. Water is mainly used as a carrier to complete fertilization by irrigation with the help of pressure system or the height difference of terrain. This system can be designed according to soil environment, fertilizer demand from crops, and growing stage-variant water demands. Through optimization of water and fertilizer proportions, efficient use of water and fertilizer as well as pinpoint management can be achieved.

Liquid digestate is rich in water-soluble nutrients which can be used as an integrated fertilizer source that is inexpensive and easily available. However, at present, the end use of integrated water and fertilizer is mainly drip irrigation or spray irrigation. The high suspended solid content in liquid digestate can easily lead to pipeline blockage, the selection and maintenance of filtration system is therefore extremely important. Furthermore, liquid digestate is rich in magnesium, ammonium and phosphate ions, which will form struvite (magnesium ammonium phosphate) precipitation and block sprinkler heads when pH increases. Therefore, the integrated use of water and fertilizer requires reforming of end units.

(5) Safety issues in application of liquid digestate to agricultural land

The safety issues in applying liquid digestate to agricultural land mainly lie in the fact that heavy metals and antibiotics contained in livestock and poultry manure will remain in the digestate, posing a threat to soil and water and further affecting the crops.

Duan et al. (2008) sampled and analyzed soil with liquid digestate applied for 6 consecutive years. The results showed that the residual of heavy metal in soil was not obvious, but Cu and Zn content increased significantly. The amount of heavy metal and veterinary drug residue is related to application of digestate fertilizer. The detection rate of veterinary antibiotic and ciprofloxacin in soil applied with digestate fertilizer were higher than that of the control group without fertilizer application. Studies have also shown that excessive application of liquid digestate can cause accumulation of heavy metals in soil (Chen 2010). For crops, the content of heavy metals in rice treated by liquid digestate was not significantly different from that in the control group (Zhao et al. 2012; Jiang et al. 2011). When the amount of liquid digestate was large, heavy metal content in Chinese cabbage was significantly increased compared with that in soil with the low liquid digestate application, but the increased content was still within the standard (Huang et al. 2012). Compared with chemical fertilizers, liquid digestate has a lower content of heavy metals, but the required amount of liquid digestate is often larger than that of chemical fertilizer, so heavy metal content of soil-crop system with liquid digestate application is slightly higher. A suitable amount of liquid digestate irrigation does not introduce significant impacts on crop safety, but a risk of heavy metal accumulation in crops still exists under high-intensity irrigation.

The risk of using liquid digestate as a foliar fertilizer lies in heavy metal residues. Xia et al. (2014) tested some samples of kidney beans that were applied with liquid digestate. Results showed that the ranges of Zn, Cu, Cr, Cd content were 31.93–38.88, 5.56–8.29, 0.64–1.29 and 0.09–0.15 mg/kg. Pb was not detected, and the overall heavy metal content did not exceed the maximum requirements of *Maximum levels of Contaminants in Foods (GB 2762-2005).*

Another important potential threat in application of liquid digestate is that substances such as nitrogen and phosphorus in liquid digestate may affect the water environment. Once liquid digestate is applied to the field, ammonia nitrogen and phosphate transfer through water in soil, which poses potential pollution risks to surrounding surface water and groundwater. After liquid digestate application, nitrogen, phosphorus, and many other substances that are not absorbed by crops may enter surface water through surface runoff, resulting in eutrophication. Ammonia nitrogen in liquid digestate will be converted into nitrate nitrogen under nitrifying bacteria in soil. Nitrate nitrogen will then flow into the lower water level along with leachate in soil, causing groundwater pollution. Studies have shown that ammonia nitrogen content increases in paddy water where liquid digestate is applied for three consecutive years (the dosage is less than 540 kg/hm^2). However, the nitrogen content in soil leachate changes only a little (Jiang et al. 2011). Li et al. (2007) found that the contents of total nitrogen, ammonia nitrogen and nitrate nitrogen in leachate were all within the safe range when the mass fraction of liquid digestate did not exceed 50% in the potted plant experiment with hybrid pennisetum grasses.

In conclusion, the application of liquid digestate will pose a certain potential threat to soils, water environment, and food safety. However, with appropriate irrigation and proper spraying intensity, the degree of threat is limited, and the risk can be controlled within standard requirements through reasonable utilization and strict supervision.

9.4.2 High Value Utilization of Liquid Digestate

High value utilization of liquid digestate is utilization by way of technologies such as filtration, compounding, complexation, and membrane separation to concentrate nutrients in liquid digestate, reduce volume, and improve the content of humus and inorganic nutrients in liquid fertilizers. As a result, commercial fertilizers can be produced based on different needs of crops.

(1) Utilization of liquid digestate as nutrient solution of soilless cultivation
Soilless cultivation is a kind of crop cultivation method in which the artificial root systems replace soil environment, and the artificial root systems can be adjusted and controlled to meet the needs of plant growth. It has advantages such as a high yield, good qualities, and no pollution, while saving water, fertilizer, and land occupation with no restriction on regions. At present, soilless cultivation mainly uses chemical synthetic solutions as nutrient solutions, and the preparation process is complex, which is mainly used in facility agriculture. Liquid digestate contains all the elements required for plant growth, and it can be used as specialized nutrient solution after dilution or treatment (Table 9.26). Soilless cultivation by liquid digestate can break the traditional concept that soilless cultivation must be realized with chemical nutrient solutions, and it has overcome the demerits of traditional soilless cultivation by chemical nutrient solutions. Moreover, soilless cultivation by liquid digestate retains the advantages of no geographical restrictions, achievable continuous cropping, effective prevention of underground pests and diseases, fertilizer economy, water economy, and high crop yield. Some studies have shown that liquid digestate does not cause excessive accumulation of heavy metals in aboveground parts of a plant, so it is feasible to carry out pollution-free crop production. Compared with the cultivation using standard inorganic nutrient solution, soilless cultivation of tomato, cucumber, celery, lettuce and eggplant using liquid digestate has significantly improved the product quality, especially in vitamin C content, soluble solid content, soluble sugar content, and decreases in nitrate content, and the yield is also improved (Wang et al. 2009; Chen et al. 2011a, b). This largely reveals the advantages of using liquid digestate as nutrient solution for hydroponic culture of plant. Soilless cultivation facilities of liquid digestate is simple and convenient by utilization of local materials and ingredients. Therefore, soilless cultivation can be promoted widely as a follow-up industry of biogas plants.

Table 9.26 Comparison of nutrients in liquid digestate and special nutrient solution of soilless cultivation (Zhou et al. 2013) unit: mg/L

Nutrient	Special nutrient solution	Original liquid digestate	Five times diluted liquid digestate
Nitrate nitrogen	189	0	0
Ammonia nitrogen	7.0	984	197
Phosphorus	45	247	49
Potassium	360	500	100
Calcium	186	590	118
Magnesium	43	161	35
Sulfur	120	68	13
Manganese	0.55	11	2.2
Zinc	0.33	1.2	0.24
Boron	0.27	0.91	0.18
Molybdenum	0.048	0.2	0.04
Copper	0.05	0.19	0.04
Iron	0.88	4.5	0.9

Liquid digestate with high levels of ammonia nitrogen and salinity does not facilitate plant growth. Therefore, such liquid digestate usually needs to be diluted prior to soilless cultivation. Before used as a nutrient solution, liquid digestate also needs to be matured and stabilized for 30 days in a storage pool. After precipitation and filtration, the liquid digestate can finally be used as a nutrient solution for soilless cultivation through dilution of 1:4 to 1:8 ratio according to nutritional requirements of different crops. The conductivity should be controlled between 2.0 and 4.0 ms/cm, and the pH of liquid digestate can be adjusted to 5.8–6.5 by using nitric acid or phosphoric acid. According to the needs of different crops, micronutrients can be added. In cultivation, liquid digestate needs to be added or replaced regularly. Adding chelate iron to maintain certain iron concentration in nutrient solution can effectively solve the deficiency of iron in liquid digestate (Zhou et al. 2013).

(2) Concentration of liquid digestate to produce fertilizer

Compared to discharge wastewater, liquid digestate has excessively high concentration of nitrogen and phosphorus, whereas a relatively low concentration of nutrients compared with fertilizers. As a result, the unit transportation cost of nutrients is rather high, and the major utilization of liquid digestate from livestock and poultry wastes is mainly limited uses in nearby farmlands. With the intensive and scaled development of livestock and poultry breeding industry, the contradiction between the consumption capacity of surrounding lands and liquid digestate output is becoming increasingly prominent. Common measures to solve this contradiction include negative pressure evaporation concentration, membrane concentration, etc.

Negative pressure evaporation concentration is a mature and effective liquid concentration technology that has simple operations, low requirements for environmental conditions and high tolerance for suspended solids. Applying this technology in liquid digestate concentration can effectively prevent the loss of effective components while achieving concentration. At a temperature of about 50–80 °C, four times of concentration can be realized. The concentrate contains certain nutrients and low toxicity to plants. Therefore, the concentrate can be used as a fertilizer, while the condensate can be discharged or recycled if national standards are met (Table 9.27), but the energy consumption is notable in evaporation concentration processes.

Membrane concentration has the characteristics of simple process, easy operation and no change of composition. Membrane concentration can be categorized into nanofiltration, osmosis, and reverse osmosis separations. Nanofiltration and reverse osmosis require high pressure, while osmosis requires sodium chloride and magnesium chloride as the draw solution. Since liquid digestate contains relatively high concentration of suspended matte that may block the membrane, a pretreatment is required before concentration by nanofiltration and reverse osmosis. Pretreatment methods include sand filtration, microfiltration, ultrafiltration and flocculation. Concentration ratio of liquid digestate by membrane separation is 3–5 times (Table 9.27).

Reverse osmosis membrane separation is most widely used for membrane concentration processes. Common membrane separation devices include the tube type, the hollow fiber type, the coil type and the plate frame type. Among them, the disk tube reverse osmosis (DTRO) process has the strongest pollution resistance and is used as a treatment process of landfill leachate, which is suitable for liquid digestate concentration. The concentration experiment of liquid digestate from chicken manure is shown in Fig. 9.3, and the concentration effect is shown in Table 9.28. COD, SS, total nitrogen, total phosphorus and ammonia nitrogen were all decently concentrated. The COD content and total phosphorus in effluent were relatively low, but the total nitrogen and ammonia nitrogen content were high.

At present, membrane separation technology has been applied in liquid digestate deriving from chicken wastes which have relatively high nutrient content. For the low nutrient liquid digestate coming from pig farm and cattle farm, further concentration processing and higher production costs are required to achieve an ideal nutrient content.

Numerous tests have demonstrated that membrane separation technology in liquid digestate concentration is feasible with satisfactory results. But there are still some problems limiting the application of this technology.

(1) Operation cost

The price of membrane component itself is high. During operation, high pressure is required for permeation through the membrane, which requires high energy consumption and operation costs. However, with the development of membrane technology and optimization of membrane concentration process,

Table 9.27 Technologies for concentrating liquid digestate for producing fertilizer

Concentration method	Concentration ratio	Concentration effect	References
Negative pressure evaporation concentration, 50 °C and 2 kPa	4 times	The ammonia nitrogen removal efficiency was 86.41%, and the plant physiological toxicity of the concentrate was lower	He et al. (2016a, b)
Osmosis, 2 mol/L NaCl as draw solution	5 times	The concentration of soluble organic matters, TP, TN, ammonia nitrogen, total potassium has been markedly increased, retention rate reaching over 80%	Xu et al. (2016a, b)
Osmosis, 2 mol/L NaCl as draw solution	5 times	The recovery efficiency of COD, humic acid and amino acid was higher than 99.5%	Lu et al. (2016)
Reverse osmosis	3–4 times	The removal efficiency of ammonia nitrogen, COD and conductivity in the permeable liquid was up to over 90%	Liang et al. (2011)
Tubular microfiltration—dish-tube reverse osmosis	3–5 times	The removal efficiency of COD, ammonia nitrogen and conductance was above 93%	Wang (2015)
Precision filter-tube reverse osmosis	4.3 times	The removal efficiency of COD, NH_3–N and TP all exceeded 90%, and the desalination efficiency was 91.2%	Zhou et al. (2016)

the concentration cost of liquid digestate becomes increasingly accepted by many farms.

(2) Pollution

Liquid digestate contains a large number of organic compounds, suspended particles, and some colloidal particles, which can cause membrane pollution

Table 9.28 Concentration effect of liquid digestate from chicken manure (Liang et al. 2011)

Characteristic	Conductivity (mS/cm)	COD (mg/L)	SS (mg/L)	TN (mg/L)	Ammonia nitrogen (mg/L)	TP (mg/L)
Liquid digestate	22.1	7467	7240	3600	3347	120
Permeate	0.778	59.1	5.0	228	197	5.2
Concentrate	66.7	26208	24145	12550	10339	420.5

and shorten the normal operation time of system. A pretreatment is a common method to overcome these problems, but it involves a cost increase. Livestock and poultry waste contains heavy metals, which enter the liquid digestate together with feces and urine. Membrane separation technology does not only concentrate nutrients in liquid digestate, but it also concentrates heavy metal elements, increasing the risk of using concentrated solutions as fertilizers. However, some research results show that the heavy metal content of concentrated solutions is within the allowable range of relevant national standards.

In addition, ammonium magnesium phosphate (struvite) precipitation technology and ion exchange resin can be used to recover nitrogen, phosphorus, humic acid and other substances in liquid digestate, but it is difficult to be widely spread in a large scale due to high cost.

9.4.3 *Application of Liquid Digestate in Aquaculture*

Liquid digestate is rich in nutrients, which can also be used in aquaculture. On one hand, organic matters in liquid digestate can be directly used as bait for fish and shrimps. On the other hand, nitrogen and phosphorus in liquid digestate can promote the growth of algae which provide bait for fish and shrimps.

(1) Liquid digestate for fish farming

Using liquid digestate for fish cultivation is also a common way of resource utilization. Liquid digestate can be used for the cultivation of grass carp, herring, mantis, carp, abalone and bat fish. Liquid digestate can provide nutrition for animals and plants in water, increase the output of animals and plants in ponds, and enrich bait for filtering-feeding fish. Thereby the use of chemical fertilizers such as urea will be reduced. It can also avoid fish diseases and losses caused by parasite ovum and germs in fresh livestock and poultry wastes. By using liquid digestate, the survival rate of fry can increase by more than 10%, and the yield can increase by more than 27%.

Liquid digestate can be used as fishpond base fertilizer and topdressing. Liquid digestate, when used as the base fertilizer of fishpond, should be applied after disinfection of fishpond and before fish seeds are placed into the pond. 3000–4500 kg of liquid digestate could be applied for each hectare of water surface, but in general it should not exceed 4500 kg. The application amount of topdressing is generally 1000–3000 kg/hm^2 (Jing et al. 2016). In addition to fish farming, liquid digestate can also be used for other aquaculture products, such as mussels.

(2) Liquid digestate for algae cultivation

Algae are a kind of low-grade plant in water. The nitrogen and phosphorus in liquid digestate can be utilized by algae. Algae can greatly reduce the content of nitrogen and phosphorus in liquid digestate and remove CO_2 in biogas. Using liquid digestate for growing algae does not only purify liquid digestate, but it also obtains high density and high quality algal bodies which can be used to produce biodiesel or feed protein. Compared with applying liquid digestate to farmland, the land area required for the cultivation of microalgae is markedly smaller (Xia and Murphy 2016).

Liquid digestate is rich in nutrients, but it is not always beneficial to the growth of algae. For example, liquid digestate is generally brown and dark, containing a lot of suspended matter. Its turbidity is high and is not desirable for light transmission. The content of ammonia nitrogen in liquid digestate is high, which will hurt the growth of algae. In an open environment, liquid digestate is easy to breed miscellaneous bacteria and algae, as well as their predators such as rotifers and algophages. In the process of liquid digestate treatment by microalgae, microalgae cells are easily affected by pests. Therefore, a pretreatment for liquid digestate is very important in the whole process of microalgae cultivation. The main purpose of the pretreatment is to reduce turbidity, adjust the nutrition structure, and prevent insect pests.

Using liquid digestate to cultivate microalgae can realize resource reutilization. Three main components can be extracted from microalgae biomass: lipids (including tri-acylglycerol and fatty acids), carbohydrates and proteins. Fats and carbohydrates are the sources of bioenergy (such as biodiesel, bioethanol, etc.). Proteins can be used as a feed for animal and fish.

At present, the harvesting, drying and separation of target products of microalgae biomass are highly energy intensive, which remains to be the urgent problems to be solved in algae resource utilization. Microalgae can accumulate a large amount of protein if there is high nitrogen content in liquid digestate. Protein as feed does not need extraction and is a more preferable way at the present stage. In addition, it would be ideal to grow algae that can produce high value-added products. For example, astaxanthin can be produced by the cultivation of haematococcus pluvialis, and spirulina can be used for producing healthcare products.

References

Cai, A.X., Q. Jiang, and Y.C. Chang. 1999. Study on the effect of biogas fertilizer on alkali soil improvement and yield increase. *Chinese Journal of Soil Science* 1: 4–6. (in Chinese).

Cao, R.K., H. Chen, and Y.Z. Zhao. 2015. Resource utilization of biogas slurry: current status and future prospects. *China Biogas* 33 (2): 42–50. (in Chinese).

Chang, P., Y. Zhang, Y.M. Li, X.Q. Zhang, and Y.H. Qu. 2010. Influence of biogas residue of artificial substrates on tomato seedling. *Northern Horticulture* 15: 134–137. (in Chinese).

Chen, Z.G. 2010. *Studies on the Effects of Vegetable Yield and Safety and Soil Security Capacity After Application of Biogas Fertilizer*. Shanghai Jiao Tong University, 2010. (in Chinese).

Chen, W. 2016. Effect of biogas fertilizer on rice output. *Horticulture & Seed* 4: 57–58. (in Chinese).

Chen, M., and Y.S. Cui. 2012. A review on the resource and bioavailability of heavy metals in biogas fertilizer from the manure of livestock. *Chinese Journal of Soil Science* 43 (01): 249–256 (in Chinese).

Chen, Y.X., H.M. Dong, X.P. Tao, and B. Shang. 2011a. Effect of irrigating winter wheat with anaerobic digested swine farm wastewater (ADSFW) on soil quality. *Chinese Agricultural Science Bulletin* 27 (3): 154–158. (in Chinese).

Chen, Y.X., B. Shang, X.P. Tao, and H.M. Dong. 2011b. Study on effects of biogas slurry from swine farms on tomato quality in soilless cultivation. *Chinese Agricultural Science Bulletin* 27 (19): 172–175. (in Chinese).

Chen, W., H.Y. Meng, and Y.J. Wang. 2014. Study on nutrient contents and security of biogas residue and fluid. *Journal of Anhui Agricultural Sciences* 23: 7960–7962. (in Chinese).

Cordovil, C.M.D.S., A.D. Varennes, R. Pinto, and R.C. Fernandes. 2011. Changes in mineral nitrogen, soil organic matter fractions and microbial community level physiological profiles after application of digested pig slurry and compost from municipal organic wastes to burned soils. *Soil Biology & Biochemistry* 43 (4): 845–852.

Cui, Y.R., Y.M. Zhao, J. Jie, S.N. Zhao, and F.X. Pang. 2015. Effects of the formula of seedling substrate based on biogas slurry on the physiological indexes of rice. *Shanxi Agricultural Economy* 9: 52–54. (in Chinese).

Deng, L.W. 2015. *Biogas Engineering*. Beijing: Science Press.

Deng, L.W., W.G. Wang, and D. Zhen. 2017. *Treatment and utilization of swine wastewater: Principle and technology*. Beijing: Science Press.

Dong, Z.X. 2015. Study on the utilization of different organic wastes as fertilizer. Doctoral dissertation of Shanxi Agricultural University. (in Chinese).

Dong, Z.X., F.W. Liu, Z. Xu, X.Z. Liu, W.P. Zhang, and Y.S. Bu. 2014. Effect of organic-inorganic fertilization blended by biogas residue on the growth and quality of Pak-choi. *Northern Horticulture* 22: 1–5. (in Chinese).

Dong, Y.Y., X.J. Nie, Q. Wang, B. Ye, Y.X. Liu, Z.B. Fan, and J. Jing 2017. Nutrient difference analysis of biogas slurry from different scale pig farm. *Journal of Zhejiang Agricultural Sciences* 58 (12): 2089–2092. (in Chinese).

Duan, R., G. Wang, S.Q. Yang, and Z.L. Yang. 2008. Reliminary Research of Potential Pollution on Farmland Soil after Using Biogas. *Journal of Jilin Agricultural University* 30 (3): 310–315. (in Chinese).

Duan, L.J., J.G. Cao, F. Xiong, and Z.Z. Yang. 2015. Experimental Study of Chicken Manure Fermentation Liquid Treatment by Struvite Precipitation. *Environmental Engineering* 33 (07): 66–71. (in Chinese).

Gan, FuD, S.Q. Wei, W.N. Tan, J.H. Li, and H.B. Jiang. 2011. Effect of biogas slurry on tabe bean quality and soil fertility. *China Biogas* 29 (1): 59–60. (in Chinese).

Ge, Z., Y.S. Wei, J.W. Liu, and Y.Z. Zhao. 2014. Characteristics of digestate and utilization: An overview. *China Biogas* 32 (03): 74–82. (in Chinese).

Gu, H.J., Y.X. Zhao, and M.Z. Yu. 2016. Application of biogas slurry from pig farm in ornamental fish pond. *Journal of Liaoning Agricultural Technical College* 18 (4): 3–4. (in Chinese).

Han, M., C.H. Feng, K.F. Liu, Y.Q. Fu, L.Q. An, S.L. Wang, and C.D. Gao. 2014. Application of coagulants on cattle biogas slurry. *Tianjin Agricultural Sciences* 20 (09): 96–101. (in Chinese).

He, N.N., Y.X. Guan, Y.H. Liang, H. Wu, Z.H. Luo, H. Zhao, and Z.L. Gu. 2016a. Simultaneous determination of four fluoroquinolones in biogas slurry by high performance liquid chromatography with fluorescence detection. *Journal of Agro-Environment Science* 35 (10): 2034–2040. (in Chinese).

He, Q.R., W.C. Wang, K. Cai, S.P. Yan, and Y.L. Zhang. 2016b. Effect of vacuum concentration on CO_2 absorption performance and phytotoxicity of biogas slurry. *Transactions of the Chinese Society for Agricultural Machinery* 47 (2): 200–207. (in Chinese).

He, N.N., Y.X. Guan, Y. Liang, T.X. Wu, H.Y. Zhao, Z.H. Luo, Z.L. Gu, and H. Wu. 2017. Simultaneous determination of 3 TCs and 6 SAs in biogas slurry by solid phase extraction and high performance liquid chromatography. *Journal of Analytical Science* 33 (3): 373–377. (in Chinese).

Huang, D.D., H.J. Luo, H.C. Ying, and K.Y. Wang. 2012. Methane and ammonia emissions patterns during digested slurry storage. *Transactions of the Chinese Society for Agricultural Machinery* 43 (S1): 190–193. (in Chinese).

Huo, C.Y., S.B. Wu, J. Guo, H.C. Cheng, H. Reng, R.J. Dong, and C.L. Pang. 2011. Analysis of phytohormone and qunolin-ketone components in anaerobic digestion effluent. *China Biogas* 29 (5): 7–10. (in Chinese).

Jian, S.W., G.H. He, B.G. Ma, Y. Zhang, M.Q. Sun, Z.Z. Zhi. 2015. Experimental analysis of biogas sludge as admixture of porous sintered wall material. *Journal of Agro-Environment Science* 34 (11): 2222–2227. (in Chinese).

Jiang, L.N., Q. Wang, D.J. Chen, A.F. Li, S.M. Qian, J.R. Fu, and J.M. Wang. 2011. Effects of paddy field disposal of biogas slurry on the rice production, soil quality and environmental safety. *Journal of Agro-Environment Science* 30 (7): 1328–1336. (in Chinese).

Jin, Y.K. 1987. Study on soil structure improvement of biogas slurry. *Chinese Journal of Soil Science* 3: 118–120. (in Chinese).

Jing, D.D., Y.L. Chen, C. Dai, P.P. Li. 2016. Research situation and prospect of biogas slurry utilization in fish culture. *Hubei Agricultural Sciences*, 55(22): 5886–5890, 5906. (in Chinese).

Jin, H.M., Z.Z. Chang, X.M. Ye, Y. Ma, and J. Zhu. 2011. Physical and chemical characteristics of anaerobically digested slurry from large-scale biogas project in Jiangsu Province. *Transactions of the Chinese Society of Agricultural Engineering* 27 (1): 291–296. (in Chinese).

Kratzeisen, M., N. Starcevic, M. Martinov, C. Maurer, and J. Müller. 2010. Applicability of biogas digestate as solid fuel. *Fuel* 89 (9): 2544–2548.

Li, X. 2016. Dynamic changes in plant nutrition's of liquid digestate and metabolism pathway of indoleacetic acid. Doctoral dissertation of China Agricultural University.

Li, Q., C.R. Yang, and C.Y. Zhang. 1992. Effect of digester residue on the improvement of soil fertility and the yield of rice and wheat. *China Biogas* 1: 13–18. (in Chinese).

Li, Y.C., X.D. Niao, and Y.B. Wu. 2007. Effects of slurry application on biomass of crossbred pennisetum and soil nutrient. *Journal of Agro-Environment Science* 26 (4): 1527–1531. (in Chinese).

Li, A.F., J. Li, X.W. Zhang, and X.C. Gong. 2011. Application techniques of biogas slurry in production of machine-transplanted single cropping late rice. *Acta Agriculturae Zhejiangensis* 23 (2): 382–387. (in Chinese).

Li, Y., L.B. Xie, J.G. Yao, S.Z. Fan, and J.F. Li. 2012. Effects of different substrate with biogas residue proportion on seedlings of eggplant. *Northern Horticulture* 3: 28–30. (in Chinese).

Li, Y.R., Y.X. Liu, C.Q. Sun, Z.L. Peng, Z.Y. Zhao, L.J. Li, Y.B. Yue, Q.Q. Tong, B. Nantakorn, and W. Sodchol. 2013. Organic hydroponics using anaerobic fermentation liquid residue of animal manures from biogas generation. *Southwest China Journal of Agricultural Sciences* 26 (6): 2422–2429. (in Chinese).

Li, H.N., Z.W. Shi, and C.X. Zhu. 2014a. Concentration of biogas slurry with forward osmosis technology. *Transactions of the Chinese Society of Agricultural Engineering* 24: 240–245. (in Chinese).

Li, T., H.M. Dong, and X.P. Tao. 2014b. Disinfection effects of ultraviolet on supernatants of flocculated digestate from swine farm. *Transactions of the Chinese Society of Agricultural Engineering* 30 (6): 165–171. (in Chinese).

Li, W.Y., X.N. Wu, X.Q. He, X.M. Shao, Z.P. Peng, and S.S. He. 2014c. Optimization of process combination on purification treatment of intensive piggery biogas slurry. *Journal of Agriculture* 4 (8): 85–87. (in Chinese).

Li, Y.Q., K. Sheng, S.J. Peng, Z.W. Meng, and Z.R. Dong. 2014d. Effects of biogas slurry on wheat yield and the physical and chemical properties of soil. *Chinese Agricultural Science Bulletin* 30 (12): 181–186. (in Chinese).

Li, Y.F., J.X. Xu, Q.P. Sun, B.S. Liu, J. Li, and J.J. Li. 2017. Effects of biogas residue application on soil nematode community structure. *Journal of China Agricultural University* 22 (8): 64–73. (in Chinese).

Liang, K.Q., Z. Yan, M. Zhu, X.W. He, Y.Q. Wei, K.J. Wang. 2011. Application research of reverse osmosis in concentrating biogas slurry from biogas projects. Journal of China University of Mining & Technology, 40(3): 470–475. (in Chinese).

Lin, B., G.H. Luo, Q.X. Xu, Q.S. Wang, and X.F. Guan. 2010. Effects of biogas residue on yield and quality of tea. *Fujian Journal of Agricultural Sciences* 25 (1): 90–95. (in Chinese).

Lin, B.S., Y.Q. He, Y.X. Huang, and F.Y. Qiu. 2015. Screening of compound bacteria for anaerobic fermentation with piggery biogas slurry and study the treatment effect of compound bacteria static test. *Journal of Northeast Agricultural University* 1: 74–78. (in Chinese).

Liu, D.C. 2013. New terpenoid compounds and biological activity from biogas slurry. Doctoral dissertation of Shanghai Ocean University. (in Chinese).

Liu, S.C., L.W. Wang, X.X. Li, Y.C. Chen, and M.M. Fu. 2014. Potential risk assessment of heavy metal in biogas slurry irrigation. *Journal of Plant Nutrition and Fertilizers* 20 (6): 1517–1524. (in Chinese).

Liu, Y.X., X. Nie, Y.Y. Dong, Q. Wang, B. Ye, Z.B. Fan, and J. Jing. 2017. Nutrient analysis of rural large-scale biogas digester under dry manure cleaning process. *Journal of Zhejiang Agricultural Sciences* 58 (11): 1997–2000. (in Chinese).

Liu, F., Z.B. Li, Q.R. Li, and Z.G. Wang. 2009. Effects of combined application of biogas digestate and chemical fertilizer on the yield and quality of sweet maize. *Chinese Journal of Soil Science* 40 (6): 1333–1336. (in Chinese).

Lu, X.F., F. Ma, H.D. Wang, S.W. Wang, S. Qiu, N.B. Zhao, and G. Zhao. 2016. Characteristics of forward osmosis technology for biogas slurry concentrating and its efficiency. *China Biogas* 34 (1): 62–67. (in Chinese).

Ma, Y., H. Li, Z.Z. Chang, Y.D. Xu, and J.Y. Zhang. 2011. Biocontrol effect and mechanism of biogas slurry on plant disease I: Primary study of growth inhibition effects and mechanism on phytopathogen fungi. *Journal of Agro-Environment Science* 30 (2): 366–374. (in Chinese).

Meng, Q.G., F.L. Zhao, Y.G. Zhang, S.Q. Wang, D.Y. Zhao, and T.H. Zhang. 2000. The determination of seven free amino acids in liquid from a methane generator by gas-liquid chromatography using pre-column derivation. *Journal of Agro-Environment Science* 19 (2): 104–105. (in Chinese).

Min, S.D. 1990. Study on the peroral infection of epidemic hemorrhagic fever virus under experimental condition. *Journal of Microbiology* (z1):86–87. (in Chinese).

Ni, L., G.H. Sun, G.G. Luo, W.Y. Shi, H. Lu, and W.Z. Ye. 2008. Effect of irrigation of biogas slurry on soil quality. *Soil* 40 (4): 608–611. (in Chinese).

Nie, X.J., J. Jin, Q. Wang, B. Ye, Y.X. Liu, and Z.B. Fan. 2017. Nutrient analysis of biogas residue from zhejiang rural large-scale biogas project. *Journal of Zhejiang Agricultural Sciences* 58 (12): 2111–2114. (in Chinese).

Nkoa, R. 2014. Agricultural benefits and environmental risks of soil fertilization with anaerobic digestates: a review. *Agronomy for Sustainable Development* 34 (2): 473–492.

Pan, Y.B. 2009. Investigation report on the harmless recycling of pollutants in large-scale pig farms in Gansu Province. *China Animal Health* 11: 99–101. (in Chinese).

Qian, J.H., C. Lin, J.H. Wang, and X.L. Wang. 2005. Effects of biogas residue on apples' quality and soil fertility. *Renewable Energy Resources* 4: 34–36. (in Chinese).

Qiao, W. 2015. Feedstock characteristics and fermentation efficiency evaluation of agricultural biogas engineering [A]. Chinese society of biogas, guangzhou institute of energy research, Chinese academy of sciences, German agricultural association, 2015 annual conference of Chinese society of biogas and BBS proceedings of sino-german biogas cooperation, Chinese society of biogas, guangzhou institute of energy research, Chinese academy of sciences, German agricultural association. (in Chinese).

Qin, F.J., L. Qi, F. Wang, B. Wang, X.L. Zhang, Y.J. Mao, and W. Jiang. 2015. Nutrient content analysis of three different fermentation materials in biogas slurry. *Journal of Zhejiang Agricultural Sciences* 56 (07): 1097–1099. (in Chinese).

Qu, M.S., N. Guo, Z.F. Liu, Y.F. Wang, G.N. Zhao, M. Jing, and Q. Chen. 2013. Nutrient and heavy metal contents in biogas slurry of large scaled biogas project in suburb of Beijing. *China Biogas* 31 (04): 37–40. (in Chinese).

Shang, C.F., L.K. Wang, Q.M. Chen, S.Z. Zhang, C.L. Huang, and W.G. Liu. 2009. Effect of fermentation time and temperature on amino acid content in biogas slurry. *Animal Husbandry & Veterinary Medicine* 41 (11): 42–44. (in Chinese).

Shen, Q.L., S.D. Shan, J.J. Zhou, and Z.R. Wang. 2014. Determination and analysis of compositions in biogas slurry produced by swine manure digestion. *China Biogas* 32 (3): 83–86. (in Chinese).

Sheng, J., G.F. Sun, and J.C. Zhen. 2015a. Pig farm-cropland configuration under typical waste treatment modes—A case study of anaerobic liquid fermentation following solid-liquid separation of wast. *Chinese Journal of Eco-Agriculture* 23 (2): 199–206. (in Chinese).

Sheng, J., G.F. Sun, and J.C. Zhen. 2015b. Pig farm-cropland configuration under typical waste treatment mode——A case study of direct anaerobic fermentation of manure. *Chinese Journal of Eco-Agriculture* 23 (7): 886–891. (in Chinese).

Shi, J.W, G.M. Lei, Y.Y. Li, G. Wang, G.Q. Zhao, X. Ma, G. Feng, Y.M. Shang. 2013. Effects of fermentation raw material and process on volatile organic acids in biogas slurry. *Journal of Henan Agricultural Sciences* 3: 55–58. (in Chinese).

Shi, H.J., S.S. Ma, K.L. Zhao, L.Q. Ye, H. Li, Y. Shen, W.M. Zhao, and Z.Q. Ye. 2017. Effect of organic materials on improvement of Carya cathayensis forest acidic soil. *Journal of Zhejiang A & F University* 34 (4): 670–678. (in Chinese).

Singh, M., D.L. Reynolds, and K.C. Das. 2011. Microalgal system for treatment of effluent from poultry litter anaerobic digestion. *Bioresource Technology* 102 (23): 10841.

Song, C.J., Y.S. Tian, J. Luo, Y. Qi, L.X. Zhao, X.L. Zhao, X. Tian. 2015. Ecological characteristics of *Festuca arundinacea* L. turf cultivated by solid residues from anaerobic digestion project. *Transactions of the Chinese Society of Agricultural Engineering* 31(17): 254–260. (in Chinese).

Song, C.F., S.D. Shan, M.X. Zhang, and X.H. Wen. 2011. Concentration and determination of composition of biogas slurry. *Transactions of the Chinese Society of Agricultural Engineering* 27 (12): 256–259. (in Chinese).

Su, T. 2017. Study on substrate formula of container seeding on betula platyphylla suk. *The Journal of Hebei Forestry Science and Technology* 3: 6–7. (in Chinese).

Sun, H.X., H.J. Zhang, Y.B. Xu, and S.S. Zhang. 2017. Determination and analysis of heavy metal elements in pig feed, excrement, biogas residue and biogas slurry. *Heilongjiang Animal Science and Veterinary Medicine* 09: 285–287. (in Chinese).

Tambone, F., V. Orzi, G. D'Imporzano, and F. Adani. 2017. Solid and liquid fractionation of digestate: mass balance, chemical characterization, and agronomic and environmental value. *Bioresource Technology* 243: 1251–1256.

Tang, W., J. Wu, B.Y. Sun, G. Yang, and Q. Yang. 2010. Effects of application amounts of biogas slurry on yield and quality of rice. *Journal of Agro-Environment Science* 29 (12): 2268–2273. (In Chinese).

Tao, H.G., X.B. Li, and T.L. Zhao. 2003. Bioliquid, biosolid and eco-agriculture. *Renewable Energy Resources* 2: 37–38. (in Chinese).

Tatsuya, 2009. *Methane Fermentation*. Ltd: Gihodo Shuppan Co.

Wan, J.B., H.Y. He, Y.M. Wu, P. Gu, and W.Y. Zhang. 2010. Study on nitrogen removal from pig farm biogas slurry in pilot-scale sequencing batch biofilm reactor. *China Water & Wastewater* 26 (21): 127–129. (in Chinese).

Wang, X.J. 2006. Research on properties of organic substrate in soilless culture based on biogas residue. Doctoral dissertation of Huazhong Agricultural University. (in Chinese).

Wang, J.J., J. Liu, R.P. Zhang, B.S. Yu, J.R. Li, S. Wang, B. Sun. 2012a. Effect of matching application of biogas residues and chemical fertilizers on rice growth, yield and quality. *Heilongjiang Agricultural Sciences* 4: 66–70. (in Chinese).

Wang, L.J. 2015. Study on concentrate treatment process of biogas slurry by using disk-tubular reverse osmosis (DTRO). Zhejiang Gongshang University. (in Chinese).

Wang, Y.Y., R.H. Liu, F. Shen, and L.J. Wu. 2008. Effect of biogas slurry as topdressing fertilizer on yield and quality of Chinese cabbage. *Jiangsu Agricultural Sciences* 1: 220–222. (in Chinese).

Wang, W., S.D. Zhu, L.Y. Yuan, X.J. Hao, R.J. Li, and J. Long. 2009. Application of biogas liquid and biogas residue in soilless culture of peppers. *Journal of Anhui Agricultural Sciences* 37 (24) 11499–11500. (in Chinese).

Wang, L., S. Wu, and C.L. Li. 2010. Heavy metal detection and safety analysis in excrement, biogas slurry and biogas residue. *Journal of Northern Agriculture* 06: 56–57. (in Chinese).

Wang, W.P., X.M. Lu, Z.H. Wei, X.C. Chen, Z.Y. Xue, Y.C. Hu, C.L. Hong, and F.X. Zhu. 2011. Influence of applying biogas slurry on yield and quality of citrus and soil environment. *Journal of Agro-Environment Science* 30 (11): 2300–2305. (in Chinese).

Wang, C., L. Gao, and W.L. Yang. 2012b. A preliminary study on the composition of biogas slurry with pig manure as the main fermentation material. *Modern Journal of Animal Husbandry and Veterinary Medicine* 8: 51–52. (in Chinese).

Wang, C., Y.Y. Wang, Y.N. Chen, H.Z. Shi, K. Liu, H.B. Zhang, and Z.F. Tong. 2017. Treatment of cattle and pig biogas slurry by magnetic bentonite. *Non-Metallic Mines* 40 (03): 93–96. (in Chinese).

Wei, D., M. Wan, R. Liu, G.R. Wang, X.D. Zhang, X.G. Wen, Y. Zhao, and L.J. Chen. 2014. Study on the quality of digested piggery wastewater in large-scale farms in Jiaxing. *Environmental Science* 35 (7): 2650–2657. (in Chinese).

Winandy, J.E., and Z. Cai. 2008. Potential of using anaerobically digested bovine biofiber as a fiber source for wood composites. *BioResources* 3 (4): 1244–1255.

Wu, D.W. 2010. Application technology of biogas slurry and residue in citrus production. *Fujian Agricultural Science and Technology* 2: 58–59. (in Chinese).

Wu, H.B. 2015. Analysis of biogas slurry active components terpenoid transform mechanisms. Doctoral dissertation of Shanghai Ocean University. (in Chinese).

Wu, Yu. 2015. Study of biomass production by *Chlorella pyrenoidosa* cultivated in poultry waste anaerobic digested effluents and carbon fixation by *Spirulina platensis*. Doctoral dissertation of China Agricultural University. (in Chinese).

Wu, L.J., R.H. Liu, and Y.Y. Wang. 2007. Analyses of biogas fermentation material and characteristics of its products in northern energy-ecology ecosystem. *Journal of Agricultural Mechanization Research* 07: 183–186. (in Chinese).

Wu, Y.Z., C.X. Shi, and M.J. Zhang. 2009. Application of biogas slurry and residue in fruit tree. *Agricultural Engineering Technology (New Energy Industry)* 7: 38–39. (in Chinese).

Wu, F.L., M.F. Ye, D.Y. Lin, and B.Q. Weng. 2011. Application technology of biogas slurry and residue in citrus production. *Fujian Agricultural Science and Technology* 26 (1): 103–107. (in Chinese).

Wu, J., J.W. Wang, L.W. Zhang, Y. Liu, B.X. Chen, and X.X. Meng. 2014. Effects of biogas slurry on yield and quality of maize. *Journal of Nuclear Agricultural Sciences* 28 (5): 905–911. (in Chinese).

Xia, A., and J.D. Murphy. 2016. Microalgal cultivation in treating liquid digestate from biogas systems. *Trends in Biotechnology* 34 (4): 264–275.

Xia, C.L., Z. Zhang, Y. Zhang, J.W. Chen, and X.Y. Cai. 2014. Effects of biogas slurry spraying on yield and heavy metals residues of kidney bean (*Phaseolus vulgaris*). *Environmental Science and Technology* 1: 7–12. (in Chinese).

Xiang, Y.S., D.F. Sun, M.R. Wang, S.Z. Hu, and W.H. Yang. 2006. Effect of different concentration of biogas slurry on spring tea. *Hubei Agricultural Sciences* 45 (1): 78–80. (in Chinese).

Xie, S.S., S.Z. Huang, S.P. Lin, and C.P. Luo. 2010. Effect of applying cow dung biogas slurry on grass and soil of tall grass family. *Contemporary Animal Husbandry* 10: 39–41. (in Chinese).

Xie, S.H., L.G. Zong, H. Chu, Z.Y. Wang, X.L. Qiu, A.J. Ma, and R.H. He. 2013. Effects of different biomass materials on the amelioration of acidic tea garden soil. *Journal of Tea Science* 3: 279–288. (in Chinese).

Xu, Y.X., X.X. Tian, D.Z. Li, W. Tian, and X.L. Xu. 2012. Comparison of nutrient contents of fermentation residues in biogas digesters with different raw materials. *Bulletin of Agricultural Science and Technology* 05: 100–102. (in Chinese).

Xu, Q.T., Z.L. Kong, and M.K. Zhang. 2016a. Comprehensive evaluation of improving effects of different organic wastes on a newly reclaimed cultivated land. *Chinese Journal of Applied Ecology* 27 (2): 567–576. (in Chinese).

Xu, M.L., X. Ye, Y.G. Li, Y.Q. Song, and F. Xiao. 2016b. Optimization of forward osmosis process for concentration of biogas slurry. *Transactions of the Chinese Society of Agricultural Engineering* 32 (2): 193–198. (in Chinese).

Yan, B.Z., W.T. Xu, and P.B. Yu. 2012. Study on improving yield and quality of red grape by biogas slurry. *China Biogas* 30 (2): 47–48. (in Chinese).

Yin, S.L., L.P. Zhang, Y.H. Xi, J.L. Ma, J.Y. Yang, G.W. Zhang, P.F. Duan, H.C. Cheng, H.W. Liu, and X.M. Liu. 2017. Effect of biogas residue application on soil micro eco-structure, enzyme activity and physicochemical property. *China Biogas* 35 (1): 72–76. (in Chinese).

Yu, H.L., B.B. Lv, K.S. Li, H. Chen, J.X. Li, X.X. Zhang, X.D. Shang, and Q. Tan. 2012. Cultivation of *Flammulina velutipes* using biogas residue. *Acta Edulis Fungi* 19 (4): 41–43. (in Chinese).

Yuan, Y. 2010. The experimental study on biogas slurry as lettuce and orange's foligage fertilizer. Doctoral dissertation of Huazhong Agricultural Univerisity. (in Chinese).

Zeng, Y., H.S. Hong, W.Z. Cao, N.W. Chen, and W.P. Wang. 2004. Economic analysis on uses of livestock manure on cropland. *Journal of Xiamen University (Natural Science)* 43 (S1): 195–200. (in Chinese).

Zhang, X.H. 1994. Application of marsh manure in controlling crop pests and diseases. *Renewable Energy Resources* 6: 23–24. (in Chinese).

Zhang, J.F. 2013. Comparative test of biogas slurry fertilizer and pigpen fertilizer applied to corn base. *Agricultural Technology & Equipment* 16: 47. (in Chinese).

Zhang, G.M. 2014. Research on pest control of maize by applying marsh fertilizer. Doctoral dissertation of Helongjiang University. (in Chinese).

Zhang, W.D., H.C. Song, Q. Ding, X.K. Wei, Z.W. Xiong, and P.L. Tao. 2001. Application of methane fermenative residues in control of crop diseases and insect pests. *Research of Agricultural Modernization* 22 (3): 167–170. (in Chinese).

Zhang, H., Z.M. Lei, X.C. Dou, B. Lu, and Q. Li. 2008. Analysis of nutritive value of biogas residues and heavy metal content in feedstuff with biogas residues and the pork fed on it. *Chinese Journal of Eco-Agriculture* 05: 1298–1301. (in Chinese).

Zhang, C.A., Y. Liu, M. Cao, Y.Q. Wang, and L. Yao. 2011a. Pricing method and application effects of biogas slurry. *Acta Ecologica Sinica* 31 (6): 1735–1741. (in Chinese).

Zhang, Y.F., L. Dong, Y. Li, Z.H. Liu, W.F. Bian, and P. Liu. 2011b. Effects of biogas slurry on soybean yield, quality and nutrient absorption and soil chemical property. *Journal of Soil and Water Conservation* 25 (4): 135–138. (in Chinese).

Zhang, C.A., Y.F. Zhang, H.T. Lin, Y.W. Shen, and L. Yang. 2012. Pricing biogas slurry based on nutrient content. *China Biogas* 30 (6): 43–46. (in Chinese).

Zhang, W.W., A.J. Gong, L.N. Qiu, and R.L. Yao. 2013. Solid state fermentation of biogas residues for production of *Bacillus thuringiensis* based bio-pesticide. *Transactions of the Chinese Society of Agricultural Engineering* 29 (8): 212–217. (in Chinese).

Zhang, Y., Y.J. Liu, and Z. Jiang. 2016. Analysis of nutrient analysis and the feasibility of agricultural residue. *Journal of Northeast Agricultural University* 47 (03): 59–63. (in Chinese).

Zhao, L., L.N. Zhou, and J.P. Qiu. 2005. Physicochemical properties of biogas slurry and its effect on nutrient composition of pollution-free vegetables. *Journal of Zhejiang Agricultural Sciences* 1 (2): 103–105. (in Chinese).

Zhao, Q.L., J. Wu, B.X. Chen, and W.L. Lv. 2012. Effect of biogas slurry on heavy metal accumulation of soil and maize. *Journal of Soil and Water Conservation* 2: 251–255. (in Chinese).

Zhao, G.H., G. Chen, and J. Xu. 2014. Distribution characteristics of main nutrients and heavy metals in biogas slurry from pig manure and urine. *Journal of Zhejiang Agricultural Sciences* 9: 1454–1456. (in Chinese).

Zheng, S.X., L. Qiu, Q.Y. Liu, and Y.L. Zahng. 2014. Fertilizer efficiency of digestate and safe & effective utilization. *China Biogas* 32 (1): 95–100. (in Chinese).

Zhou, Y.F., L. Qiu, Z.L. Li, Y. Zhang, and Q.P. Sun. 2013. Progress of biogas slurry soilless culture and optimization technology. *Journal of Agricultural Mechanization Research* 5: 224–227. (in Chinese).

Zhou, Y.Y., Y.C. Luo, H.M. Ruan, R.Y. Hu, and J.N. Shen. 2016. Research on the treatment of biogas slurry in farm by using a dish tube reverse osmosis system. *Agriculture and Technology* 36 (4): 132–134. (in Chinese).

Zhu, Q.W. 2014. Distribution characteristics of heavy metals in feeds. pig manures and biogas slurry. *Research of Soil and Water Conservation* 21 (6): 284–289. (in Chinese).

The manufacturer's authorised representative in the EU is Springer Nature Customer Service Centre GmbH, Europaplatz 3, 69115 Heidelberg, Germany. If you have any concerns regarding our products, please contact ProductSafety@springernature.com

Printed and bound by CPI Group (UK) Ltd, Croydon, CR0 4YY
05/10/2026
02228124-0016